Colloid Science

Colloid Science

Physical Chemistry of Colloidal Dispersions

Albert P. Philipse

Great Clarendon Street, Oxford, OX2 6DP,
United Kingdom

Oxford University Press is a department of the University of Oxford.
It furthers the University's objective of excellence in research, scholarship,
and education by publishing worldwide. Oxford is a registered trade mark of
Oxford University Press in the UK and in certain other countries.

© Oxford University Press 2026

The moral rights of the author have been asserted.

All rights reserved. No part of this publication may be reproduced, stored in a retrieval system, transmitted, used for text and data mining, or used for training artificial intelligence, in any form or by any means, without the prior permission in writing of Oxford University Press, or as expressly permitted by law, by licence or under terms agreed with the appropriate reprographics rights organization. Enquiries concerning reproduction outside the scope of the above should be sent to the Rights Department, Oxford University Press, at the address above.

You must not circulate this work in any other form
and you must impose this same condition on any acquirer.

Links to third party websites are provided by Oxford in good faith and
for information only. Oxford disclaims any responsibility for the materials
contained in any third party website referenced in this work.

Published in the United States of America by Oxford University Press
198 Madison Avenue, New York, NY 10016, United States of America

British Library Cataloguing in Publication Data
Data available

Library of Congress Control Number: 2025947852

ISBN 9780198850663

Printed and bound by CPI Group (UK) Ltd., Croydon, CR0 4YY

The manufacturer's authorized representative in the EU for product safety is
Oxford University Press España S.A. of Parque Empresarial San Fernando de Henares,
Avenida de Castilla, 2 – 28830 Madrid (www.oup.es/en or product.safety@oup.com).
OUP España S.A. also acts as importer into Spain of products made by the manufacturer.

Preface

Colloid science is the science of finely dispersed matter in the form of colloids. It is the interdisciplinary branch of learning *par excellence* as it employs and studies ideas, theories, and techniques from a diverse range of fields including chemistry, physics, formulation science, biology, and materials science. Lyklema's encyclopaedia *Fundamentals of Interface and Colloid Science (FICS)* (Academic Press, 2005) showcases the large variety of subjects that are united under the colloid flag. The implication is that in an introductory book like this, some large topics are unavoidably left out or discussed only in a succinct manner. If you miss in this Primer a subject of your particular interest, I must apologize—and point you to *FICS* for an in-depth treatment of the issue.

The topics that *are* included in this Primer have in common that I have lectured about them—in physical chemistry or colloid science courses—or that they played a role, sometimes even as the subject of investigation, in my own research projects. Despite this somewhat personal choice of topics, I believe this Primer to be a representative and illustrative introduction to colloid science, one that provides a stepping stone to more comprehensive treatments in *FICS* or in any other of the references at the end of the chapters.

During my time (1991–2023) at the Utrecht Van 't Hoff Laboratory for Physical and Colloid Chemistry I had the pleasure of working together on various projects with Jan Dhont, Ben Erné, Willem Kegel, Bonny Kuipers, Henk Lekkerkerker, Haran Pathmamanoharan, Dominique Thies-Weesie, Agienus Vrij, and Gert Jan Vroege. I would like to thank all these 'Van 't Hoff' colleagues, who each contributed their own considerable expertise in colloid science, from which I have learned a lot.

In writing this text I gratefully made use of the instructive (unpublished) lecture notes *Colloidal Analysis Techniques* of Ben Erné. Bonny Kuipers is thanked for his careful proofreading of earlier versions of various chapters.

I also would like to acknowledge the pleasant collaboration with OUP editors Kehinde Batmus and Hayden Merrick, and in particular with Brooke James, whose competent input and dedicated support were very helpful, if not indispensable, in getting this Primer completed.

A. P. P., Utrecht, 2026

Brief Table of Contents

Preface — v

1 The Colloidal State of Matter — 1

2 Colloidal Nucleation, Growth, and Ripening — 12

3 Brownian Motion — 30

4 Van 't Hoff's Law, Donnan Equilibria, and the Depletion Force — 50

5 Charged Colloids and Electrical Double Layers — 75

6 Colloidal Kinetics — 103

7 Sedimentation, Flow in Porous Media, and Random Particle Packings — 120

8 Dispersion Flow and Rheology — 139

9 Magnetic Dispersions — 161

10 Association Colloids and Emulsions — 180

11 Colloid Characterization by Light Scattering and Electrophoresis — 193

Answers to Chapter Exercises — 217

Glossary — 255

Index — 263

Detailed Table of Contents

Preface   v

1 The Colloidal State of Matter   1
- 1.1 Introduction   1
- 1.2 Dispersions and their applications   3
- 1.3 Rise and fall of colloidal dispersions   6
- Summary   8
- References   8
- Exercises   9

2 Colloidal Nucleation, Growth, and Ripening   12
- 2.1 Introduction   12
- 2.2 Thermodynamics of homogeneous precipitation   13
- 2.3 Precipitation kinetics   16
- 2.4 Particle growth and polydispersity   18
- 2.5 Particle solubility and Ostwald ripening   20
- 2.6 Catalysed, heterogeneous nucleation   23
- Summary   25
- References   25
- Appendix 2A: Inhomogeneous particle growth—and gelation   26
- Exercises   27

3 Brownian Motion   30
- 3.1 Introduction   30
- 3.2 Timescales in colloidal dispersions   32
- 3.3 The random walk   37
- 3.4 Fick's first law and the diffusion coefficient   40
- 3.5 The Einstein equations for Brownian motion   41
- 3.6 Quadratic displacements from the diffusion equation   46
- Summary   48
- References   48
- Exercises   48

4 Van 't Hoff's Law, Donnan Equilibria, and the Depletion Force   50
- 4.1 Introduction   50
- 4.2 Osmotic pressure from osmotic equilibrium   52
- 4.3 Van 't Hoff's law from particle kinetics   55
- 4.4 A membrane pressure probe   56

	4.5	The Donnan equilibrium	59
	4.6	Depletion forces	66
		Summary	69
		References	70
		Appendix 4A: The Derjaguin approximation	70
		Appendix 4B: Hyperbolic functions	71
		Exercises	72

5 Charged Colloids and Electrical Double Layers — 75

	5.1	Introduction	75
	5.2	Ion distributions near a charged surface	77
	5.3	The Poisson–Boltzmann equation	80
	5.4	A single, flat EDL	82
	5.5	Two interacting EDLs	86
	5.6	The DLVO potential	93
		Summary	96
		References	97
		Appendix 5A: The Debye cube	97
		Appendix 5B: Van der Waals attraction between two parallel plates	98
		Exercises	100

6 Colloidal Kinetics — 103

	6.1	Introduction	103
	6.2	Diffusion to a target sphere	104
	6.3	Rapid early-phase flocculation	106
	6.4	Slow flocculation	109
	6.5	Kinetics of late-stage coagulation	113
	6.6	Diffusional growth	114
	6.7	Flow-induced flocculation	116
		Summary	117
		References	117
		Exercises	118

7 Sedimentation, Flow in Porous Media, and Random Particle Packings — 120

	7.1	Introduction	120
	7.2	Single-particle sedimentation	121
	7.3	Sedimentation–diffusion equilibrium	123
	7.4	Flow in particle packings and Darcy's law	126
	7.5	The Kozeny–Carman scaling relation	128
	7.6	Random particle packings	131
		Summary	134

	References	134
	Exercises	135

8 Dispersion Flow and Rheology — 139

8.1	Introduction	139
8.2	Shear flow, Newton's law, and the Reynolds number	140
8.3	The Stokes equation for viscous flow	143
8.4	Flow in simple geometries	144
8.5	Diffusion versus convection: the Péclet number	147
8.6	Concentration-dependent viscosity	149
8.7	Non-Newtonian dispersions	152
	Summary	156
	References	156
	Appendix 8A: The Krieger–Dougherty relation	157
	Exercises	158

9 Magnetic Dispersions — 161

9.1	Introduction	161
9.2	Magnetism in solids and solutions	162
9.3	Ferrofluids	164
9.4	Equilibrium magnetization of superparamagnetic dispersions	167
9.5	Interactions between dipolar magnetic colloids	171
9.6	Magnetizable dispersions	174
	Summary	176
	References	176
	Appendix 9A: Small-scale ferrofluid synthesis	177
	Exercises	178

10 Association Colloids and Emulsions — 180

10.1	Introduction	180
10.2	Interfacial tension and the Laplace pressure	181
10.3	Stabilized emulsions	183
10.4	Micellization	187
	Summary	190
	References	191
	Exercises	191

11 Colloid Characterization by Light Scattering and Electrophoresis — 193

11.1	Introduction	193
11.2	Static light scattering	194
11.3	Dynamic light scattering	200
11.4	Electrophoresis	203

Summary	212
References	212
Appendix 11A: Apparent light-scattering radii	213
Exercises	214
Answers to Chapter Exercises	217
Glossary	255
Index	263

1 The Colloidal State of Matter

1.1 Introduction

What do important everyday materials such as blood, ink, milk, paint, and toothpaste have in common? At first sight these substances are quite disparate: they have different colours, they have different textures and flow properties, they vary in chemical composition, and they are dissimilar in terms of mass density, refractive index, and many other physical properties. Alongside these differences, however, these materials share one crucial characteristic: they all contain structural entities and particles that are known as **colloids**—minuscule objects that, by definition, have at least one linear dimension in the size range from several nanometres to a few microns (Figure 1.1).

Matter in the colloidal state has significant features that are a direct consequence of the minute size of its colloidal components. In the first place, small particle sizes entail that colloidal systems harbour a large **specific surface area** A_s. This quantity is defined as the ratio between particle surface S and particle volume V:

$$A_s = \frac{S}{V}, \tag{1.1}$$

which for colloidal spheres of radius a modifies to

$$A_s = \frac{3}{a}. \tag{1.2}$$

As an example, when a glass marble with radius $a = 1$ cm is divided into small glass ('silica') colloids with radius $a = 5$ nm, the specific surface area A_s of the glass increases by a staggering factor of two million! Owing to such large areas, surface effects are important for colloidal systems; these effects include adsorption of ions or polymers at the particle surface, and large capacities for catalysis and for the uptake of impurities in the process of water purification. Large surface areas also entail dynamic properties such as the viscous forces exerted by solvent on the surface of moving colloidal particles.

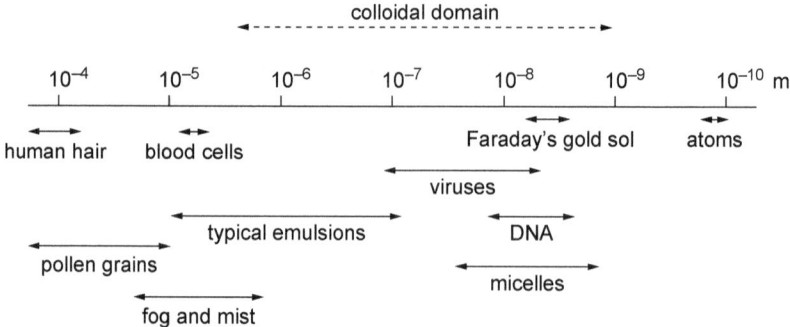

Figure 1.1 The colloidal domain encompasses particles that have at least one linear dimension in the size range from a few nanometres to several microns.

Another consequence of the inherent large specific area of small particles is the significant role in colloidal systems played by Van der Waals attractions. Any two pieces of matter attract each other via two forces: Newton's gravitational force and the **Van der Waals force**. The relative magnitude of the two strongly depends on particle size: for planets the gravitational force totally dominates, whereas for water molecules the Van der Waals attraction completely dwarfs gravity. It is the Van der Waals attraction that makes water molecules adhere to each other to form liquid water; it is also the Van der Waals attraction that makes colloids stick together into aggregates. Gravity, of course, has an impact on colloidal systems as a whole: ink and milk, for example, flow under their own weight. And gravity may also induce individual colloids to migrate towards the Earth's surface—the process of sedimentation that is investigated in Chapter 7. But there is nil gravitational attraction between colloids: they are totally unaffected by each other's mass.

Colloids have, in addition to the importance of surface chemistry and Van der Waals attractions, a third characteristic feature that is entailed by their small size: they are in everlasting **thermal motion**. The mass of colloids is sufficiently small for them to participate in the thermal movements of surrounding solvent molecules, in contrast to macroscopic objects such as glass marbles that remain motionless on the bottom of the vessel. How exactly particle mass and particle diffusion are related is explained in Chapter 3 on **Brownian motion**—the common name for the thermal diffusion of colloids.

How does matter attain the colloidal state? That may happen along at least three different routes, which we will briefly review in arbitrary order. In the process of **precipitation**, molecules associate or polymerize in solution to colloidal particles. Precipitation reactions occur when two water-soluble substances form an insoluble material. For example, when a pale-yellow solution of iron chloride and a colourless sodium hydroxide solution are mixed, almost instantaneously brownish agglomerates settle in the test tube. What happens is that the very soluble iron (III) ions and hydroxyl ions combine to give the poorly soluble iron hydroxide, according to the net reaction:

$$Fe^{3+}(aq) + 3\,OH^-(aq) \rightarrow Fe(OH)_3(s)\downarrow. \tag{1.3}$$

Precipitation can also be initiated by lowering the solute solubility via a change in temperature, pH, or solvent composition. For example, when water is added to a sulphur solution in ethanol, sulphur particles precipitate because water is a very poor solvent for sulphur. A change in pH is involved in the preparation of colloidal silica. A transparent water-glass solution (i.e. an aqueous solution of sodium silicate) has a pH of about 11. A slow decrease in pH triggers precipitation of discrete, small silica particles, owing to the marked reduction of silicate solubility at decreasing pH.

In precipitation methods, colloids are formed from molecular species, whereas in a **dispersion method** the colloidal size range is reached by dividing a macroscopic phase into progressively smaller parts. A well-known example is the formation of **emulsions**, which comprises the dispersion of one liquid in another (see Chapter 10, Section 10.3). The dispersal of inorganic materials, a process also referred to as **comminution**, consists of mechanical fracture in a ball mill. Such a mill is a rotating cylindrical vessel in which minerals are pulverized between colliding ceramic balls. In this way, for example, the mineral magnetite (Fe_3O_4) can be dispersed into colloidal magnetite, an instance of the magnetic colloids that are the subject of Chapter 9.

A third route molecules may follow to form colloidal structures is that of **self-assembly**. This route is a spontaneous one, in the thermodynamic sense, meaning that molecules associate by themselves to form colloids without external energy input; compare this to the dispersion method, which is a process forced by input of mechanical energy. Examples of self-assembled colloids are viruses that result from spontaneous association of proteins, and micelles, which are aggregates that form in solutions of surface-active molecules—an instance of the association colloids that are studied in Chapter 10.

1.2 Dispersions and their applications

A high specific surface area is, of course, not unique for small solid (S) particles or gas (G) bubbles dispersed in a liquid (L). A large interfacial area is characteristic for a much larger class of disperse systems (i.e. fine dispersions of one phase in another, continuous phase; see Table 1.1). Note that not all dispersed phases exhibit Brownian motion, as in the case of a solid dispersion of small

Table 1.1 Examples of disperse systems

S in S	L in S	G in S
Polycrystalline materials, opals, ruby glass, bone	Solid emulsions: ice cream, tar	Solid foams: foam rubber, porous glass
S in L	**L in L**	**G in L**
Paints, inks, clay, toothpaste, ferrofluids, silica sols	Emulsions, milk, margarine, mayonnaise	Beer foam, soap bubbles
S in G	**L in G**	**G in G**
Aerosols of solid particles: smoke	Aerosols of droplets: fog, spray, clouds	No examples exist

gold particles in glass which gives ruby glass its characteristic red colour. There are many other examples of structures containing large surface areas which make them colloidal in nature. For example, blood is a dispersion of, among other corpuscles, cells and blood platelets in an aqueous electrolyte solution (serum). Other biological structures may be mentioned, such as bone, in which calcium phosphate is dispersed in collagen. Incidentally, it should be noted that dispersions of gases in gases do not exist.

Dispersions of one liquid phase in another are the emulsions, which we encountered earlier. This type of colloidal dispersion is developed and applied in many fields, including food and dairy products, cosmetics and personal care products, paints, pharmaceuticals, and agrochemicals.

Let us review some applications of *inorganic* colloids. Metallic colloids such as cobalt and iron particles, attached to porous carriers, are employed in heterogeneous catalysis. Colloidal silver, owing to its anti-microbial properties, is incorporated in burn and wound dressings, hospital clothing, and fitness outfits; water can be disinfected by silver colloids, and aqueous colloidal silver dispersions are marketed for their alleged 'universal antibiotic' properties. Gold colloids have widespread biological applications, including their use as sensors in pregnancy tests, and labelling (their attachment to biomolecules so that these can be located on electron micrographs).

Abundant in our environment are the inorganic colloids in the form of clay particles in soil; here 'clay' denotes soil particles with diameters below a few microns. Inorganic colloids in the form of dispersions of clays or milled powders are applied on a large scale for the fabrication of traditional ceramics such as earthenware and bathroom equipment. A variety of oxide as well as non-oxide powders (SiC, SiN, Al_2O_3, etc.) are processed in the form of colloidal suspensions to manufacture technical ceramics, including ceramic membranes and high-temperature-resistant materials. Dispersions of plate-like clay minerals such as kaolinite (see Figure 1.2) and montmorillonite are also employed in large volumes as oil-well drilling fluids, which are pumped into well bores to facilitate the drilling process by suspending cuttings and providing cooling and lubrication.

Colloidal processing underlies the very pages on which this text is printed. Paper-making starts with the degradation of wood chips to an aqueous suspension of cellulose fibres, with a significant percentage of fibres and fibre fragments with cross-sectional dimensions in the colloidal size range. Inorganic particles, in the form of silica or bentonite sols, are added to improve the quality and rate of paper-making, a process which comprises the filtering and drying of the mixture of fibres and colloids on a wire mesh. Carbon colloids function as pigments, giving ink its black appearance. Many other pigments in printing ink, paints, and plastics are found in the form of finely ground inorganic oxides or hydroxides. Iron oxides such as red haematite (α-Fe_2O_3), dark-brown maghemite (γ-Fe_2O_3), and black magnetite (Fe_3O_4) are in some of the most practically important pigments.

Virtually all dispersions mentioned so far comprise colloids that are **polydisperse**—that is, the colloids have a broad distribution in size or shape, or in both: a telling example of the latter are the magnesium hydroxide colloids

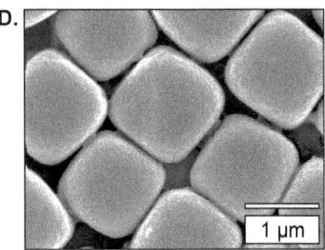

Figure 1.2 Examples of anisotropic colloids. **(A)** Haematite spindles (length 500 nm); **(B)** gibbsite platelets (diameter 160 nm); **(C)** kaolin clay platelets; and **(D)** silica cubes.

Source: (A) Thies-Weesie et al. (1995); (B) Wierenga et al. (1998); (C) van Olphen (1977); (D) Rossi et al. (2011).

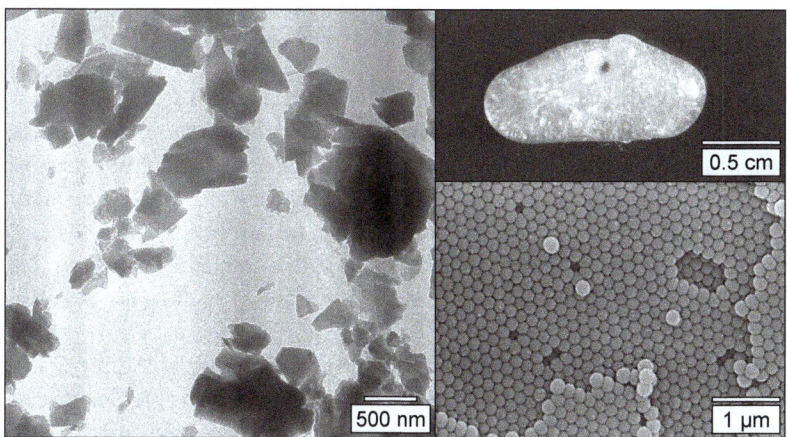

Figure 1.3 Left: Example of a polydisperse system with a wide distribution of aggregate size and shape; TEM-image taken from a dispersion of magnesium hydroxide Mg(OH)$_2$ colloids. **Lower right**: SEM image of an ordered packing of monodisperse silica spheres (radius 200 nm), slightly sintered together to form (**upper right**) a synthetic, solid opal that diffracts visible light.

Source: Images taken by the author; see Philipse (1989).

depicted in Figure 1.3. Size polydispersity is an inevitable consequence of precipitation methods, for reasons that will become clear in our study of precipitation kinetics in Chapter 2. Also, the process of self-assembly may produce a distribution in colloid size, as is the case for the formation of micelles, as discussed in Chapter 10, Section 10.4. For most applications, polydispersity does not present a problem; on the contrary, it can even be beneficial: dispersions of polydisperse spheres flow more easily than those of mono-sized spheres—as we will see in Chapter 8 on dispersion rheology. And ceramic materials made from polydisperse colloids have a much higher mechanical strength than ceramics based on monodisperse colloids; the latter tend to form ordered structures (see the opal structure in Figure 1.3) that, on firing in an oven, sinter to materials that harbour large cracks that do not form in structures of randomly packed, polydisperse particles.

Polydispersity, however, *is* a problem in the quantitative study of colloidal dispersions; particle-sizing methods, such as the ultracentrifugation treated in Chapter 7 and the light-scattering techniques considered in Chapter 11, cannot be applied to dispersions with a broad size or shape distribution. That is why, for more fundamental research, model dispersions are investigated that comprise colloids of uniform size and shape. Classic examples of such monodisperse colloids are latex and silica spheres. Monodisperse non-spherical colloids have also become available (Figure 1.2). Incidentally, monodisperse silica spheres are also found in nature: the iridescence of opals is due to the crystalline arrangement of uniform colloidal silica spheres (Figure 1.3) which gives rise to Bragg reflections of visible light.

1.3 Rise and fall of colloidal dispersions

Colloids composed of insoluble substances such as metals, metal oxides, and hydroxides are termed **lyophobic**, derived from the Greek verbs λύω 'I loosen' and φοβέω 'I fear'. There are, of course, also substances whose molecules love to loosen themselves in a liquid, and they are called **lyophilic**, where '-philic' stems from φιλέω 'I love'. Examples are polyelectrolytes and proteins, many of which spontaneously disperse in water, and a variety of surfactants that may dissolve as well-defined aggregates, such as the already-mentioned micelles.

When we look at the preparation of lyophobic colloids via the precipitation method, it seems at first sight quite obvious that particles precipitate out of solution, solvent fearing as they are. However, in many cases solvent fear does *not* lead to visible, macroscopic de-mixing or precipitation. Carbon, for example, is a very insoluble material, yet we are familiar with precipitate-free dispersions of carbon particles ('carbon black') in water, better known as inks. And haematite and titania are water-insoluble oxides but, nevertheless, red and white paints contain, respectively, haematite and titania pigment particles that remain in stable dispersion; that is to say, the particles do not settle as precipitates. Window glass consists of dense amorphous silica that—how fortunate!—does not dissolve in rainwater; yet colloidal silica may stay dispersed in water over extensive periods of time.

When a precipitate, for example that of the iron hydroxide in reaction equation (1.3), is investigated under an electron microscope, it turns out that it is actually composed of large aggregates of primary, small crystallites with sizes typically in the range from a few to several tens of nanometres. So apparently lyophobic colloids come into existence as very small particles that rapidly stick together (by Van der Waals attractions) into larger structures that settle under gravity. To prevent or very much slow down this particle aggregation, a repulsion between the colloids must be operative—for example, the electrical double-layer repulsion that is the subject of Chapter 5. A repulsion between colloids can also be induced by polymers adsorbed on the colloid surface: the carbon particles in ink are wrapped in water-soluble polymers (coming from 'gum arabic') that prevent their aggregation.

The inherent tendency of lyophobic colloids to aggregate signifies that they are unstable, in the thermodynamic sense of the word. This instability reveals itself as the spontaneous tendency of the colloids to lower their **Gibbs free energy** G. This lowering occurs predominantly by the reduction of surface area, because molecules at the surface are in a higher energy state than molecules in a bulk phase, which is the basic reason why in the dispersion method energy input is needed to bring a macroscopic phase into the colloidal state. In addition, when colloids are made via the precipitation method, the Gibbs energy rises because of the formation of small particles with large specific surface areas. In the Gibbs free-energy landscape in Figure 1.4, we see that the **nucleation** event (marked as *Nu*) corresponds to a rise in Gibbs energy with

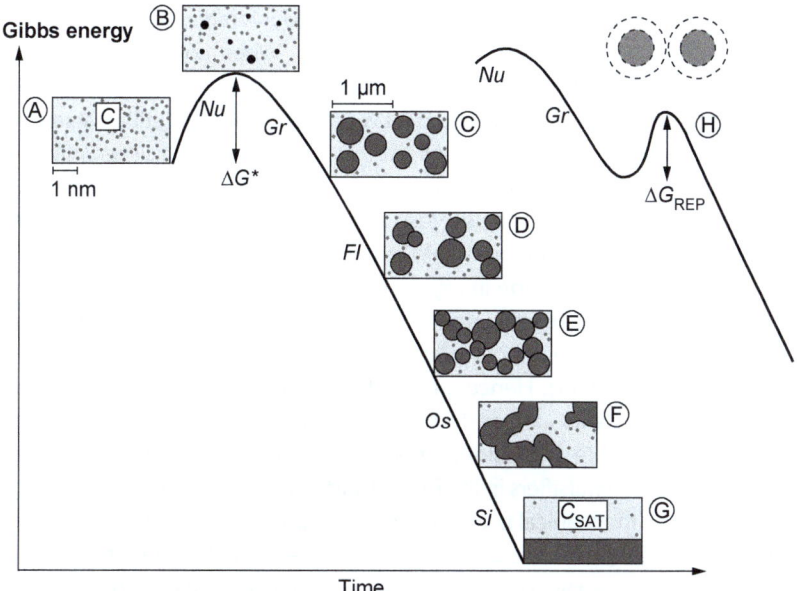

Figure 1.4 In a supersaturated solution A with monomer concentration $C > C_{SAT}$, nucleation Nu of small particles occurs, requiring a Gibbs energy rise ΔG^* to form the critical nuclei in B. The Gibbs energy falls along various steps (further explained in Section 1.3), until colloids end up in the macroscopic bulk phase G. Note that the monomer concentration in solution during the whole sequence A–G monotonically decreases, from the supersaturation C to the saturation concentration C_{SAT} in equilibrium with the bulk in G. A repulsive energy barrier ΔG_{REP} in H, caused by polymers or electrical charges on the colloid surface, traps colloids in C in a kinetically stable dispersion.

an amount ΔG^*—an amount to be evaluated in Chapter 2. A growth phase marked as Gr follows, with a fall in Gibbs energy owing to spontaneous formation of bulk phase, leading to a dispersion of discrete colloids in box C. From that stage on, the Gibbs energy keeps on declining due to spontaneous surface-area reduction via the processes of **flocculation** (D and E), **Ostwald ripening**, and **coalescence** (F), until the colloids have merged into the bulk phase in box G.

The time it takes for colloids to traverse the Gibbs free-energy landscape in Figure 1.4 varies enormously. Nucleation may occur in a split-second, or it could take *very* much longer. And coarsening in Figure 1.4 from box C to G may occur swiftly, as in the case of bare oil droplets in water that quickly merge into a macroscopic oil phase. Inorganic precipitates, in contrast, may remain in box E almost indefinitely, owing to their solidity and low solubility that obstruct ripening and coalescence. A repulsion between colloids due to, for example, electrical surface charges, retards **aggregation** to the clusters in boxes D and E; if the inter-colloid repulsion is strong enough, dispersions may remain forever in a state of kinetic, colloidal stability.

An inter-colloid repulsion, incidentally, only counteracts aggregation; it does *not* inhibit coarsening by Ostwald ripening. When colloids are not in contact or part of an aggregate, ripening may still proceed via monomer diffusion through the continuous phase. This way the colloids can still coarsen in time, and go downhill in Figure 1.4. As a practical example of this scenario we mention here catalyst nanoparticles that are fixed on a substrate such that they cannot aggregate; they nevertheless lose surface area in time due to monomer transport through the continuous liquid or gas phase.

A last note on the landscape in Figure 1.4: the dispersion method brings a material *uphill* in Gibbs energy into the dispersed stage C in Figure 1.4, from where the fractured colloidal particles will tend to go downhill again via aggregation and coarsening. Hence, at the start of the milling process suitable polymers ('deflocculants') are added which adsorb on the fractured particles and keep them in stable dispersion. When the macro-phase in G is an oil, to be dispersed in water, emulsifiers in the form of surfactant molecules are mixed in; they facilitate the mechanical break-up of oil into droplets and retard coarsening once the oil-droplet dispersion in box C has been reached. Kinetic stabilization of emulsions can also be achieved by emulsifiers in the form of proteins or small solid particles—a topic that is further discussed in Chapter 10.

Summary

- Many materials in nature, technology, and daily life have among their constituents particles or multi-molecular structures with one or more linear dimensions that are in the colloidal domain.
- Two results of colloids' high specific surface area are their high capacity for surface chemistry, and their strong susceptibility to surface forces such as Van der Waals attractions.
- Colloidal masses are small enough for colloids to make significant displacements due to their thermal motion; this Brownian motion is in essence a slower version of molecular diffusion.
- Besides via polymerization in solution, and dispersion methods such as emulsification and comminution, colloidal structures may also form by spontaneous molecular self-assembly.
- Lyophobic colloids, being composed of poorly soluble substances, are thermodynamically unstable, as attested by their spontaneous flocculation and surface-area reduction via Ostwald ripening or coalescence.

References

R. Baars et al. (2012), 'Morphology-controlled functional colloids by heterocoagulation of zein and nanoparticles', *Colloids and Surfaces A* **483**, 245101.

C. J. Brinker and G. W. Scherer (1990), *Sol-Gel Science*. Academic Press, San Diego.

H. Darley and G. Gray (1988), *Composition and Properties of Drilling and Completion Fluids*, fifth edn. Butterworth-Heinemann, Woburn.

R. K. Iler (1979), *The Chemistry of Silica: Solubility, Polymerization, Colloid and Surface Properties and Biochemistry*. Wiley, New York.

J. Lyklema (ed.) (2005), Fundamentals of Interface and Colloid Science: *vol. 1 Fundamentals; vol. 2 Solid-Liquid Interfaces; vol. 3 Liquid-Fluid Interfaces; vol. 4 Particulate Colloids; vol. 5 Soft Colloids*. Elsevier, Amsterdam.

H. van Olphen (1977), *An Introduction to Clay Colloids*. Wiley, New York.

A. Philipse (1989), 'Solid opaline packings of colloidal silica spheres', *J. Mat. Sci. Letters* **8**, 1371.

L. Rossi et al. (2011), 'Cubic crystals from cubic colloids', *Soft Matter* **7**, 4139–4142.

U. Schwertmann and R. M. Cornell (1991), *Iron Oxides in the Laboratory; Preparation and Characterization*. VCH, Weinheim.

T. H. Tadros (ed.) (2009), Colloid and Interface Society Series, *vol. 3 Colloidal Stability and Application in Pharmacy; vol. 4 Colloids in Cosmetics and Personal Care; vol. 5 Colloids in Agrochemicals; vol. 6 Colloids in Paints*. Wiley-VCH, Weinheim.

D. Thies-Weesie et al. (1995), 'Preparation of sterically stabilized silica-hematite colloids', *J. Coll. Interface Sci.* **174**, 211.

A. Wierenga et al. (1998), 'Aqueous dispersions of colloidal gibbsite platelets', *Colloids and Surfaces A* **134**, 355

Exercises

1.1 When water is added to a solution of the maize protein zein in ethanol, spherical zein colloids are formed (see R. Baars et al., 2012). Can you provide an explanation?

1.2 Silica sols are commercialized as stable, concentrated dispersion of silica particles (diameter about 10 nm) in alkaline solution. When about equal volumes of a silica sol and ethanol are mixed, the easily flowing sol abruptly turns into a stiff, space-filling gel. What is the explanation?

1.3 Instead of equation (1.1) that defines it as the area per unit volume, specific surface area is sometimes also defined as the surface area per unit weight:

$$A_s^* = \frac{S}{V\delta},$$

where S is the area of an object with volume V and mass density δ.

(a) Work this out for spheres of radius a.

(b) Calculate A_s^* (in $m^2 g^{-1}$) for the silica colloids in Exercise 1.2, for a silica mass density of $\delta = 2$ g ml^{-1}.

1.4 A cube with sides of 1 dm is divided into cubes with sides of 1 nm. By which factor does the surface area increase?

1.5 A spherical droplet with radius R and mass density δ is divided into N smaller droplets with radius r.

    (a) By which factor does the *specific* droplet area increase?

    (b) For a given liquid–air surface tension γ, what is the increase in surface energy due to the division into N droplets?

1.6 Calculate (derive a formula for) the specific surface areas A_s of the following particle shapes:

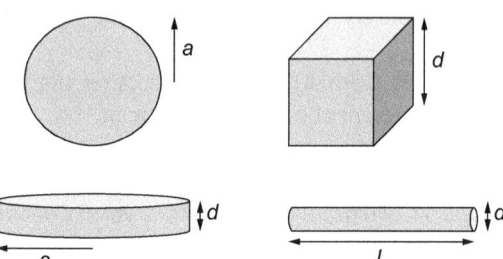

Show that for very thin discs and very thin rods, A_s only depends on the smallest dimension d.

1.7 A sphere of radius R is composed of molecules with diameter d. Show that the fraction ϕ of molecules that are at the sphere's surface is to a good approximation given by $\phi = 3d/R$. Evaluate ϕ for sphere radii $R = 1\,\text{mm}, 1\,\mu\text{m}, 10\,\text{nm}$, and $1\,\text{nm}$. Assume molecules are spheres with diameter $d = 0.1\,\text{nm}$.

1.8 A glass half-sphere (with area $O = 1\,\text{m}^2$ and mass density $\delta_{glass} = 2\,\text{g mL}^{-1}$) hangs suspended in water with mass density $\delta_{water} = 1\,\text{g mL}^{-1}$. A second glass half-sphere of 752 kg is brought to a surface-to-surface distance h:

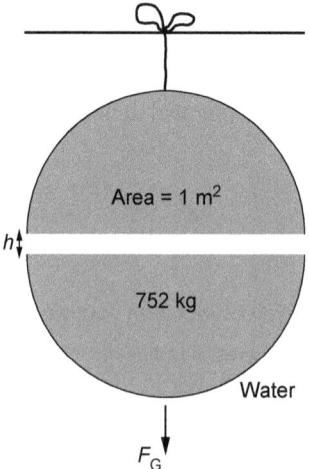

In Appendix 5B, it is shown that the Van der Waals energy of attraction (per unit area) between two flat surfaces at distance h is

$$A_{att} = -\frac{A_H}{12\pi h^2} \quad (Jm^{-2}).$$

Here A_H is the Hamaker constant, for the system glass–water–glass, equal to $A_H \approx 4 \times 10^{-21}$ J.

(a) Express the volume V_h of a half-sphere of radius R in terms of its base area O.

(b) Calculate the buoyant weight of (i.e. the net gravitational force on) the lower glass half-sphere.

(c) What is the expression for the Van der Waals *force* between the two surfaces?

(d) Calculate the magnitude of this Van der Waals force between the two half-spheres at distances $h = 1$ nm, 10 nm, and 100 nm.

Find the solutions to these exercises at the end of the book.

2 Colloidal Nucleation, Growth, and Ripening

2.1 Introduction

In our exploration of the colloidal life cycle sketched in Figure 1.4, we made acquaintance with colloid formation via precipitation in a **supersaturated** solution. This precipitation is an instance of a **phase transition**, in which one phase, here a solution of molecules, spontaneously transforms into another one, here a solid composed of those same molecules. Colloid precipitation exemplifies a phase transition that occurs along the route of nucleation and growth: small nuclei of the new phase come into existence and these 'embryos' subsequently grow further by incorporation of molecules from the surrounding solution. This chapter explores the energetics and the kinetics of this colloidal nucleation and growth process.

This exploration is of much wider relevance than only for the birth of a colloidal dispersion. In nature, household and industry examples are abundant of phase transformations that, just as with precipitating colloids, follow a nucleation-and-growth route. Everyday instances are the nucleation of air bubbles in boiling water, the lime formation in a water cooker (triggered by the decreased solubility of $CaCO_3$ at elevated temperature), and water droplet nucleation on cold bathroom mirrors. Industrial crystallization processes involve the large-scale production of crystalline products that form in supersaturated solutions. The sugar industry, for example, employs sugar crystallization from aqueous solutions thereof, by temperature elevation of supersaturated sugar solutions or syrups. An ancient, still widely applied method for crystalline salt formation makes use of solar evaporation of seawater. Rain and fog are important examples of nucleation from a water vapour phase. When vapour ascends into the sky it becomes saturated: the partial pressure of water vapour in the air becomes equal to the water vapour pressure at the given temperature. Then the condensation of water vapour to liquid becomes thermodynamically favourable (though cloud formation requires droplet condensation on airborne nuclei such as sea-salt particles and particulates stemming from industrial combustion).

Supersaturated solutions have in common that they are **metastable** systems because the new phase is a more stable state of lower Gibbs free energy, a state reached by crossing a nucleation free-energy barrier. When embryos of a new phase are formed completely within the bulk of metastable fluid, we speak of **homogeneous nucleation**. Homogeneous nucleation is actually a pretty rare phenomenon because foreign sites, such as suspended impurities, and microscopic cracks and crevices on container walls, may greatly speed up the formation of the new phase—in a process that is called **heterogeneous nucleation**. A familiar example here is the rapid nucleation of carbon dioxide gas bubbles on a glass surface, observed on a freshly poured glass of beer or soft drink. You may wish to verify the drastic effect of adding extra nucleation sites in the form of sugar or sand grains.

This chapter proceeds in Section 2.2 with an examination of the energetics of homogeneous colloid nucleation. The thermodynamics of this nucleation does not give any clue about the rate at which it actually occurs: to address nucleation kinetics we have to incorporate in our modelling—as we will do in Section 2.3—the rate at which molecules from a supersaturated solution reach newly born colloids by diffusion. Next the colloids enter a region of irreversible particle growth, which is the subject of Section 2.4. One important consequence of this growth for the colloids is their decreasing solubility. As a consequence of that size-dependent solubility, colloids have the spontaneous tendency to decrease their joined surface area; this is the **Ostwald ripening** that is explained in Section 2.5. Up to this point we have only dealt with colloids that form by homogeneous precipitation, so Section 2.6 addresses the pronounced catalytic effect of a substrate on the kinetics of colloidal nucleation.

2.2 Thermodynamics of homogeneous precipitation

Classical nucleation theory

Particle nucleation in a supersaturated solution is usually described via the so-called classical nucleation theory (CNT, see Abraham, 1974). The essential assumption in CNT is that particles nucleating in a supersaturated solution are small pieces of bulk matter, with a surface tension equal to that of a macroscopic bulk. So, for example, a small silica particle in an aqueous silicate ('waterglass') solution has a surface tension equal to that of a macroscopic glass–water interface. Starting from this assumption, the thermodynamics of particle nucleation is elaborated as follows. Suppose in a supersaturated solution, N solute molecules form a new colloidal particle. The particle's formation Gibbs energy change comprises two contributions:

$$\Delta G = \Delta G(\text{bulk}) + \Delta G(\text{surface}). \quad (2.1)$$

The bulk term is negative, as it is the driving force for nucleation:

$$\Delta G(\text{bulk}) = -N|\Delta \mu|; \quad N = \frac{V}{v_m}. \quad (2.2)$$

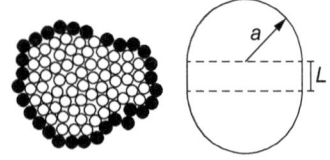

Figure 2.1 In classical nucleation theory a nucleus (**left**) is modelled by a droplet composed of bulk molecules and surface molecules which have a higher free energy per molecule than the bulk. The nucleus is not necessarily spherical and is modelled here (**right**) by a spherocylinder (the subject of Exercise 2.3).

Here v_m is the contribution per molecule to bulk volume V, and $\Delta\mu$ is the chemical potential difference between bulk phase and supersaturated solution—which is to say that $\Delta\mu$ is the reversible (minimal) work required to transfer *one* molecule from solution to the bulk. Surface area A is proportional to $(Nv_m)^{2/3}$, with a proportionality constant, called the shape parameter β, with a magnitude that depends on the shape of the nucleus. The Gibbs energy increase required to create surface area A is the product of A and surface tension γ:

$$\Delta G(\text{surface}) = \gamma A = \gamma\beta(v_m N)^{2/3}, \qquad (2.3)$$

where $v_m^{2/3}$ is the contribution per molecule to the particle area. Combination gives the total Gibbs formation free energy of a cluster composed of N molecules:

$$\Delta G = \gamma\beta(v_m N)^{2/3} - N|\Delta\mu|. \qquad (2.4)$$

For relatively small clusters the surface area term dominates, whereas ΔG as a function of N only starts to decrease due to the bulk term beyond a critical value N^* (Figure 2.2). This critical cluster size follows from the condition $d\Delta G/dN = 0$, for $N = N^*$:

$$(N^*)^{1/3} = \frac{2\gamma\beta v_m^{2/3}}{3|\Delta\mu|}, \qquad (2.5)$$

which can be used to rewrite the formation Gibbs energy of a particle in (2.4) as

$$\Delta G = A\gamma\left[1 - \frac{2}{3}\left(\frac{N}{N^*}\right)^{1/3}\right]. \qquad (2.6)$$

The expression for ΔG in this form is independent of the shape of the cluster (the shape parameter β has dropped out of the equation) and equally holds, for example, for cubes (Exercise 2.2) and spheres.

Activation free energy

The height of the maximum (see Figure 2.2) in the Gibbs energy

$$\Delta G^* = \frac{1}{3}A^*\gamma; \quad A^* = \beta(v_m N^*)^{2/3} \qquad (2.7)$$

is the activation barrier in the formation of colloidal particles by homogeneous nucleation in a supersaturated solution. Note that the (reversible) work needed to form the surface of the critical cluster equals $A^*\gamma$, but that the maximum in ΔG is only one third of $A^*\gamma$; this reduction is due to the spontaneous bulk formation that lowers the net Gibbs energy formation of the cluster.

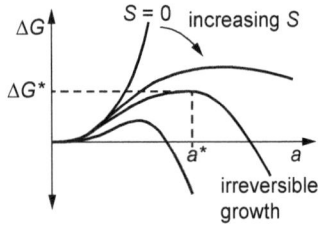

Figure 2.2 Sketch of (2.13) for nucleation and growth of a spherical precipitate of radius a in a solution with supersaturation ratio S.

We are now in a position to understand the effect of increasing the solution's supersaturation: a higher $|\Delta\mu|$ in equation (2.5) pushes the critical size N^* to smaller values which, according to equation (2.7), also lowers the activation barrier (see also Figure 2.2). Hence, enlarging supersaturation induces a more rapid nucleation of smaller particles. A decrease in the interfacial tension γ between colloids and solution, which could be caused by adsorption of surfactant molecules at the colloid–solution interface, brings about the same trend: both critical size and activation barrier, according to (2.5) and (2.7), then diminish: for smaller surface tensions the spontaneous bulk formation 'takes over' in an earlier stage of the nucleation process.

Degree of supersaturation

The **saturation concentration** c(sat) is defined as the solubility of a flat bulk phase, that is to say, c(sat) is the concentration of a solution in thermodynamic equilibrium with the bulk. Supersaturated solutions have concentrations $c > c(sat)$ and their degree S of supersaturation is specified by:

$$S = \frac{c}{c(sat)}. \qquad (2.8)$$

The chemical potential difference $\Delta\mu$ between solution and bulk can be expressed in terms of S as follows. For ideal, non-interacting solute molecules the chemical potential μ_{sol} is defined by:

$$\mu_{sol} = \mu_0 + k_B T \ln \frac{c}{c_0}, \qquad (2.9)$$

where c is the solute concentration and μ_0 is a reference chemical potential for a solution with reference concentration c_0. Further, T is the absolute temperature and k_B the **Boltzmann constant**. For a saturated solution we have:

$$\mu_{sat} = \mu_0 + k_B T \ln \frac{c(sat)}{c_0}. \qquad (2.10)$$

Since the saturated solution is in equilibrium with the bulk, the chemical potential of the latter equals μ_{sat}. Hence the chemical potential difference $\Delta\mu$ between supersaturated solution and bulk equals:

$$\Delta\mu = \mu_{sat} - \mu_{sol} = k_B T \ln \frac{c(sat)}{c} = -k_B T \ln S, \qquad (2.11)$$

with the supersaturation S as defined in (2.8). On substitution of (2.11) in (2.4) we obtain for the formation Gibbs energy of a cluster of N molecules:

$$\Delta G = \gamma \beta (v_m N)^{2/3} - N k_B T \ln S. \qquad (2.12)$$

This result, it should be noted, is less general than (2.4) because ideality of solute molecules is assumed in the chemical potentials (2.9) and (2.10).

Spherical nuclei

We will now work out (2.12) for the nucleation of a non-crystalline, amorphous sphere of radius a. The sphere area equals $4\pi a^2 = \beta(v_m N)^{2/3}$ and the sphere contains $N = (4/3)\pi a^3/v_m$ molecules. Hence the formation Gibbs energy of the spherical nucleus is

$$\Delta G = 4\pi a^2 \gamma - (4/3)\pi a^3 \frac{k_B T \ln S}{v_m} = 4\pi a^2 \gamma \left(1 - \frac{2a}{3r_k} \ln S\right). \tag{2.13}$$

Here we took the opportunity to introduce the characteristic length r_k, also referred to as the Kelvin length, which is defined as

$$r_k = \frac{2v_m \gamma}{k_B T}. \tag{2.14}$$

The physical significance of r_k is that when particle radii approach this length the particle solubility significantly increases. We note here in passing that the Kelvin length scale is a molecular one: surface tension approximately equals thermal energy divided by the surface area per molecule. Thus $\gamma \sim k_B T/v_m^{2/3}$ such that the Kelvin length is about $r_k = 2\gamma v_m/k_B T \sim 2v_m^{1/3}$, where $v_m^{1/3}$ is a typical distance between two neighbouring molecule centres.

The Gibbs energy maximum (Figure 2.2) for a spherical nucleus is:

$$\Delta G^* = \frac{4}{3}\pi (a^*)^2 \gamma; \quad a^* = \frac{r_k}{\ln S}. \tag{2.15}$$

On substitution of (2.14) in (2.15) we observe for the Gibbs energy maximum the following proportionality:

$$\Delta G^* \propto \frac{\gamma^3}{(\ln S)^2}. \tag{2.16}$$

Thus a modest change in surface tension γ has a pronounced impact on the barrier height owing to the cubic dependence γ^3; this modest change has, as we will see in Section 2.3, an even bigger effect on the kinetics of nucleation.

2.3 Precipitation kinetics

In the precipitation kinetics of colloids in a supersaturated solution, we can distinguish two regimes (Figure 2.2). Colloids significantly larger than the critical size are in the regime of irreversible growth. In the initial, nucleation regime small particles struggle with their own solubility to pass the activation barrier ΔG^*. This passage is called a **nucleation** event, which for simplicity we will model as the capture of one or several molecules by a critical cluster, assuming that after this capture the cluster enters the irreversible growth regime from which a new colloid is born. This model, of course, neglects the finite probability that critical clusters may also dissolve. For an order-of-magnitude assessment of the nucleation rate, however, this simple picture suffices.

Homogeneous nucleation rate

Our definition of a nucleation event entails that the number I of colloids which per second come into existence is proportional to c_m and c^*:

$$I = kc_m c^*, \qquad (2.17)$$

where k is a rate constant; c_m and c^* are, respectively, the concentrations of single, unassociated molecules and critical clusters. To quantify I, we first evaluate the frequency at which molecules encounter a nucleus of radius a by diffusion. The stationary, time-independent flux J of molecules diffusing through a spherical envelope of radius r in the direction of the origin (see Figure 2.3) is:[1]

$$J = 4\pi r^2 D \frac{dc(r)}{dr}. \qquad (2.18)$$

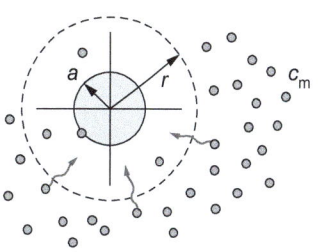

Figure 2.3 Molecules diffuse to a growing nucleus of radius a to which they irreversibly attach upon arrival at the sphere surface. A stationary diffusion flux develops from the bulk solution with constant molecule concentration c_m, to the zero concentration at $r = a$.

Here $c(r)$ is a radial concentration profile and D is the molecular diffusion coefficient relative to the sphere positioned at the origin at $r = 0$. Each molecule reaching the nucleus irreversibly attaches so the concentration of *free* molecules at $r = a$ is zero. Further, we assume that the concentration c_m of molecules in the liquid far away from the sphere remains constant: $c(r = a) = 0$; $c(r \to \infty) = c_m$. For these boundary conditions, solution of (2.18) yields the stationary diffusion flux (Exercise 2.4):

$$J = 4\pi D a^* c_m. \qquad (2.19)$$

We now identify the nucleation rate I as this flux J multiplied by the concentration c^* of critical nuclei with radius a^*:

$$I = Jc^* = 4\pi D a^* c_m c^* \ [s^{-1} m^{-3}]. \qquad (2.20)$$

Comparing (2.17) and (2.20) we see that the reaction constant defined by (2.17) equals $k = 4\pi D a^*$. Next we approximate concentration c^* as follows. If we assume that an equilibrium distribution of clusters with radius a is established, this distribution should satisfy:

$$c(a) \propto \exp\left[-\Delta G(a)/k_B T\right]. \qquad (2.21)$$

Here $\Delta G(a)$ is the reversible work (the minimal work) needed to form a cluster of radius a. In principle, no work is associated with the formation of a single, monomeric molecule, so the constant of proportionality in (2.21) should equal the monomeric number concentration c_m in the bulk of the supersaturated solution. Hence

$$c(a) = c_m \exp\left[-\Delta G(a)/k_B T\right]. \qquad (2.22)$$

---

[1] (2.18) is an instance of Fick's first diffusion law, which is further explained in Section 3.4.

This is the Boltzmann distribution for the concentrations of clusters, with the formation free energy $\Delta G(a)$ given by (2.13). Applying this distribution to clusters with a critical size a^* we have

$$c^* = c_m \exp\left[-\Delta G^*/k_B T\right], \qquad (2.23)$$

where ΔG^* is the height of the nucleation barrier. On substitution of (2.23) in (2.20) we find for the nucleation rate:

$$I = 4\pi D a^* c_m^2 \exp\left[-\Delta G^*/k_B T\right]; \quad \Delta G^* = (4\pi/3)(a^*)^2 \gamma. \qquad (2.24)$$

Since we have seen in (2.16) that $\Delta G^* \propto \gamma^3$, the conclusion here is that any uncertainty in γ is strongly amplified by the exponent in (2.24) and as surface tensions of solid–liquid interfaces are not accurately known, a prediction for the nucleation flux from (2.24) is at best an order-of-magnitude estimate.

Kinetic pre-factor

Equation (2.24) shows that the nucleation rate is extremely sensitive to the value of a^* and, thus, to the supersaturation via (2.15). The pre-exponential kinetic factor in (2.24) is of the order:

$$4\pi D a^* c_m^2 \sim \frac{k_B T}{\eta} c_m^2, \qquad (2.25)$$

as follows from substitution of the Stokes–Einstein diffusion coefficient $D = k_B T / 6\pi\eta a^*$, where we neglect the size difference between molecules and critical clusters. For an aqueous solution at room temperature with a molar concentration $c_m = 10^{-3}$ M, we find an astronomical pre-factor of order $10^{30}\,\mathrm{m}^{-3}\,\mathrm{sec}^{-1}$. This pre-factor is the reason why homogeneous nucleation is an explosive event, as illustrated in Figure 2.4: the Boltzmann factor $\exp[-\Delta G^*/k_B T]$ in equation (2.24) increases within a small range of supersaturation from zero to a small finite value—which then multiplies the large pre-factor, resulting in an outburst of particle nucleation events.

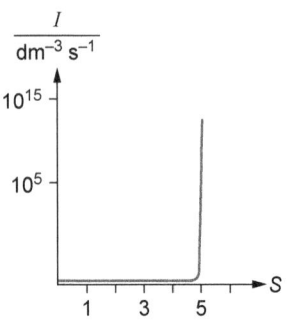

Figure 2.4 Homogeneous nucleation is an all-or-nothing event: within a narrow range of supersaturation S, the nucleation rate I jumps from practically zero to astronomic numbers. Shown here is the order-of-magnitude estimate for the nucleation rate of silica particles in the silicate solution from Exercise 2.5.

2.4 Particle growth and polydispersity

Precipitation from a supersaturated solution usually produces polydisperse colloids because nucleation of new particles and further particle growth overlap in time. This overlap is a consequence of the statistical nature of the nucleation process: near the critical size particles may grow as well as dissolve. To narrow down the initial size distribution as much as possible, nucleation should take place in a short time, followed by equal growth of a constant number of particles.

La Mer's scheme

To separate nucleation from growth, one should rapidly create the critical supersaturation required to initiate homogeneous nucleation, after which particle growth lowers the saturation sufficiently to suppress new nucleation events.

This so-called La Mer's scheme rests on the extreme sensitivity of homogeneous nucleation rates to supersaturation. An instance of La Mer's scheme is found in the double-jet precipitation of silver halide colloids, in which $AgNO_3$ and NaBr solutions are simultaneously added to an agitated gelatine solution. Here, the number of newly formed silver bromide (AgBr) crystals quickly reaches a constant value and further addition of reagents causes only further growth of fairly monodisperse cubic crystals.

Narrowing distributions

A fortunate consequence of particle growth is that in many cases the size distribution is *self-sharpening*. We will illustrate this effect for colloidal spheres of radius a, which irreversibly grow by the uptake of molecules from a solution according to the rate law

$$\frac{da}{dt} = k_0 a^n, \qquad (2.26)$$

where k_0 and n are constants. This growth equation leads to either spreading or sharpening of the relative size distribution, depending on the value of n, as can be demonstrated as follows. Consider at a given time t any pair of spheres with arbitrary size from the population of independently growing particles. Let $1+\varepsilon$ be their size ratio such that $a(1+\varepsilon)$ and a are, respectively, the radii of the larger and smaller sphere (see also Figure 2.5). The larger sphere grows according to:

$$\frac{d}{dt} a(1+\varepsilon) = k_0 a^n (1+\varepsilon)^n, \qquad (2.27)$$

which can be combined with growth equation (2.26) for the smaller sphere to obtain the time evolution of the size ratio:

$$\frac{d\varepsilon}{dt} = k_0 a^{n-1} \left[(1+\varepsilon)^n - (1+\varepsilon) \right]; \varepsilon \geq 0. \qquad (2.28)$$

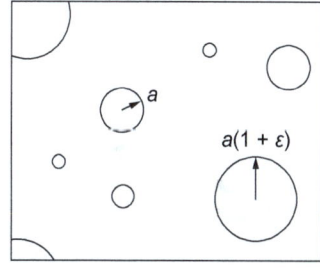

Figure 2.5 When particle radius a grows in time according to $da/dt \propto a^n$ with $n<1$, the relative size difference ε between any pair of particles decreases in time t; hence the particle size distribution sharpens in time.

Clearly, the relative size difference ε increases with time for $n>1$, in which case particle growth broadens the distribution. For $n=1$ the size ratio ε between two spheres remains constant, whereas for $n<1$ it monotonically decreases in time. Since this decrease holds for any pair of particles in the growing population, it follows that for $n<1$ the relative size distribution is self-sharpening.

Reaction limited growth

The requirement $n<1$, for self-sharpening, is in practice a realistic one. For example, when the growth rate is completely determined by a slow reaction of molecules at the sphere surface, we have

$$\frac{da^3}{dt} = k_0 a^2, \qquad (2.29)$$

implying that da/dt is a constant, so $n=0$ and self-sharpening of the distribution takes place.

Diffusion-limited growth

The opposite limiting case is growth governed by the rate at which molecules reach a colloid by diffusion. The diffusion flux for molecules with a diffusion coefficient D, relative to a sphere centred at the origin at $r=0$, is given by (2.18). We assume that the saturation concentration is maintained at the particle surface, neglecting the influence of particle size on $c(\text{sat})$ (the Kelvin effect, see Section 2.5), and also suppose that the bulk concentration of molecules is constant:

$$c(r=a) = c(\text{sat}); \; c(r \to \infty) = c(\infty). \tag{2.30}$$

For these boundary conditions, the stationary (i.e. r-independent) flux towards the sphere follows from Fick's law (2.18) as

$$J = 4\pi Da\left[c(\infty) - c(\text{sat})\right], \tag{2.31}$$

showing that the rate at which the colloid intercepts diffusing molecules is proportional to its radius and not to its surface area. Suppose every molecule contributes a volume v_m to the growing colloid; then for a homogeneous sphere the volume increases at a rate

$$\frac{d}{dt}\frac{4}{3}\pi a^3 = Jv_m, \tag{2.32}$$

which on substitution of (2.31) leads to

$$\frac{da}{dt} = Dv_m\left[c(\infty) - c(\text{sat})\right]a^{-1}, \tag{2.33}$$

with the time-dependence $a^2 \sim t$ that is typical for a diffusion-controlled process. Thus, the exponent in (2.26) for diffusion-controlled growth is $n = -1$, and consequently the relative width of the size distribution decreases in time.

2.5 Particle solubility and Ostwald ripening

The particles that have come into existence via a nucleation and growth procedure remain thermodynamically active in the sense that they attempt to reduce their total surface area, as illustrated in Figure 1.4. Recall that surface molecules embody a higher free energy than bulk molecules; hence surface area decrease is a spontaneous process with a negative ΔG. In the case of liquid droplets (e.g. of oil in water) this surface area reduction can occur via the direct merger ('coalescence') of droplets. Solid particles must employ a different route for surface area decrease and that is through exchange of material dissolved in the continuous phase—a phenomenon referred to as **Ostwald ripening**. This ripening is an important ageing effect, which may occur in any polydisperse system of sufficiently small particles. It is observed in emulsions and aqueous dispersions, as well as for colloidal metal catalysts in a high-temperature gas where ripening (referred to in catalysis as 'sintering') decreases catalytic activity in time. Let us first take a closer look at the particle solubility that underlies the Ostwald ripening process.

The Gibbs–Kelvin effect

For colloidal spheres of radius a and surface tension γ, there is one solute concentration $c(a)$ at which the colloids have a critical size and reside at the Gibbs energy maximum in Figure 2.2. This concentration, $c(a)$, called the equilibrium solubility of the colloids (see Figure 2.6), follows from the critical radius in (2.15), and is in the context of particle solubility (the Gibbs–Kelvin effect) usually written as

$$c(a) = c(\text{sat}) \exp\left(\frac{r_k}{a}\right); \quad r_k = \frac{2\gamma v_m}{k_B T}. \tag{2.34}$$

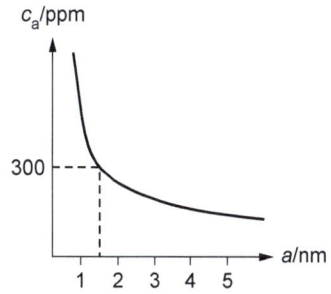

Figure 2.6 Equilibrium solubility of a sphere with radius a according to the Gibbs–Kelvin equation (2.34). For silica particles with a radius around $a = 1.5$ nm, the solubility in neutral water is about 300 ppm.

In this result, also known as the **Gibbs–Kelvin equation**, $c(\text{sat})$ is the equilibrium solubility of a flat surface. Note that (2.34) is valid only for dilute solutions: in its derivation, the chemical potential for ideal solutes in equation (2.9) is employed. The increase in solubility for small spheres (or in the vapour pressure of small droplets) is equivalent to an increasing Laplace pressure difference $\Delta p = 2\gamma/a$ across a radius of curvature a (see Chapter 10, equation (10.8)). In terms of the Laplace pressure, the Gibbs–Kelvin equation (2.34) can be rewritten to

$$c(a) = c(\text{sat}) \exp\left[\frac{v_m \Delta p}{k_B T}\right]. \tag{2.35}$$

The physical picture associated with this equation is an assembly of small particles or droplets, internally pressurized by a high Laplace pressure that forces them to release molecules into solution, giving rise to higher equilibrium solubility.

Ostwald ripening

Returning to the Gibbs energy maximum in Figure 2.2, we note that it presents an unstable equilibrium, which can be maintained only for critical particles of exactly the same size. For polydisperse particles (with the same surface tension), there is no single, common equilibrium solubility; particles either grow or dissolve. Clearly, the largest particles have the strongest tendency to grow owing to their low solubility. An obvious consequence of Ostwald ripening is loss of specific surface area, which may proceed quite rapidly for small, highly soluble particles. Illustrative examples are aqueous sols of nanometre-sized silica particles (Figure 2.7), which immediately after preparation undergo a rapid decrease in surface area on a timescale of hours to days, followed

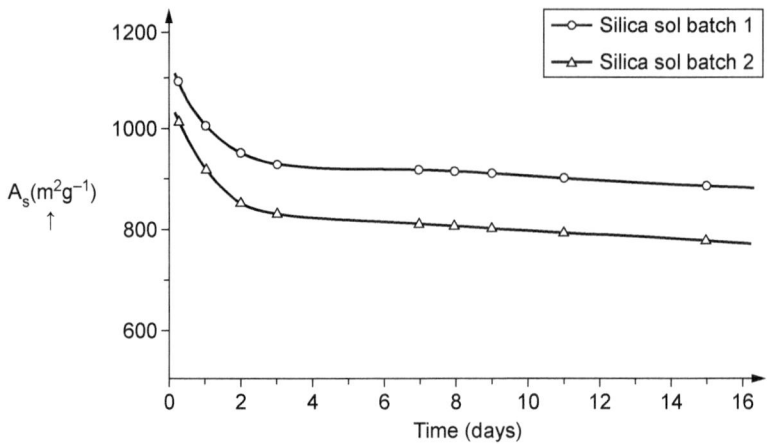

Figure 2.7 Aqueous sols of silica nanoparticles, freshly prepared by acidifying a waterglass solution, exhibit a rapid, initial decrease in specific surface area A_s, measured with the Sears method (Iler, 1979) due to Ostwald ripening. On a timescale of months, the surface area saturates at a value typically in the range of 700–800 m²g⁻¹.

Source: Courtesy of K. Larsson and B. Larsson, EKA Chemicals, Sweden.

by a much slower decay, which may continue to stabilize at 700–800 m²g⁻¹ after several months. The behaviour of small, highly soluble particles, as in Figure 2.7, illustrates that in the condensation method there is actually no sharp distinction in time between 'sol preparation' and 'ageing'. In the initial precursor solution, the specific surface area decreases as soon as precipitates start to grow.

Ripening kinetics

We will examine Ostwald ripening kinetics here only for the simple, but illustrative, case of spherical particles that gradually disappear via deposition of dissolved matter onto a flat substrate. A practical instance is silica nanoparticles in alkaline solution in the vicinity of a glass container wall. The bulk solute concentration equals $c(sat)$: the concentration that is in equilibrium with the flat substrate. The solubility $c(a)$ of the small spheres with radius a is given by the Gibbs–Kelvin equation (2.34). Assuming that in the immediate vicinity of the sphere the solute concentration remains at the value $c(a)$, a flux J will develop of monomers that diffuse from concentration $c(a)$ at distance $r = a$ from the sphere centre at $r = 0$, to the lower concentration $c(sat)$ in the bulk. Integration of Fick's first law (2.18) yields for the stationary flux J of monomers diffusing away from a dissolving sphere:

$$J = \frac{4\pi}{3v_m}\frac{da^3}{dt} = 4\pi Da[c(sat) - c(a)] \tag{2.36}$$

where we have applied the boundary conditions $c(r \to \infty) = c(sat)$ and $c(r = a) = c(a)$. For a sphere radius $a \gg r_k$, we can linearize the Gibbs–Kelvin equation to $c(a) = c(\text{sat}) \exp(r_k/a) = c(\text{sat})[1 + (r_k/a) + \ldots]$, which on substitution in (2.36) leads to:

$$\frac{da^3}{dt} = -3Dc(\text{sat})r_k v_m, \quad \text{for } a \gg r_k. \tag{2.37}$$

So in this case of Ostwald ripening in a solution of soluble spheres in contact with a flat substrate, the sphere volume decreases linearly in time and the radius of a dissolving sphere decreases in time as $a \sim -t^{1/3}$.

LSW theory

To evaluate the time evolution of a certain size distribution of growing and dissolving spheres is a demanding task dealt with in the theory of Lifshitz, Slezov, and Wagner—the LSW theory reviewed in Dunning (1973). The assumptions in this theory are the same as those underlying equation (2.37): there is only transport by diffusion and the sphere solubility is low enough to allow linearization of the Gibbs–Kelvin equation. The LSW theory predicts for the increase in average sphere radius $\langle a \rangle$, for the Ostwald ripening in a late stage, the asymptotic result

$$\frac{d\langle a \rangle^3}{dt} \sim \frac{4}{9} Dc(\text{sat}) r_k v_m. \tag{2.38}$$

Thus in the late stage of the ripening process, the average particle volume increases linearly in time and the average radius increases as $\langle a \rangle \sim t^{1/3}$. Further, the supersaturation correspondingly falls as $t^{-1/3}$ and the number of spheres as t^{-1}. The LSW theory appears to work well for emulsions; for inorganic particles, a comparison with experimental data is less straightforward. It is, in any case, not correct to use the $t^{1/3}$ scaling as the *general* hallmark for Ostwald ripening in view of the restrictive validity of the LSW theory. For example, close to the nucleation stage with many highly soluble particles, linearization of the Gibbs–Kelvin equation will be invalid and we can no longer rely on the outcome in (2.38).

2.6 Catalysed, heterogeneous nucleation

So far, we have assumed that particles nucleate and grow in a solution of one solute only. In practice, this homogeneous precipitation is difficult to realize (as already mentioned in the Introduction) because of the omnipresence of contaminants, dust motes, and irregularities on the vessel wall, which may act as nucleation sites for the new phase. Heterogeneous nucleation is an important tool in colloid synthesis as it provides a convenient way to decrease size polydispersity. The classical example here is the preparation of monodisperse gold colloids starting from the extremely fine Faraday gold sol as a seed solution (Zsigmondy and Thiessen, 1925). The seed can also differ chemically from the precipitating material, leading to the formation of core–shell colloids.

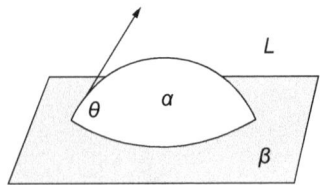

Figure 2.8 Heterogeneous nucleus of substance α in the form of a spherical cap on a planar seed β immersed in liquid L. The precipitating phase α partially wets the seed with a contact angle θ.

Examples are the heterogeneous growth of silica on gold cores and other inorganic particles, reviewed in Caruso (2001).

Catalysed nucleation

The efficiency of seeds or a container wall to catalyse nucleation is due to the reduction of the interfacial Gibbs energy of a precipitating particle. As an illustrative example we consider a phase α, which nucleates as a spherical cap of radius a on a flat seed substrate β immersed in a liquid L. The cap wets the substrate with a contact angle θ as shown in Figure 2.8. As in the case of homogeneous nucleation in Section 2.1, it is assumed here that a nucleus is a macroscopic piece of structureless bulk matter to which equilibrium thermodynamics can be applied.

When we evaluate the bulk and surface contribution (in Exercise 2.6), the total Gibbs energy ΔG_{het} for the nucleation of the cap in Figure 2.8 turns out to be

$$\Delta G_{het} = f(\theta) \Delta G_{hom}. \tag{2.39}$$

Here ΔG_{hom} is the Gibbs energy in equation (2.13) for homogeneous nucleation of a sphere of radius a, and $f(\theta)$ is a geometrical factor which equals the ratio between sphere cap volume V_{cap} and the total sphere volume (Exercise 2.6):

$$f(\theta) = \frac{V_{cap}}{(4/3)\pi a^3} = \frac{1}{4}(2 - 3\cos\theta + \cos^3\theta); \quad 0 \leq f(\theta) \leq 1. \tag{2.40}$$

So the Gibbs energy of heterogeneous nucleation is merely the result for homogeneous nucleation multiplied by the fraction $f(\theta) < 1$—a factor that only depends on the contact angle θ. This multiplication does not change the critical radius of curvature:

$$a^*_{het} = a^*_{hom} = \frac{r_k}{\ln S}, \tag{2.41}$$

but only reduces the nucleation energy barrier: the surface in Figure 2.8 acts as a true catalyst for particle nucleation. Note that for complete de-wetting, when $\theta = 180°$ and $f(\theta) = 1$, the Gibbs energy maximum equals ΔG^*_{hom}; then precipitation proceeds as if no seeds or substrates are present. For any contact angle in the range $0 \leq \theta < 180°$, the substrate lowers the activation energy for nucleation because $0 \leq f(\theta) < 1$. When the contact angle is nearly zero, the nucleation barrier is virtually absent and it will be impossible to maintain any supersaturation in the presence of the substrate.

Heterogeneous nucleation rate

In view of the strong dependence of the nucleation rate on the activation energy in (2.24), it is clear that seeds may speed up precipitation kinetics considerably. The heterogeneous nucleation rate I_{het} will have a form similar to that of the homogeneous rate I_{hom} in (2.24), and the pre-exponential factor (2.24) will remain the same in order of magnitude. Thus (Exercise 2.9):

$$I_{het} \sim I_{hom} \exp\left[(1-f(\theta))\frac{\Delta G^*_{hom}}{k_B T}\right]. \qquad (2.42)$$

For contact angles $\theta \leq 30°$, the function $f(\theta)$ in (2.42) has a value of practically zero, with the consequence that the nucleation rate is enhanced by the large factor of $exp(\Delta G^*_{hom}/k_B T)$.

Summary

- In homogeneous precipitation, small, soluble clusters nucleate by overtaking an activation free-energy barrier, after which particles enter a regime of spontaneous, irreversible growth.
- Increasing the solution's supersaturation S lowers the activation barrier for nucleation and causes faster nucleation of smaller particles.
- Particle growth-rate equations entail that size distributions of growing colloids sharpen in time. Separation of nucleation and growth time regimes also contributes to a narrow spread in size of precipitating particles.
- Homogeneous nucleation is all-or-nothing: within a narrow S window, nucleation rates jump from practically zero to astronomic values.
- Heterogeneous precipitation on a substrate is a catalysed process that merely lowers the nucleation barrier; nucleation rates are particularly strongly enhanced by substrates that are wetted by the newly formed phase.
- In Ostwald ripening, larger particles grow at the expense of smaller ones—a specific-surface-area reduction implied by the Gibbs–Kelvin equation, which says that the smaller the particles are, the more soluble they become.

References

F. F. Abraham (1974), *Homogeneous Nucleation Theory*. Academic Press, London.

F. Caruso et al. (2001), 'Multilayer Assemblies of Silica-Encapsulated Gold Nanoparticles on Decomposable Colloid Template', *Adv. Mater.* **13**, 1090–1094.

J. W. Dunning (1973), 'Ripening and Ageing Processes in Precipitates', in *Particle Growth in Suspensions* A. L. Smith (ed.),. Academic Press, London, pp. 3–28.

R. K. Iler (1979), *The Chemistry of Silica*. Wiley, New York.

A. P. Philipse and A. M. Wierenga (1998), 'On the Density and Structure Formation in Gels and Clusters of Colloidal Rods and Fibers', *Langmuir* **14** (1), 49–54.

R. Zsigmondy and P. A. Thiessen (1925), *Das Kolloide Gold*. Akademische Verlaggesellschaft, Leipzig.

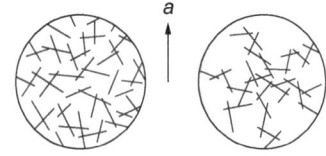

Figure 2.9 Schematic of radial clusters with outer radius a that grow by the uptake of rod-like particles from their surroundings. **Left**: particle centres are homogeneously distributed leading to a growing cluster of constant mass density. **Right**: growth of heterogeneous, fractal-like structure with an internal density profile.

Appendix 2A: Inhomogeneous particle growth—and gelation

We have studied in Chapter 2 growth of colloidal spheres that were homogeneous objects with a spatially constant mass density. Often, however, diffusional growth produces heterogeneous structures with an internal density profile (see Figure 2.9 for a schematic example). A classical case is silica precipitation at low pH, where ramified clusters are formed rather than the fully condensed SiO_2 particles at alkaline pH, owing to the low condensation rate of silanol groups at acid pH which obstructs cluster densification (Iler, 1979). Eventually growing clusters form a space-filling, rarefied network, also referred to as a **colloidal gel**. In particular, fractal clusters composed of high-aspect-ratio particles lead to remarkably low gel densities (Philipse and Wierenga, 1998).

Suppose a monomer volume fraction profile $\varphi(x)$ is present in the growing colloid, where x is the distance to its centre. Then, the rate of growth is, instead of (2.33), given by

$$\varphi(a)\frac{da}{dt} = Dv_m\left[c(\infty)-c(sat)\right]a^{-1}, \qquad (2.43)$$

because each monomer contributes a volume $v_m/\varphi(a)$ to the growing colloid upon arrival at its surface at $x=a$. When this volume contribution increases with the colloid radius (i.e. when the average mass density of the colloid decreases), the large particles in the size distribution have a gain in growth rate. This scenario will occur for the fractal clusters produced by **diffusion-limited aggregation** (DLA). Consider monomers with volume p^3 which diffuse towards a single spherical cluster with total radius a. The number of monomers, N, in the cluster scales as

$$N \sim \left(\frac{a}{p}\right)^{d_f}, \qquad (2.44)$$

where d_f is the fractal dimensionality. The average monomer volume fraction in the cluster is accordingly

$$\langle\varphi\rangle \sim \frac{Np^3}{a^3} = \left(\frac{a}{p}\right)^{d_f-3}, \qquad (2.45)$$

assuming the monomers are spheres. However, clusters may also be composed of randomly oriented fibres (Figure 2.9) or platelets. They significantly reduce the average density of a cluster, but do not necessarily change its fractal dimension. The local volume fraction at a distance x from the cluster centre is

$$\varphi(x) = \frac{d_f}{3}\left(\frac{x}{p}\right)^{d_f-3}, \qquad (2.46)$$

as can be checked (Exercise 2.12) by substitution of (2.46) into the definition of the average density:

$$\langle \varphi \rangle = \frac{3}{4\pi a^3} \int_0^a \varphi(x) 4\pi x^2 dx. \tag{2.47}$$

Thus, the volume fraction at the edge of the cluster is

$$\varphi(a) \sim \frac{d_f}{3}\left(\frac{a}{p}\right)^{d_f - 3} = \frac{d_f}{3}\langle \varphi \rangle, \tag{2.48}$$

which on substitution into (2.43) leads to the following scaling of the growth rate of the outer radius of the cluster:

$$\frac{da}{dt} \sim a^{2-d_f}; \quad a \sim t^{1/(d_f - 1)}. \tag{2.49}$$

For $d_f = 3$, we recover the square-root time dependence $a \sim t^{\frac{1}{2}}$ of diffusional growth of a homogeneous sphere—which we already met in equation (2.33). A fractal dimensionality $d_f < 3$ enhances the growth rate, but as long as $d_f > 1$, the form of da/dt is such that self-sharpening will occur. Since for three-dimensional DLA the fractal dimensionality certainly exceeds unity, it follows that self-sharpening will occur for diffusional growth of both homogeneous and heterogeneous clusters.

In the above we have examined clusters growing independently in a bulk solution. In a later stage of growth, clearly aggregation of fractal clusters themselves becomes the kinetically dominating event in the formation of large aggregates and space-filling gels.

Exercises

2.1 Estimate the Kelvin length for water molecules at room temperature.

2.2 Demonstrate that for a colloidal cube of side d the formation Gibbs energy of a critical nucleus is given by:

$$\Delta G^* = \frac{1}{3}\Delta G^*(\text{surf}) = \frac{1}{3}A^*\gamma.$$

Here A^* is the surface area of a critical cube nucleus. Discuss whether or not this result is typical for a cubical nucleus.

2.3 Consider a cylinder of length L, capped at both ends by a hemisphere of radius a (Figure 2.1). The number of molecules in the sphero-cylinder with volume V equals $N = V/v_m$, where v_m is the molecular volume.

(a) Show that the Gibbs energy for the formation of the amorphous sphero-cylinder is

$$\Delta G = 4\pi a^2 \gamma \left[\left(1+\frac{L}{2a}\right) - \frac{2}{3}\frac{a}{r_k}\left(1+\frac{3L}{4a}\right)\ln S\right].$$

(b) Show that the sphero-cylinder can only spontaneously grow when its radius a exceeds the critical value:

$$a^* = \frac{2v_m \gamma}{k_B T}\frac{1}{\ln S}.$$

2.4 Refer to Section 2.3: show that the stationary diffusion flux J of monomers from a bulk concentration c_m towards a sphere with radius a equals $J = 4\pi Dac_m$.

Discuss the assumptions underlying this result. When are they questionable?

2.5 (a) Derive that—in order of magnitude—the stationary nucleation flux equals

$$I \sim \frac{k_B T}{\eta} c_m^2 \exp\left[-\Delta G^*/k_B T\right]$$

(b) Estimate I for silica (waterglass) solutions with c_m = 200 ppm, 500 ppm, and 1,000 ppm; here 1 ppm = 1 part per million in weight = 1 mg/kg.

Take for the silica–water surface tension γ = 0.05 N m⁻¹. Further, $S = c_m/c_{sat}$; c_{sat} = 100 ppm, and V_m = 27.2 cm³/mol. Assume the solutions of 'soluble silica' consist of Si(OH)$_4$ molecules for which M = 96 g/mol.

2.6 (a) Derive the Gibbs energy formation in equation (2.39) of a spherical cap on a substrate (Figure 2.8). Make use of Young's equation $\gamma_{\beta L} = \gamma_{\alpha\beta} + \gamma_{\alpha L}\cos\theta$ for the three interfacial tensions involved.

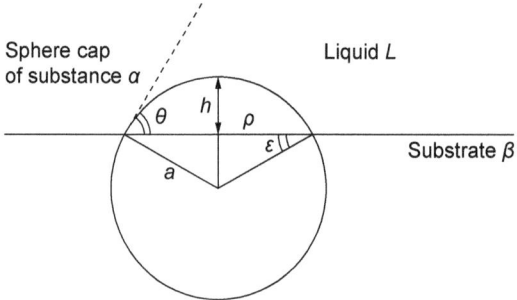

(b) Sketch the factor $f(\theta)$ by which the rate of heterogeneous nucleation differs from homogeneous nucleation, as function of the wetting angle θ. Explain the physical meaning of your sketch.

2.7 Consider the dissolution of a silica sphere with radius a, near a flat glass wall, a case of Ostwald ripening.

(a) Show how solution of Fick's law (2.18) leads to (2.37):

$$\frac{da^3}{dt} \approx -3Dc(\text{sat})r_k v_m, \quad a \gg r_k.$$

(b) Estimate da/dt for a silica sphere of radius $a = 100$ nm. Assume the sphere is dissolving in a solution of $Si(OH)_4$ molecules; $c(sat) = 100$ ppm, $V_m = 27.2$ cm^3 mol^{-1}, $M = 96$ g mol^{-1}. The typical value of the diffusion coefficient of a small molecule in water is $D \sim 10^{-5}$ cm^2 s^{-1}.

2.8 Suppose a colloidal sphere with radius a grows, as discussed in Section 2.4, at a rate given by:

$$\frac{da}{dt} = ka^n; \quad k = \text{rate constant}.$$

(a) Show that for any $n < 1$, the size distribution of a sphere dispersion sharpens in time.
(b) What is the value for n for, respectively, diffusion-controlled growth and growth governed by a reaction taking place at the sphere surface?

2.9 Verify (2.42).

2.10 The increased solubility of small particles according to equation (2.34), also referred to as the Gibbs–Kelvin effect, is not easy to quantify for sols of inorganic particles as their surface tension is difficult to determine experimentally. For an order-of-magnitude estimate of the Gibbs–Kelvin effect, we consider the case of amorphous silica in water (Figure 2.6). For the silica–water surface tension the range $\gamma = 0.05$–0.1 N/m is reported in Iler (1979). Estimate the corresponding range in equilibrium radii at $T = 298$ K, for a molar silica volume of $V_m = 27.2$ cm^3/mol, and a supersaturation of $c/c_{sat} = 3$.

2.11 Evaluate the shape parameter β in equation (2.3) for a sphere of radius a, a cube with side d, and a thin rod of diameter d and length $L \gg d$.

2.12 Verify that the profile $\varphi(x)$ from (2.44) indeed leads to the average volume fraction $\langle \varphi \rangle$ in (2.45).

Find the solutions to these exercises at the end of the book.

3 Brownian Motion

3.1 Introduction

The mass density of glass is about twice that of water, so it will come as no surprise that glass marbles quickly settle in an aqueous solution, to remain at rest once they have arrived at the bottom. But if we first fragment these marbles into tiny, sub-micron glass particles, and only then disperse them into solution, these particles would not settle but, instead, remain distributed throughout the whole solution. This is, at first sight, a puzzling state of affairs: the total glass weight has not changed, and since the glass colloids have the same mass density as the initial marbles, why would they not sink in water as well? A view through a powerful microscope would resolve the issue: glass colloids do not sediment but, instead, move and tumble around in random directions. The serenely settling marbles have been pulverized into small glass bits that participate in the dynamic pandemonium that is referred to as **Brownian motion**. It is this motion that keeps colloids distributed in a dispersion.

How long will this Brownian motion endure? Imagine that the bottle containing the colloidal glass dispersion is hermetically sealed and stored in a safe for, say, 5,000 years—about the age of the Great Sphinx of Giza. Suppose your (very distant) descendant opens the bottle and looks at it through a microscope; she will see the same pandemonium as you did: in their splendid isolation the colloids have continued to dance through the ages. Such a timescale is, however, insignificant for the air bubbles that have been observed in water inclusions in quartz; here the erratic bubble dance has endured since the Jurassic period—from about 200 million years ago, when dinosaurs were roaming our planet.

So, Brownian motion is a *very* persistent phenomenon—and that raises the question: where does the energy come from that seems to power these never-ending Brownian movements? To keep, for example, bikes or buggies in motion, we must pedal and push—that is, we must invest energy in the form of mechanical work. When we stop doing that, bikes and buggies grind to a halt, because of the resistance between internal parts of a vehicle, and friction between the moving vehicle and its surroundings. This friction involves the dissipation of mechanical energy as heat—that is, the *irreversible* distribution of energy over vast numbers of molecules in the surroundings. But that is the situation in the macroscopic world that we live in. The microscopic domain of molecules and colloids, in contrast, is an agitated realm where particles never come to a standstill. The particles *do* experience viscous friction, as a result of which they transfer kinetic energy to their environment. However, now the energy donation is *reversible*: in thermodynamic equilibrium particles gain on average in time just as much energy as they lose. The never-ending self-motion is the manifestation of a system's temperature—which gives rise to the term **thermal motion**, and the expression **thermal energy** for the energy associated with this 'heat motion' of particles. For molecules the spontaneous thermal motion is referred to as diffusion (from the Latin verb *diffundere* 'to scatter, pour out') whereas for colloids it is usually called Brownian motion; the difference between the two terms is nominal as they both denote thermal motion.

The particle size below which Brownian motion becomes significant is about a few microns. For comparison, the thickness of a human hair (Figure 3.1) is around 50 microns, and pollen grains have diameters in the range 50–100 microns. Hairs and grains are examples of **athermal** or **granular** particles that exhibit no Brownian motion.

When it comes to effects and consequences of particle size, it is instructive to compare (here and in later chapters) the four reference particles that are listed in Table 3.1: molecular *M*-sphere, nano *N*-sphere, colloidal *C*-sphere, and granular *G*-sphere. The four, jointly denoted as the particle quartet, span seven decades in particle size. The smallest quartet member is the molecular *M*-sphere that models solvent molecules as spheres of radius $R = 0.1$ nm—about the collision radius of a water molecule. The size of the *M*-sphere also represents in order of magnitude radii of ions in aqueous solutions. Nanoparticles are exemplified by the nano *N*-sphere with radius $R = 5$ nm, a characteristic dimension for metal and metal-oxide nanoparticles. For the colloidal domain the reference is the colloidal *C*-sphere with $R = 100$ nm. The largest quartet member is the granular *G*-sphere with a radius of one millimetre; it stands for

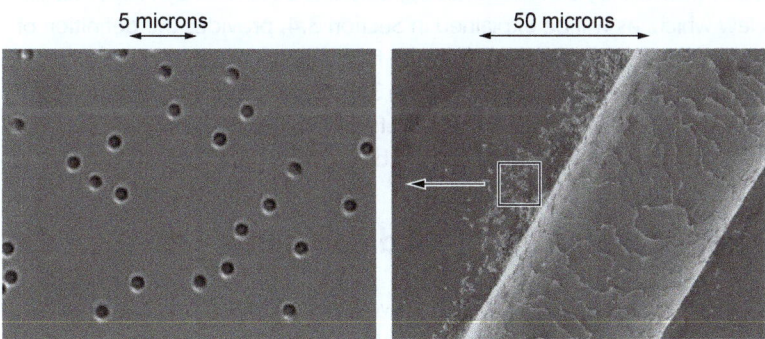

Figure 3.1 Transmission electron microscope (TEM) image (**right**) of a human hair sided by colloidal silica spheres of about one micron (to its left), that are (**left**) observed under an optical microscope. A hair is too massive to displace itself by its thermal energy, whereas the silica colloids are small enough to exhibit significant Brownian displacements.

Source: TEM image courtesy B. Erné; optical microscope image courtesy L. Rossi.

Table 3.1 Properties of the particle quartet

Unless stated otherwise, particles are immersed in water with a viscosity of $\eta = 0.89$ mPa s and temperature $T = 298$ K.

	R* nm	δ_p** g ml^{-1}	M*** g mol^{-1}
Molecular *M*-sphere	0.1		18
Nano *N*-sphere	5	2.0	6.10^5
Colloidal *C*-sphere	10^2	2.0	5.10^9
Granular *G*-sphere	10^6	2.0	5.10^{21}

* Sphere radius; 0.1 nm is about the collision radius of a water molecule.
** Mass density.
*** Molar mass; 18 g mol^{-1} is that of water.

granular matter composed of large, visible particles such as sand granules or rice grains. *M*-spheres have the same molar mass as water, and for the other quartet members, molar masses follow from a mass density equal to 2.0 g mL^{-1} —the mass density of the glass marbles that we started this Introduction with.

We will start our exploration of Brownian motion in Section 3.2 by examining the various timescales that underlie the process. This examination is needed, among other things, to understand why diffusive displacements are independent of the colloid mass—would you not have expected that more massive particles diffuse around more slowly? In fact, it is only at very short times that the colloid makes minuscule 'ballistic steps' with a length set by the colloid mass m. But after making numerous of these randomly oriented steps, the colloid enters the diffusive time regime where, as we will see in Section 3.2, it has lost all memory of its mass. The tortuous path tracked by colloids in Brownian motion is analysed via the random walk model in Section 3.3. The joint diffusive motion (the 'flux') of particles induced by a concentration gradient is described by Fick's first diffusion law which, as will be explained in Section 3.4, provides the definition of a particle's **diffusion coefficient**. The two fundamental equations that quantify Brownian motion are the subject of Section 3.5; they are referred to as *Einstein I*, which is the expression for the diffusion coefficient D, and *Einstein II*, which quantifies how the mean-squared displacement by Brownian motion grows in time.

3.2 Timescales in colloidal dispersions

When watching a video of particles in Brownian motion, the first question that might come to mind is, what are we actually looking at? Do we see forces at work that make particles change their direction of motion? And do these particles perhaps jitter around at speeds dictated by their kinetic energy? According to the equipartition principle, colloids, molecules, or any other particles for that matter all have —in thermal equilibrium—on average the same kinetic energy E_{kin}:

$$\langle E_{kin} \rangle = \left\langle \frac{1}{2}mu^2 \right\rangle = \frac{3}{2}k_B T. \tag{3.1}$$

Here angular brackets denote the average of kinetic energies of particles with mass m and speed u; k_B is the Boltzmann constant. For the typical colloidal C-sphere from Table 3.1 the microscopic speed of the colloid, as measured by its root-mean-squared (RMS) speed, is about $<u^2>^{1/2} \approx 4$ cm s^{-1}. That is not the speed at which particles move around in the Brownian motion movie. We would only observe under a microscope this RMS speed for colloids when they are present as a dilute gas, instead of being dispersed in a solution. In the gas state a colloid traverses at RMS speed the straight, free path between two collisions with other colloidal gas particles. However, when immersed in a solvent composed of densely packed molecules, the RMS speed of the colloid can only be discerned on a spatial scale that is much too small, and for times much too short, to be accessible via microscopic imaging.

Thus, to understand what is observed in Brownian motion we have in the first place to examine the timescales that are experienced by colloids in a dispersion. That examination starts with the assessment that these timescales fall into two classes (see also Figures 3.2 and 3.7): the first is that of the characteristic times that are determined by particle *mass*; the second is that of diffusive timescales that are governed by particle *size*. Let us first consider the mass-related timescales.

Molecular collision time τ_c

The fastest process in a colloidal dispersion is collision of solvent molecules with each other—and with a colloid. Since solvent molecules are closely packed together, molecules will collide when they travel a distance that is about equal to the molecular radius R. From (3.1) we can infer that the time τ_c it takes to traverse that distance R follows from

$$R \sim \tau_c \left<u^2\right>^{1/2}. \quad (3.2)$$

So, the collision time τ_c for molecules with radius R is of the order:

$$\tau_c \sim \frac{R}{\sqrt{k_B T/m}}. \quad (3.3)$$

Here the symbol ~ indicates an estimate in order of magnitude. For molecular M-particles (Table 3.1) at room temperature we find from (3.1) an RMS speed of $<u^2>^{1/2} = 370$ m s^{-1}. Taking $R = 0.1$ nm we then obtain for the molecular collision time $\tau_c \sim 2 \times 10^{-13}$ s. The colloid, which is completely at rest on this timescale, gets bombarded by solvent molecules with staggering frequencies of order $1/\tau_c \sim 10^{13}$ s^{-1}. This high frequency implies that on a timescale $t \gg \tau_c$ the colloid experiences a *continuous* fluid rather than a collection of discrete molecules. In such a fluid, the motion of a colloid is damped by the Stokes friction, and it is this viscous energy dissipation by which a colloid 'relaxes' its momentum.

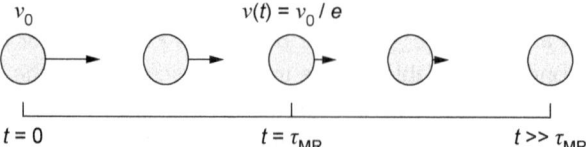

Figure 3.2 The speed $v(t)$ of a sphere decays exponentially in time due to viscous kinetic energy dissipation. At the momentum relaxation time τ_{MR} the sphere speed has decreased by a factor e from its initial value v_0. Only a drift speed of the sphere imparted by an external force will fully decay to zero. The momentum of a particle in thermal motion will only partially decay as the average speed stays at the equilibrium value determined by equation (3.1).

Momentum relaxation time τ_{MR}

Suppose a colloid is given an initial 'drift' speed v_0, and an initial momentum of magnitude $P = mv_0$ at time $t = 0$ (see also Figure 3.2). Let us compute the time τ_{MR} needed for this colloid to come to a standstill—that is, the time needed by the colloid to lose its initial momentum due to viscous energy dissipation to the solvent. Assuming the solvent is a viscous continuum, the viscous force on the sphere equals $fv(t)$, where f is the **Stokes friction factor**. Newton's second law reads:

$$F_{tot} = \frac{dP(t)}{dt}, \quad (3.4)$$

where $P(t) = mv(t)$ is the magnitude of the particle's momentum at time t. Since the total force on the sphere equals $F_{tot} = -fv(t)$ we obtain the following differential equation for the speed $v(t)$ for a particle with constant mass m:

$$m\frac{d}{dt}v(t) = -fv(t), \text{ for } t \gg \tau_C, \quad (3.5)$$

from which we find that the instantaneous speed $v(t)$ decays with time t as

$$v(t) = v_0 \exp\left[-\frac{ft}{m}\right] = v_0 \exp\left[-\frac{t}{\tau_{MR}}\right]. \quad (3.6)$$

The exponential decrease of the colloid's momentum (see also Figure 3.2) is set by a decay time, also referred to as the momentum relaxation time:

$$\tau_{MR} = \frac{m}{f} = \frac{2}{9}\frac{\delta_p}{\eta}R^2, \quad (3.7)$$

for a sphere with mass $(4/3)\pi\delta_p R^3$ and friction factor $f = 6\pi\eta R$. The appearance of the ratio m/f in (3.7) is understandable: when the colloid mass m increases it will take the colloid a longer time to expel its momentum; a viscosity increase has the opposite effect as it makes the colloid's motional energy dissipate faster. The distance $l(t)$ travelled by the sphere during the momentum-relaxation process equals:

$$l(t) = \int_0^t v(t')dt' = v_0 \tau_{MR}\left[1 - \exp(-t/\tau_{MR})\right]. \quad (3.8)$$

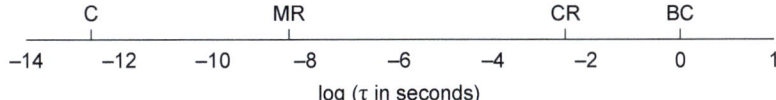

Figure 3.3 Illustration of the broad range in timescales harboured by a dispersion of the Brownian C-spheres from Table 3.1. Indicated are the molecular collision time C, the momentum relaxation time MR, the time CR needed for a change in colloid configuration, and the typical time BC for spheres to encounter each other via Brownian motion in a dispersion that contains one per cent spheres by volume.

For times much smaller than the momentum relaxation time we find from (3.8):

$$l(t) = v_0 t, \quad \text{for } t \ll \tau_{MR}. \tag{3.9}$$

This is the ballistic displacement by a sphere moving at uniform speed v_0. When the sphere has dissipated all its momentum it has, according to equation (3.8), travelled a total distance of

$$l = v_0 \tau_{MR}, \quad \text{for } t/\tau_{MR} \to \infty. \tag{3.10}$$

It should be noted that this limit only applies to an initial momentum provided by an external force, which provides a directed drift velocity that fully decays (see also Figure 3.3). Brownian particles receive thermal shocks from the environment that make them change direction: the colloids only probe an initial part of the velocity decay in (3.6), such that in thermal equilibrium their average speed remains at its equipartition value, given by (3.1).

Hydrodynamic decay time τ_{HD}

A moving colloid disturbs the surrounding fluid in two ways: it causes pressure waves that travel at the speed of sound, and the colloid motion initiates a shear wave (a flow pattern of fluid layers moving at different speeds). When a liquid layer moving in the x-direction contacts a slower layer, it transfers x-momentum to the slower layer. This momentum transfer is discussed further in Chapter 8 on viscous flow, where we will conclude that the time τ_{HD} needed for momentum to travel via a shear wave a distance R is, in order of magnitude:

$$\tau_{HD} \sim \frac{\delta}{\eta} R^2 \sim \tau_{MR}. \tag{3.11}$$

Here δ is the mass density of the fluid. This hydrodynamic decay time is comparable to the moment relaxation time τ_{MR} in (3.7) needed for a colloid to dissipate its momentum; that makes sense because viscous dissipation is primarily losing momentum via shear waves. Also, the propagation of pressure (sound) waves occurs on a timescale similar to τ_{MR}. In water, for example, the velocity of sound is 1,500 m/s so it takes about 10^{-9} s for a pressure disturbance to travel a distance of $R = 100$ nm.

Diffusive or Brownian timescale $t \gg \tau_{MR}$

When a colloid has performed many momentum-exchanging steps, it enters the diffusive time regime $t \gg \tau_{MR}$. In this regime the colloid tracks a tortuous path composed of many randomly directed, moment-exchanging steps (Exercise 3.4). The resulting net diffusive displacements will be addressed in Section 3.3 on the random walk.

Configurational relaxation time τ_{CR}

For colloids to alter their positions—their **configuration**—they must diffuse at least a distance comparable to their own radius R, as in Figure 3.4. Accordingly, the relaxation time τ_{CR} is defined as the time needed for sphere centres to inscribe an area of order R^2 by diffusion. This time is of order $\tau_{CR} \sim R^2/D$; the diffusion coefficient of the sphere depends on thermal energy $k_B T$ and solvent viscosity η as $D \sim k_B T / \eta R$ (we ignore here any constants). So, the configurational relaxation time is in order of magnitude:

$$\tau_{CR} \sim \frac{R^2}{D} \sim \frac{\eta R^3}{k_B T}. \tag{3.12}$$

Relaxation of particle configurations is extremely slow in comparison to moment relaxation. For the C-particles from Table 3.1 with $R = 100$ nm and $\ell \approx 0.1$ nm, the difference is six orders of magnitude: $\tau_{CR} \sim 5.10^{-3}$ s versus $\tau_{MR} \sim 5.10^{-9}$ s. To put this wide time span into human perspective: if colloids needed one minute to relax their momentum, they would require more than one year to change their configuration.

Brownian collision time τ_{BC}

Colloids encounter each other by Brownian motion on a timescale τ_{BC} which can be estimated as follows. Suppose a tracer sphere (Figure 3.4) with radius R diffuses in a dilute dispersion with low colloid number density ρ. The tracer

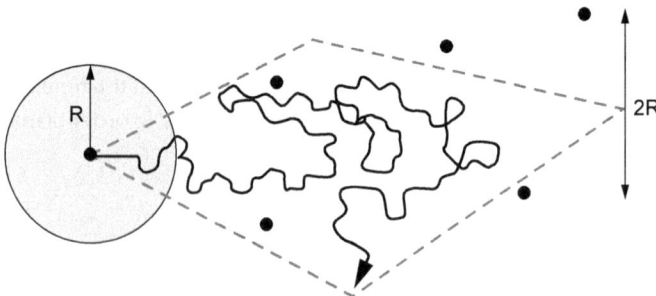

Figure 3.4 Illustration accompanying the estimate of the Brownian collision time τ_{BC} from equation (3.13). A sphere inscribes by diffusion an area of $2Dt$ per second during which it collides with other spheres that have their centres (black dots) in a volume of order DRt.

sweeps by diffusion in *t* seconds an area of the order of *Dt* square metres. Since spheres collide at a centre-to-centre distance 2*R*, the tracer scoops up in *t* seconds a 'collision volume' 2*DRt* in which it encounters about ρDRt other spheres. Therefore, the typical time between two such encounters is of the order:

$$\tau_{BC} \sim \frac{1}{\rho DR}. \tag{3.13}$$

Substituting the sphere diffusion coefficient $D = k_B T / 6\pi \eta R$ and the sphere volume fraction $\varphi = \rho(4/3)\pi R^3$ we have:

$$\tau_{BC} \sim \frac{\eta R^3}{k_B T \varphi} = \frac{\tau_{CR}}{\varphi}, \tag{3.14}$$

where τ_{CR} is the configurational relaxation time from (3.12). The *R*-dependence in (3.14) stems from the conversion of number density to volume fraction: if for a given φ we reduce the particle radius, the number density increases and particles collide at higher frequency. The timescale τ_{BC} determines the coagulation kinetics of colloids—to which we will return in Chapter 6.

3.3 The random walk

In its Brownian motion a colloid executes a sequence of uncorrelated steps (Figure 3.5) that are taken in random directions. This sequence can be modelled by a so-called random walk, also known as the drunken man's walk—like an intoxicated person attempting to reach home by stumbling in random directions. We will consider the one-dimensional case of a random walk consisting of jumps on a straight line that occur with equal probability to the left and to the right in Figure 3.6.

If the particle in Figure 3.6 makes steps at a frequency ϕ (unit of s⁻¹) then the total number of steps *N* in time interval *t* equals $N = \phi t$. The displacement after *N* steps is given by the vector:

$$\vec{x} = \ell\hat{\delta}_1 + \ell\hat{\delta}_2 + \ell\hat{\delta}_3 + \ldots \ell\hat{\delta}_N = \ell\sum_{j=1}^{N}\hat{\delta}_j. \tag{3.15}$$

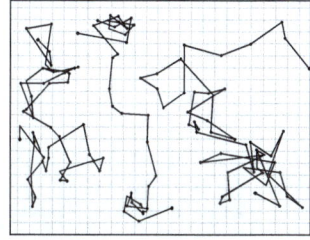

Figure 3.5 Brownian motion observed by Jean Perrin for mastic spheres (radius 0.53 microns) in water. Particle positions were marked every 30 seconds. The side of a grid square is about 3 microns.

Source: Perrin (1916).

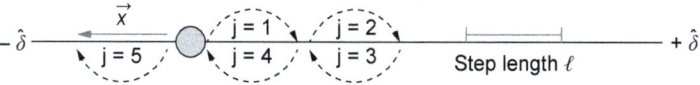

Figure 3.6 In a one-dimensional random walk a particle makes steps to the left and right with equal probability. Here $N = 5$ steps of length ℓ lead according to equation (3.15) to a net particle displacement $\vec{x} = -\ell\hat{\delta}$. Unit vectors in the +*x* and −*x* directions are, respectively, $+\hat{\delta}$ and $-\hat{\delta}$.

The unit vectors $\hat{\delta}_j$ are in either the plus or the minus x-direction (Figure 3.6), and occur with equal probability: $\hat{\delta}_j = +\hat{\delta}$ or $-\hat{\delta}$. The average value of many displacements is zero:

$$\langle \vec{x} \rangle = \left\langle \ell \sum_{j=1}^{N} \hat{\delta}_j \right\rangle = \ell \sum_{j=1}^{N} \langle \hat{\delta}_j \rangle = 0, \tag{3.16}$$

because for a random walk the average value of the unit vector $\hat{\delta}_j$ is (by definition) zero. The dot product of a vector with itself equals the square of the vector's length: $\vec{x} \cdot \vec{x} = x^2$. Hence, the average of the quadratic displacement is:

$$\langle \vec{x} \cdot \vec{x} \rangle = \langle x^2 \rangle = \ell^2 \left\langle \sum_{j=1}^{N} \sum_{k=1}^{N} \hat{\delta}_j \cdot \hat{\delta}_k \right\rangle. \tag{3.17}$$

In the double summation in (3.17) we can recognize dot products $\hat{\delta}_j \cdot \hat{\delta}_k$ with equal indices ($j = k$) and dot products with non-equal indices ($j \neq k$) such that we can split (3.17) into the following two terms:

$$\langle x^2 \rangle = \ell^2 \left\langle \sum_{k=1}^{N} \sum_{k=1}^{N} \hat{\delta}_k \cdot \hat{\delta}_k \right\rangle + \ell^2 \left\langle \sum_{j \neq k}^{N} \sum_{k=1}^{N} \hat{\delta}_j \cdot \hat{\delta}_k \right\rangle. \tag{3.18}$$

Since the dot product $\hat{\delta}_j \cdot \hat{\delta}_k$ equals 1 or −1 with equal probability, the $j \neq k$ double summation in (3.18) adds up to zero. Because $\hat{\delta}_k \cdot \hat{\delta}_k = 1$, the $j = k$ double summation equals the total number of steps, N, taken by the random, 'drunk' walker:

$$\left\langle \sum_{k=1}^{N} \sum_{k=1}^{N} \hat{\delta}_k \cdot \hat{\delta}_k \right\rangle = \sum_{k=1}^{N} \hat{\delta}_k \cdot \hat{\delta}_k = N. \tag{3.19}$$

So, the **mean-squared displacement (MSD)** equals $<x^2> = \ell^2 N$, which on substitution of $N = \phi t$ leads to the following time dependence for the MSD:

$$\langle x^2 \rangle = \ell^2 \phi t. \tag{3.20}$$

Taking the step length ℓ in Figure 3.6, equal to the momentum relaxation step l from (3.10), we have:

$$\ell^2 = l^2 \sim (v_0 \tau_{MR})^2 \sim \frac{k_B T \tau_{MR}^2}{m}, \tag{3.21}$$

and taking for the step frequency $\phi = 1/\tau_{MR}$, we find for the MSD:

$$\langle x^2 \rangle \sim \frac{k_B T}{m} \tau_{MR} t. \tag{3.22}$$

Since the momentum relaxation time equals $\tau_{MR} = m/f$, the MSD turns out to be proportional to:

$$\langle x^2 \rangle \sim \frac{k_B T}{f} t. \qquad (3.23)$$

Two important features of this MSD should be noted. First, the colloid mass m has dropped out of the equation; m has no effect at all on diffusive displacements. Consequently, one cannot determine a particle mass from tracking diffusive displacements (under a microscope or via the dynamic light scattering treated in Chapter 11): all we can determine is the particle's friction factor f, which for a sphere then yields the sphere radius. Second, the MSD by Brownian motion grows linearly in time, in contrast to the case of uniform ballistic particle motion for which the squared displacement increases *quadratically* in time, as shown in Figure 3.7. The difference is due to the circumstance that in ballistic motion all steps are added up, whereas for particles in Brownian motion steps are not only added but just as frequently subtracted, with the result that a Brownian particle migrates slower than a ballistic object.

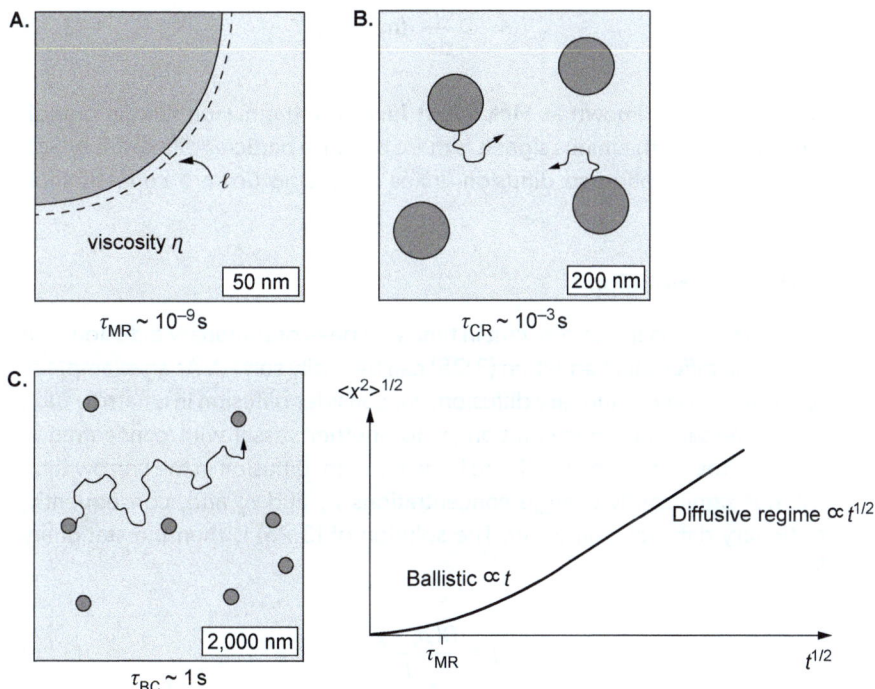

Figure 3.7 Above: In **(A)**, C-spheres (Table 3.1) relax their momentum over a molecular distance ℓ, on a timescale at which the solvent already behaves as a viscous continuum. After many relaxation steps C-spheres enter the diffusive regime where first configurations change in **(B)**, followed at longer times by Brownian encounters in **(C)**. **Below**: The RMS displacement due to momentum relaxation in **(A)** grows linearly in time; upon entering the diffusive regime the time dependence of the RMS displacement modifies to a square root of time.

We recognize in the proportionality constant in (3.23) the translational diffusion coefficient $D = k_B T / f$. The correct factor multiplying time t in (3.23) is actually $2D$, as shown in Section 3.5 and Section 3.6.

3.4 Fick's first law and the diffusion coefficient

The diffusion coefficient for a Brownian particle or molecule is defined as follows. When a particle concentration ρ changes in the x-direction, a concentration gradient $d\rho/dx$ is present which induces an x-directed, net diffusive flux j_d. The diffusion coefficient D is defined as the quotient of this flux and the gradient that drives it:

$$D = -\frac{j_d}{d\rho/dx} \quad (\mathrm{m^2\,s^{-1}}). \tag{3.24}$$

Rewritten to

$$j_d = -D\frac{d\rho}{dx} \quad (\mathrm{m^{-2}\,s^{-1}}), \tag{3.25}$$

we have what is known as **Fick's first law** for the diffusion flux (in one dimension). Note the minus sign: it is there because particles diffuse from high to low concentration so diffusion fluxes always go down a concentration gradient.

Stationary diffusion

When a diffusion flux is constant in time we speak of stationary diffusion—for which the differential equation (3.25) can be easily solved. As an example of one-dimensional stationary diffusion, we consider diffusion in a narrow tube from a vessel with concentration ρ_A to another vessel with concentration $\rho_B < \rho_A$. Provided both vessels are large enough, diffusion in the narrow tube will not significantly change concentrations ρ_A and ρ_B and, consequently, stationary diffusion will occur. The solution of (3.25) is then the stationary flux:

$$j_d = \frac{D(\rho_A - \rho_B)}{L}. \tag{3.26}$$

Here L is the tube length that runs from $x = 0$ to $x = L$. Measurement of a stationary diffusion flux for known concentrations ρ_A and ρ_B allows the determination of the diffusion coefficient D. A much more convenient and rapid method for measuring colloidal diffusion coefficients is the dynamic light scattering technique—which will be dealt with in Chapter 11.

Scent of a sphere

As an example of *radial* stationary diffusion, we consider a porous sphere containing pleasant fragrance molecules that are radially released in all directions by diffusion to the surroundings. The diffusion flux of odour molecules at a distance r from the centre of the sphere is

$$j_d = -D\frac{d\rho}{dr} \ (m^{-2} \ s^{-1}). \tag{3.27}$$

The total flux J of odour molecules through a spherical envelope with radius r is

$$J = 4\pi r^2 j_d = -4\pi r^2 D\frac{d\rho}{dr} \ (s^{-1}). \tag{3.28}$$

In a stationary state, this total flux J is independent of the distance r: the total number of fragrance molecules that leave the surface of the fragrant sphere at $r = R$ will pass through any spherical envelope with radius r. Solving equation (3.28) yields the stationary diffusion flux (Exercise 3.11):

$$J = 4\pi DR(\rho_0 - \rho_\infty). \tag{3.29}$$

Here we have employed the boundary condition that the concentrations of fragrance molecules at the surface of the porous sphere and far away from the sphere are constant in time:

$$\rho(r = R) = \rho_0; \ \rho(r \to \infty) = \rho_\infty; \ \rho_0 > \rho_\infty. \tag{3.30}$$

Note that if the bulk concentration exceeded the odour concentration in the sphere ($\rho_0 < \rho_\infty$), flux in (3.29) would change sign: the sphere would then adsorb diffusing odour molecules instead of releasing them.

3.5 The Einstein equations for Brownian motion

The trajectory of a Brownian particle is an erratic curve (Figure 3.5) with the characteristic feature that the observed distance in a time interval Δt depends on the magnification of the microscope. Thus, one cannot differentiate this distance unambiguously with respect to time to obtain a particle speed. Instead, we must consider the time dependence of the net *displacement* of a particle, defined as the shortest distance between two positions of the colloid at some time t. The MSD and the magnitude of the diffusion coefficient D were derived by Einstein, who showed that both D and the MSD can be found via Fick's first diffusion law, as will be explained in what follows.

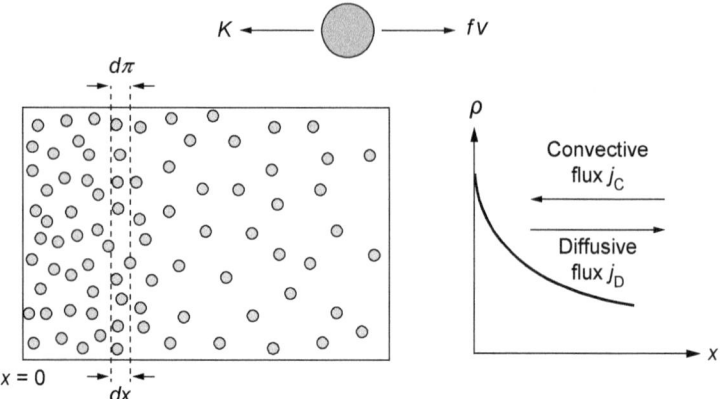

Figure 3.8 An equilibrium concentration profile of colloids resulting from an external force K on individual colloids that is balanced by an opposite osmotic force due to a concentration gradient. Force K induces a convective particle flux in the $-x$-direction, counteracted by an opposite diffusive flux. Together force and flux balance entail expression (3.38)—'Einstein I'—for the diffusion coefficient D of particles of arbitrary size and shape.

Einstein I: The diffusion coefficient

In a concentration profile of colloids that is in equilibrium the colloids experience no net force, and there is no net particle displacement. So, equilibrium can be described via a force balance as well as a flux balance—and young Einstein's ingenious insight showed that from those two balances, an expression for the diffusion coefficient can be deduced.

Force balance

Suppose a time-independent force K acts on each solute particle in the negative x-direction (Figure 3.8); then for a particle number density ρ, the force per unit dispersion volume is ρK. As ρ increases in the negative x-direction, an osmotic force develops in the opposite direction that, per unit volume, equals $-d\pi/dx$. In equilibrium the sum of forces must be zero, which implies:

$$\rho K = -\frac{d\pi}{dx}. \qquad (3.31)$$

When the colloids are ideal particles that obey Van 't Hoff's osmotic pressure law $\pi = \rho k_B T$ the force balance becomes:

$$\rho K = -k_B T \frac{d\rho}{dx}. \qquad (3.32)$$

Flux balance

Let force K impart to a particle a stationary speed v; then, according to **Stokes' law**,

$$K = fv, \qquad (3.33)$$

where f is the particle's Stokes friction factor. Thus, in the negative x-direction (Figure 3.8), under the influence of force K, a **convective flux** of particles is present with a magnitude equal to

$$j_C = \rho v = \rho \frac{K}{f} \quad (\text{m}^{-2}\,\text{s}^{-1}). \tag{3.34}$$

The diffusive flux in the positive x-direction is given by Fick's first law:

$$j_D = -D \frac{d\rho}{dx} \quad (\text{m}^{-2}\,\text{s}^{-1}). \tag{3.35}$$

In equilibrium there can be no net particle displacement, so the oppositely directed fluxes j_C and j_D must have the same magnitude:

$$\frac{\rho K}{f} = -D \frac{d\rho}{dx}. \tag{3.36}$$

Eliminating ρK from (3.36) and (3.32) we obtain

$$-\frac{k_B T}{f}\frac{d\rho}{dx} = -D\frac{d\rho}{dx}, \tag{3.37}$$

which entails that the diffusion coefficient is given by:

$$D = \frac{k_B T}{f}. \tag{3.38}$$

This derivation of the diffusion coefficient makes no assumptions on particle size or shape, nor on the type of medium in which particles are diffusing—these are features accounted for by the friction factor. For example, f could be that of a protein diffusing in a gel matrix, a clay platelet migrating in soil, or a colloid performing Brownian motion in a viscous fluid. For a sphere of radius R in a solvent with viscosity η, we can substitute the Stokes friction factor $f = 6\pi\eta R$ to obtain

$$D = \frac{kT}{6\pi\eta R}, \tag{3.39}$$

which is generally referred to as the Stokes–Einstein diffusion coefficient.

Einstein II: The mean-squared displacement

No microscopic picture of particle diffusion is assumed in the derivation of the diffusion coefficient: we merely evaluate an equilibrium colloid profile in terms of zero net force and zero net flux, from which $D = k_B T / f$ then follows. The evaluation of the *displacement*, however, requires a microscopic picture of diffusion, namely a process in which particles change positions in a totally unpredictable manner.

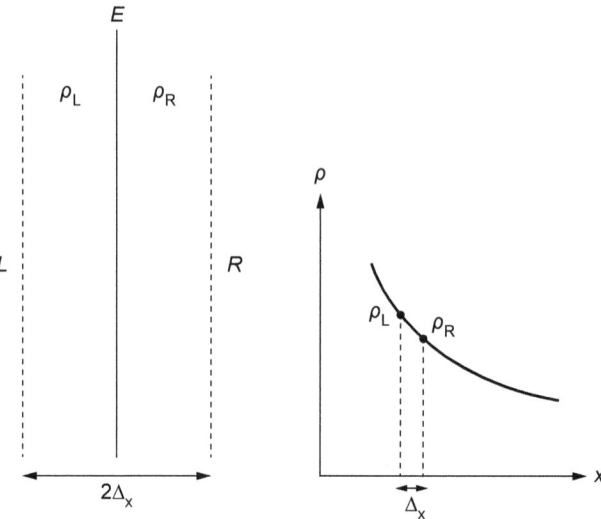

Figure 3.9 Particles migrate through plane E by diffusive jumps from region L to R and vice versa. From the net diffusion flux through E, expression (3.46)—'Einstein II'—for the mean-squared displacement is deduced in Section 3.5.

Imagine particles diffusing in the x-direction in a time interval t so short that solute concentrations hardly change. Suppose Δ_x is a typical particle displacement in the x-direction in time t, a displacement that is just as frequently positive as negative, in accordance with the just-mentioned microscopic image of diffusion. In region L left from plane E (Figure 3.9) only solutes that are within a distance Δ_x from plane E can reach it by diffusion. Since only half of the solutes in region L move towards E, the number of particles that diffuse across plane E with area A is

$$\frac{1}{2}\rho_L \Delta_x A, \tag{3.40}$$

where ρ_L is the mean solute number density in region L. Likewise, the number of particles diffusing from region R, with average number density ρ_R, in time t through E equals $(1/2)\rho_R \Delta_x A$, so the net number of particles diffusing from L to R in t seconds is:

$$\frac{1}{2}\Delta_x (\rho_L - \rho_R) A. \tag{3.41}$$

The gradient in colloid concentration is (as illustrated in Figure 3.9)

$$\frac{\rho_R - \rho_L}{\Delta_x} = \frac{d\rho}{dx}, \tag{3.42}$$

so the number of particles diffusing through plane E can also be expressed as:

$$-\frac{1}{2}\Delta_x^2 \frac{d\rho}{dx} A. \tag{3.43}$$

Consequently, the number of particles diffusing per second per unit area is given by:

$$-\frac{1}{2}\frac{\Delta_x^2}{t}\frac{d\rho}{dx} \quad (m^{-2}\,s^{-1}). \tag{3.44}$$

This diffusive particle flux due to a concentration gradient must equal Fick's first law (3.35); hence the factor multiplying $d\rho/dx$ equals the diffusion coefficient:

$$D = \frac{1}{2}\frac{\Delta_x^2}{t}. \tag{3.45}$$

Since the displacement steps are distributed, Δ_x^2 should be replaced by the average $\langle \Delta_x^2 \rangle$. Hence the Einstein II equation for the MSD is

$$\langle \Delta_x^2 \rangle = 2Dt. \tag{3.46}$$

We note here in passing that a more precise evaluation of Einstein II takes the distribution of displacements into account from the start—an evaluation that is made in Section 3.6. Since an analysis of diffusive displacements in the y- and z-directions would yield the same MSD as for the x-direction in (3.46), we can immediately write down the MSD for Brownian motion in three dimensions:

$$\langle r^2 \rangle = \langle \Delta_x^2 \rangle + \langle \Delta_y^2 \rangle + \langle \Delta_z^2 \rangle = 6Dt. \tag{3.47}$$

Here r is the magnitude of the three-dimensional position vector for the centre of the diffusing colloid. The root of the MSD (RMSD) for diffusion in three dimensions is

$$r_{rmsd} = \langle r^2 \rangle^{1/2} = (6Dt)^{1/2}. \tag{3.48}$$

The RMSD is a distance that increases in time and, therefore, we could think of differentiating this distance with respect to time to obtain an effective speed u_{eff} of diffusion:

$$u_{eff} = \frac{d}{dt} r_{rmsd} = \left(\frac{3D}{2t}\right)^{1/2}. \tag{3.49}$$

This expression, however, makes no physical sense since u_{eff} rises unlimitedly when time t approaches zero. What is going wrong here is that the expression for the MSD in (3.47) cannot be extrapolated to $t=0$, as it is only applicable on the diffusion timescale $t \gg \tau_{MR}$ (see also Figure 3.7). So Brownian motion does not yield any information on the speed of colloids—the microscopic RMS speed that follows from their average kinetic energy.

3.6 Quadratic displacements from the diffusion equation

In Section 3.5 we derived the MSD by considering particles that undergo diffusive displacements Δ_x, replacing the squared displacement Δ_x^2 in (3.45) by its average $\langle \Delta_x^2 \rangle$, to arrive at Einstein II, equation (3.46). A more precise evaluation of Einstein II takes the distribution of displacements into account from the start—a distribution that is determined by the **diffusion equation**. The derivation here is essentially that of Einstein's 1905 paper on Brownian motion (Fürth, 1956); one difference is that we do not explicitly solve the diffusion equation (3.54) for the probability density $P(x,t)$. All we need, as will become clear in what follows, are the asymptotic values that this density for physical reasons must reach.

Consider a particle diffusing for a time t to reach a (positive or negative) displacement x with respect to the particle position at $t = 0$. We assume that no external force acts on the colloid, so positive and negative displacements occur with equal probability. The average for many particles, also referred to as the *ensemble average*, of the displacement is therefore

$$\langle x \rangle = \int_{-\infty}^{+\infty} P(x,t) x \, dx = 0, \tag{3.50}$$

where $P(x,t)dx$ is the probability that after t seconds, a particle displacement is in the interval between x and $x + dx$. The function $P(x,t)$ is a probability *density* (with unit 1/m) normalized via

$$\int_{-\infty}^{+\infty} P(x,t) \, dx = 1, \tag{3.51}$$

which expresses that the probability of finding a particle *somewhere* equals one. The average of the quadratic displacement is calculated as follows. The probability of finding a particle at a certain location x is proportional to the particle concentration $\rho(x,t)$ at that location:

$$P(x,t) \propto \rho(x,t). \tag{3.52}$$

This concentration is the solution of the diffusion equation, an equation that for diffusion in the x-direction reads:

$$\frac{\partial}{\partial t} \rho(x,t) = D \frac{\partial^2}{\partial x^2} \rho(x,t), \tag{3.53}$$

where D is the diffusion coefficient. Substitution of (3.52) yields for the probability density:

$$\frac{\partial}{\partial t} P(x,t) = D \frac{\partial^2}{\partial x^2} P(x,t). \tag{3.54}$$

The MSD follows from:

$$\langle x^2 \rangle = \int_{-\infty}^{+\infty} P(x,t) x^2 dx; \text{ for } t \gg \tau_{MR}. \qquad (3.55)$$

This MSD applies for colloids that have entered the diffusive time regime; the time t is much larger than the momentum relaxation time τ_{MR} discussed in Section 3.2. We are interested in the change of the MSD with time on this diffusive timescale:

$$\frac{d}{dt}\langle x^2 \rangle = \int_{-\infty}^{+\infty} \frac{\partial}{\partial t} P(x,t) x^2 dx = D \int_{-\infty}^{+\infty} \left[\frac{\partial^2}{\partial x^2} P(x,t) \right] x^2 dx. \qquad (3.56)$$

For physical reasons the distribution function $P(x,t)$, sketched in Figure 3.10, must have the following properties. Since steps in $+x$ and $-x$ directions are equally probable, the function is symmetric such that $P(+x,t) = P(-x,t)$. In addition, moving away from the origin both the probability density $P(x,t)$ and its first and second derivatives will asymptote monotonically to zero in the limits $x \to \pm\infty$. Given these properties of $P(x,t)$, two integrations by parts of (3.56) yield (Exercise 3.7):

$$\frac{d}{dt}\langle x^2 \rangle = 2D, \qquad (3.57)$$

which results in Einstein's equation II for the average quadratic displacement:

$$\langle x^2 \rangle = 2D \int_{t_0}^{t_0+\Delta t} dt = 2D\Delta t, \text{ for } t_0 \gg \tau_{MR}. \qquad (3.58)$$

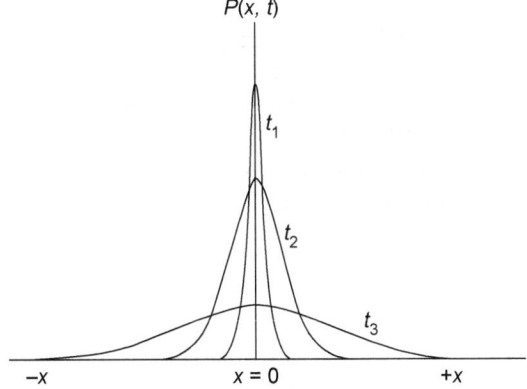

Figure 3.10 Sketch of the distribution function $P(x,t)$ at times $t_3 > t_2 > t_1$, for particles diffusing in the $+x$ and $-x$ directions, starting from position $x = 0$ at $t = 0$. $P(x,t)$ and its derivatives asymptote to zero in the limits $x \to \pm\infty$; these properties suffice to evaluate the mean-square displacement $<x^2>$.

Here Δt is a time interval in the diffusive time regime—that is, we start to monitor a particle at a time $t_0 \gg \tau_{MR}$. For a randomly diffusing colloid there is no distinction between directions x, y, and z, so $<x^2> = <y^2> = <z^2>$. Thus, the MSD for a colloid diffusing in any direction $\vec{r}$ from a central point is given by:

$$<r^2> = 6D\Delta t; \quad r^2 = x^2 + y^2 + z^2. \tag{3.59}$$

Summary

- All particles carry, irrespective of their mass, the same average kinetic energy set by temperature T; for colloids the associated speeds are high enough to make them diffuse.
- In the ballistic time regime, colloids exchange motional energy with the surrounding solvent and make low-frequency ballistic, random steps of molecular length.
- After many ballistic steps colloids enter the diffusive regime, where they track a totally erratic, non-differentiable path that can be modelled by a random walk.
- A colloid's diffusion coefficient is the ratio of the thermal energy that drives diffusion to the Stokes viscous friction that retards it.
- The root-mean-square displacement by diffusion has a characteristic $\sqrt{t}$ time dependence which entails slow spreading in comparison with ballistic or convective motion.

References

J. K. G. Dhont (1996), *An Introduction to Dynamics of Colloids*. Elsevier, Amsterdam.

R. Fürth (ed.) (1956), *Albert Einstein, Investigations on the Theory of the Brownian Motion*. Dover Publications, New York. For annotated English translations of Einstein's papers on Brownian motion.

J. Perrin (1916), *Atoms* (transl. D. L. Hammick). Constable & Company Ltd, London.

A.P. Philipse (2018), *Brownian Motion: Elements of Colloid Dynamics*. Springer Nature, Cham, Switzerland.

Exercises

3.1 **(a)** Calculate how far a marble with radius $r = 1$ cm diffuses in water over the course of a century ($T = 298$ K).

(b) One often reads that, in 1827, Robert Brown observed the Brownian motion of pollen grains. Pollen grains have diameters in the range 50–100 microns. Suppose you monitored these grains in water under a microscope for one hour: how far do the grains diffuse in that hour? Do you think Brown could have seen pollen grains diffusing around?

3.2 (a) Using equipartition, calculate the RMS speed of a sphere with radius 100 nm and mass density $\delta_p = 1.5$ g cm^{-3} at $T = 298$ K.
(b) How large is the distance the sphere would traverse in one second with this velocity in uninterrupted linear motion?
(c) How large is the RMSD in one second in the case of Brownian motion of the sphere (in the x-direction in water with viscosity $\eta = 0.89$ mPa s)? Explain the difference from the outcome in (b).

3.3 Compute for all four particles in Table 3.1 (molecular M-sphere, nano N-sphere, colloidal C-sphere, and granular G-sphere) the configurational relaxation time from equation (3.12).

3.4 (a) Estimate how many momentum relaxation steps a C-sphere has made within the time the sphere has diffused a distance equal to its radius.
(b) What is the total distance covered by these relaxation steps?

3.5 Calculate the work w done by the friction force $fv(t)$ on a sphere with mass m during momentum relaxation. Hint: recall that work w is the integral of a force (here the frictional force) over a distance—here the distance traversed by the slowing sphere.

3.6 Compute the time τ_{CR} taken by spheres of radius $R = 10, 10^2, 10^3$, and 10^4 nm to diffuse (in water at room temperature) an MSD equal to R^2.

3.7 Solve the integral in (3.56) to obtain (3.57).

3.8 Verify the limiting cases (3.9) and (3.10) of the expression (3.8) for the distance $l(t)$.

3.9 Calculate the moment relaxation time τ_{MR} for the standard colloidal C-sphere (radius 100 nm).

3.10 (a) Verify the integration in equation 3.8 and provide an estimate for the step length ℓ a colloid makes during relaxation of its moment.
(b) Estimate ℓ for a C-sphere. Comment on the outcome.

3.11 Verify that (3.29) is the solution of differential equation (3.28).

Find the solutions to these exercises at the end of the book.

4 Van 't Hoff's Law, Donnan Equilibria, and the Depletion Force

4.1 Introduction

Colloidal dispersion and solutions contain thermally agitated particles, forever dancing around in the process of Brownian motion that we studied in Chapter 3. The presence of thermal solute particles has two other consequences for a solution, and those are the **osmotic pressure** the solution exerts due to the thermal solutes, and the propensity of the solution to draw in pure solvent—an instance of solvent migration known as **osmosis**. First we look at the latter consequence, for the case of water, which will in the large majority of practical cases be the solvent that we are dealing with.

Osmosis is of the greatest biological importance: water intake by plants and trees occurs by osmosis, and osmotic forces are also the prime determinant of water distribution in our body. The free passage of water through all cell membranes entails that body fluids are in osmotic equilibrium. Osmosis is also crucial for the water management of biological cells. If, for example, the osmotic pressure inside a red blood cell (RBC) is much higher than outside, the cell may rupture—as happens for RBCs dispersed in pure water. If, on the other hand, the RBCs are immersed in a solution of high salinity, they may shrivel due to water osmosis from the cell's interior to the surroundings. Incidentally, osmotic dehydration by a salt solution is also applied to shrivel cucumbers and tomatoes in brine, and to increase the shelf life of fruit and vegetables, since the main cause of their perishability is their high water content.

One way to realize osmosis experimentally is to bring solutions into contact with pure water via so-called **semi-permeable membranes** (porous sheets of material that allow water molecules to pass but that block solute particles). Since water molecules will have the spontaneous tendency—in the thermodynamic sense that they act wholly by themselves—to migrate from zones of high water concentration to regions of lower ones, an osmotic water flow will arise, from pure water into the solution. Now when water forces its way into a solution, the hydrostatic pressure in that solution increases, and this increase forces water molecules in reverse out of solution, thereby counteracting osmosis. In osmotic equilibrium the two tendencies cancel and no net water displacement

COLLOID SCIENCE   51

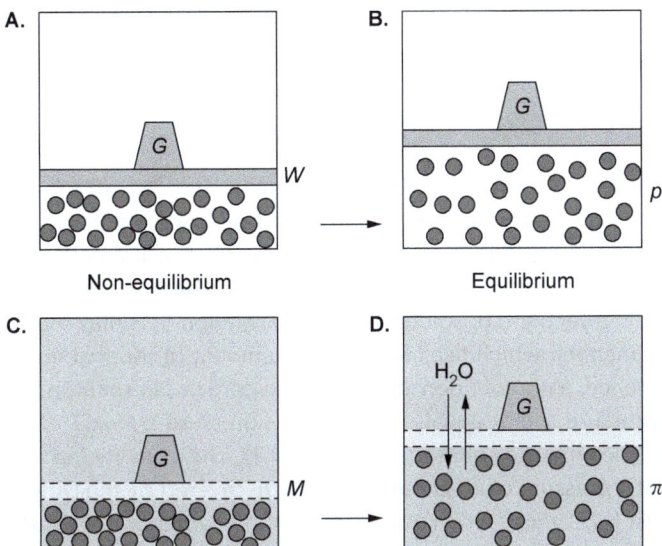

Figure 4.1 **(A)** Thermal particles in a gas push up a solid piston W until **(B)** gas pressure p equals the pressure exerted by weight G on the movable piston. **(C)** The same thermal particles in an aqueous solution force a membrane M (permeable to water) upwards, again until **(D)** osmotic pressure π equals the pressure from weight G. Going from **C** to **D**, water migrates into the solution until thermodynamic water equilibrium is reached **(D)** between pure water and solution. Additionally, two thermodynamic equilibria exist in situations **B** and **D**, namely a homogeneous particle distribution far away from the membrane, and a Boltzmann distribution of particles near that membrane. All three equilibria entail Van 't Hoff's osmotic pressure law.

occurs. Take a look, for example, at Figure 4.1C, showing a solution that is pressurized by a weight placed on the semi-permeable membrane. Osmosis will take place, and being a spontaneous process it can deliver work—here by lifting the weight G against gravity. In the equilibrium situation in Figure 4.1D, the effects of osmosis and external pressure precisely cancel. In the osmotic equilibrium sketched in Figure 4.1D, there is also mechanical equilibrium: the external pressure from the weight equals the osmotic pressure executed by the solute particles—just as in Figure 4.1B the weight on a piston equals (per unit area) gas pressure p.

With respect to osmotic pressure, we anticipate that it rises in proportion to the thermal energy $k_B T$ of the solutes; let us work out the proportionality as follows. Suppose the pressure of a system (either a solution or a gas) increases by an amount π upon addition of N ideal particles. Since the potential energy of interaction between ideal particles is, by definition, negligible, the energy associated with the N particles only comprises kinetic energy. Since the latter is proportional to $k_B T$ per particle, the energy increase of the system is proportional to $N k_B T$. Pressure is an energy density, that is, an energy per volume;

hence the pressure increase is proportional to the energy increase Nk_BT per volume V:

$$\pi \propto \rho k_B T; \quad \rho = \frac{N}{V}. \tag{4.1}$$

Here ρ is the number density of solute particles. **Van 't Hoff's law** for osmotic pressure states that $\pi = \rho k_B T$, so the constant of proportionality in (4.1) actually equals unity. For the argumentation leading to the proportionality in (4.1), it should be noted, it is irrelevant whether thermal particles are gas molecules in a vacuum, or solute particles in solution. It is only the temperature that matters, which fixes the average amount of thermal energy per particle. Hence, a number density ρ of solute particles in solution exerts an osmotic pressure π that equals the ideal-gas pressure $p = \rho k_B T$ exerted by those particles in a gas phase (see Figure 4.1). The equality between gas and osmotic pressure of ideal particles was discovered by the Dutch physical chemist J. H. Van 't Hoff (1852–1911), who now lends his name to the ideal pressure law.

In Section 4.2 we will derive this law from the osmotic equilibrium discussed above and depicted in Figure 4.1D. We find the pressure law here via a thermodynamic route along which the kinetics of colloids, and the forces they exert, remain totally out of sight. Those forces are addressed in Section 4.3, which shows how Van 't Hoff's law follows from applying Newton's second law to colloids ballistically stepping around on very short timescales. A third route to Van 't Hoff's law employs the semi-permeable membrane that blocks colloids that, consequently, undergo a repulsive force which, as detailed in Section 4.4, entails the osmotic pressure.

Van 't Hoff's law is applied in Section 4.5 to evaluate maximal pressures exerted by charged colloids; salt addition decreases this pressure from charged particles—a decrease to be studied in Section 4.5 via the so-called **Donnan equilibrium**. Differences ('gradients') in pressure correspond to net osmotic forces. Section 4.6 examines an instance of such a net force—the **depletion force** that is operative between colloids dispersed in a polymer solution.

4.2 Osmotic pressure from osmotic equilibrium

Spontaneous osmosis

The expansion of the colloidal dispersion going from Figure 4.1C to Figure 4.1D requires osmosis, the migration of water through the membrane into solution. We evaluate the change in chemical potential μ_s for water molecules undergoing this osmosis as follows. For a solution of ideal particles, the chemical potential of the *solvent* is by definition:

$$\mu_s = \mu_s(x_s = 1) + RT \ln x_s, \tag{4.2}$$

where R is the **gas constant**, x_s is the solvent mole fraction and $\mu_s(x_s = 1)$ the chemical potential of pure solvent. So, upon osmosis into the solution, the solvent chemical potential changes with an amount:

$$\Delta\mu_s = \mu_s(x_s = 1) + RT \ln x_s - \mu_s(x_s = 1) = RT \ln x_s. \quad (4.3)$$

Since $x_s < 1$ we have $\ln x_s < 0$, so the chemical potential decreases, $\Delta\mu_s < 0$, which signals that osmosis is a spontaneous process—one that occurs by itself without any external assistance.

Hydrostatic pressure increase

Solvent molecules migrating in the set-up of Figure 4.1C from solvent to solution experience an increase in hydrostatic pressure, caused by the weight G placed on the membrane. The associated change $\Delta\mu_p$ of the solvent chemical potential is as follows. Solvent and solution are incompressible substances: their volume V is independent of the applied pressure. Imagine we increase pressure p on an incompressible solution from an initial value p_i to a final state with value p_f following the gravitational route sketched in Figure 4.2. Weight G_i is brought from height $h = 0$ to height h, on a piston resting on the incompressible solvent. The required work equals the increase in potential energy of the weight in the gravity field:

$$w_i = G_i h = p_i A h = p_i V, \quad (4.4)$$

where A is the piston area and V the solution volume. Next, the final state is established by lifting a weight G_f to height h, which involves the work

$$w_f = G_f h = p_f V. \quad (4.5)$$

So the reversible work needed to go from initial to final state is

$$w_p = w_f - w_i = (p_f - p_i)n\bar{V}; \quad V = n\bar{V} \quad (4.6)$$

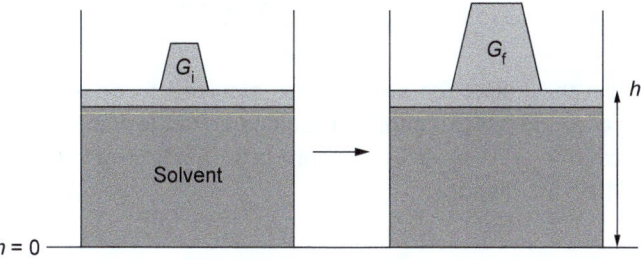

Figure 4.2 Work done to increase the pressure on an incompressible solvent is equivalent to the work needed to lift extra weight $G_f - G_i$ to height h. The corresponding chemical potential increase of the solvent is given by equation (4.7).

for n moles of incompressible solution with molar volume $\bar{V}$. Hence the increase $\Delta\mu_p$ of the solvent's chemical potential (reversible work per mole) is

$$\Delta\mu_p = \frac{w_p}{n} = \bar{V}(p_f - p_i). \tag{4.7}$$

In the membrane set-up in Figure 4.1 the pressure difference $p_f - p_i$ between solution and solvent reservoir is the hydrostatic pressure $p_{hydro} = G/A$ exerted by weight G on membrane area A. Hence:

$$\Delta\mu_p = \bar{V} p_{hydro}. \tag{4.8}$$

Solvent migration to regions of higher pressure is evidently a non-spontaneous, forced process since the chemical potential increases: $\Delta\mu_p > 0$.

Osmotic equilibrium

The total chemical potential change for water moving (as in Figure 4.1C) from pure water into solution is the sum of equations (4.3) and (4.8):

$$\Delta\mu_{tot} = \Delta\mu_p + \Delta\mu_s = \bar{V} p_{hydro} + RT \ln x_s. \tag{4.9}$$

When, as in Figure 4.1D, osmotic equilibrium has been reached—that is, when pure water is in thermodynamic equilibrium with water in the solution—then $\Delta\mu_{tot} = 0$. Moreover, in equilibrium p_{hydro} equals the solution's osmotic pressure π. Upon substitution of $\Delta\mu_{tot} = 0$ and $p_{hydro} = \pi$ in (4.9) we obtain for the osmotic pressure:

$$\pi = -\frac{\ln x_s}{\bar{V}} RT = -\frac{\ln(1-x_p)}{\bar{V}} RT. \tag{4.10}$$

Here x_p is the mole fraction of solute particles:

$$x_p = \frac{n_p}{n_p + n_s} = \frac{n_p}{n_s}, \quad \text{for } n_p \ll n_s. \tag{4.11}$$

Since from the outset the solution was assumed to be very dilute, the number n_p of solute moles is negligible compared to the number n_s of solvent moles. Because $x_p \ll 1$ we can linearize the logarithmic term to $\ln(1-x_p) \sim -x_p$ such that (4.10) modifies to

$$\pi = \frac{n_p}{n_s \bar{V}} RT = \frac{n_p}{V} RT = \rho k_B T, \tag{4.12}$$

which is Van 't Hoff's law for a number density ρ of ideal, non-interacting solute particles in a solution with volume V and temperature T.

Intermezzo: reverse osmosis

We have evaluated the hydrostatic pressure on the solution in Figure 4.1 required to stop water osmosis into the solution such that osmotic equilibrium is established. What happens if the solution is put under a pressure that exceeds π? Then water will be forced to migrate from solution to solvent—a process for obvious reasons referred to as **reverse osmosis** (RO). The RO process is employed to extract potable water from seawater, using specially designed polymeric membranes that retain salt molecules. The osmotic pressure exerted by the saline water from the North Sea is about 27 bar—so very high pressures are required for desalination, which makes RO an expensive and energy-consumptive separation method. It is nevertheless an important method, since millions of people depend for their drinking water on seawater desalination performed by coastal RO plants.

4.3 Van 't Hoff's law from particle kinetics

The thermodynamic route that we followed in Section 4.2 unfortunately says nothing about the particle kinetics that underlie the pressure. We just had to figure out how heavy the weight in Figure 4.1 should be to stop osmosis; that weight per area equals the osmotic pressure—then we know the pressure executed by a number density ρ of solute particles: Van 't Hoff's law. To bring particle kinetics into the picture, we have to recall from Chapter 3 on Brownian motion that the typical timescale on which a colloid, so to speak, displays its momentum (mass × speed) is the momentum-relaxation timescale τ_{MR}. Colloids with mass m and friction factor f make ballistic steps that are of molecular size, on a timescale $\tau_{MR} = m/f$. For example, for the colloidal C-sphere from Table 3.1, step length ℓ and relaxation time are in order of magnitude $\ell \sim 0.1$ nm and $\tau_{MR} \sim 10^{-9}$ s. In thermal equilibrium, the homogeneously distributed colloids in the bulk make these ballistic steps in all directions with equal probability. Thus for every colloid stepping in direction $L \to R$ through area A in Figure 4.3, one colloid steps in direction $R \to L$, with on average the same but opposite x-component of the velocity. The change ΔP_x in x-directed momentum due to one colloid going from L to R and one going in reverse is:

$$\Delta P_x = mv_x - (-)mv_x = 2mv_x. \quad (4.13)$$

If the momentum exchange lasts Δt seconds, the associated average force in this time interval is, according to Newton's second law,

$$F_{L \to R} = \frac{\Delta P_x}{\Delta t} = \frac{2mv_x}{\Delta t}. \quad (4.14)$$

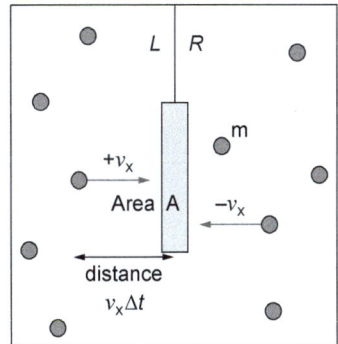

Figure 4.3 Colloids make tiny ballistic steps through a suspended wire from left to right and vice versa, resulting in a force on both sides of area A that per unit area is given by Van 't Hoff's osmotic pressure law.

The number of colloids, pacing in direction $L \to R$ with velocity component v_x, that reach the wire within Δt seconds is

$$\frac{1}{2}\rho A v_x \Delta t, \qquad (4.15)$$

where ρ is the colloid number density; the factor ½ appears because only half of the colloids step in direction $L \to R$. The pressure exerted by the colloids going from L to R equals total force per unit area:

$$\pi_{L \to R} = \frac{1}{2}\rho A v_x \Delta t \times F_{L \to R} \times \frac{1}{A} = \rho m v_x^2. \qquad (4.16)$$

In equilibrium the pressure $\pi_{L \to R}$ equals the opposite pressure $\pi_{R \to L}$ so the subscript $L \to R$ in (4.16) is superfluous. In equilibrium, moreover, particle velocities adopt the Maxwell–Boltzmann distribution, so v_x^2 should be replaced by its equilibrium average (that you are asked to compute in Exercise 4.17):

$$<v_x^2> = \frac{k_B T}{m}. \qquad (4.17)$$

Hence, equation (4.16) entails the pressure exerted by ideal, non-interacting particles:

$$\pi = \rho k_B T. \qquad (4.18)$$

Particle ideality, it should be noted, is assumed in (4.16), which comprises a linear sum of forces exerted by non-interacting particles.

4.4 A membrane pressure probe

The spontaneous tendency of solute particles to spread—via Brownian motion—into the solution can be prevented by an impervious wall, such as the membrane in Figure 4.1, that exerts a force on the colloids. That force, per unit area, equals in magnitude the solutes osmotic pressure. The use of a wall to measure osmotic pressure is analogous to employing a (vertically mobile) piston to gauge gas pressure (see Figure 4.1): if the weight on the piston prevents gas expansion, then the weight, per unit area, equals the gas pressure.

Suppose a membrane of thickness $x = 2\delta$ has its centre plane located at $x = 0$ (Figure 4.4). Colloids that arrive by diffusion at location $x \approx \delta$ experience a steeply rising repulsive force. Moving away from the membrane the repulsion exerted by the membrane decays, to vanish at distances $x \gg \delta$. The equilibrium distribution of colloids diffusing in the membrane's force field is the Boltzmann distribution

$$\rho(x) = \rho \exp\left[-V_{MF}(x)/k_B T\right]. \qquad (4.19)$$

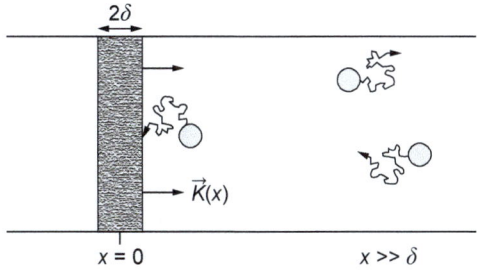

Figure 4.4 A semi-permeable membrane exerts a force $K(x)$ on colloids that is steeply repulsive for colloids approaching the membrane within a distance $x \approx \delta$, and that vanishes sufficiently far away from the membrane. These two features of $K(x)$ suffice to calculate in Section 4.4 the osmotic pressure exerted by the colloids.

Here $\rho(x)$ is colloid number density at distance x from the membrane, ρ is the bulk number density at $x \to \infty$, and $V_{MF}(x)$ is the potential of mean force (i.e. the reversible work needed to bring a colloid from infinity to position x, averaged over all configurations of surrounding molecules). The boundary conditions on this potential of mean force are

$$\frac{V_{MF}(x \to \infty)}{k_B T} = 0; \quad \frac{V_{MF}(x \leq \delta)}{k_B T} = \infty. \tag{4.20}$$

Associated with the potential $V_{MF}(x)$ is a mean force defined by:

$$K_{MF}(x) = -\frac{dV_{MF}(x)}{dx}. \tag{4.21}$$

The force F_{MF} on colloids in a slab with thickness dx and area A at a distance x from the membrane is:

$$F_{MF} = K_{MF}(x)\rho(x)A\,dx. \tag{4.22}$$

From equations (4.19), (4.21), and (4.22) we find for the total force on all colloids due to the membrane:

$$F_{tot} = A\int_0^\infty K_{MF}(x)\rho(x)\,dx = -A\rho\int_0^\infty \exp[-V_{MF}(x)/k_B T]\,dV_{MF}(x). \tag{4.23}$$

On performing—in Exercise 4.18—the integration in (4.23) and applying the boundary conditions (4.20) we obtain:

$$F_{tot} = A\rho k_B T. \tag{4.24}$$

According to Newton's third law, the total force (4.24) exerted by the membrane on the colloids directed into the solution equals in magnitude the total force exercised by the colloids on the membrane. Therefore the excess pressure π exerted by Brownian particles on the membrane equals:

$$\pi = \frac{F_{tot}}{A} = \rho k_B T. \qquad (4.25)$$

It should be noted that the potential of mean force $V_{MF}(x)$ brought about by the membrane remains unspecified; all we need to know are its boundary values in (4.20). The membrane's only function is to impose a force field on the colloids that probes their pressure.

Diffusion coefficient

Interestingly, this force field entails not only that colloid pressure equals $\rho k_B T$ but also that the colloid diffusion coefficient equals $D = k_B T/f$, for particles with friction factor f in a viscous fluid. This follows from the stationary convective colloid flux, imparted by the force $K_{MF}(x)$ in (4.21), as investigated in Exercise 4.20.

Second virial coefficient

The 'membrane pressure probe' from Figure 4.4 not only entails Van 't Hoff's law for ideal non-particles but also yields the first correction term to Van 't Hoff's law, the term that takes the interaction between two colloids into account. In the set-up in Figure 4.4, N colloids could, for the derivation of (4.3), access the whole dispersion volume V. Now suppose that each colloid centre creates an excluded volume V_{ex} that cannot be accessed by other colloid centres. Then instead of volume V, the N colloids diffuse in a smaller available volume $V - (1/2)NV_{ex}$ where they exert a pressure:

$$\frac{\pi}{k_B T} = \frac{N}{V - (1/2)NV_{ex}} = \frac{\rho}{1 - (1/2)V_{ex}\rho} \sim \rho + (1/2)V_{ex}\rho^2, \text{ for } V_{ex}\rho \ll 1. \qquad (4.26)$$

Here the colloid number density ρ is assumed to be low enough such that only simultaneous interaction of two colloids occurs: a pair interaction that gives rise to the quadratic term ρ^2, in (4.26). The coefficient multiplying this term is the **second virial coefficient** B_2:

$$\frac{\pi}{k_B T} = \rho + B_2 \rho^2; \quad B_2 = \frac{1}{2}V_{ex}, \qquad (4.27)$$

where the **excluded volume** is defined as

$$V_{ex} = 4\pi \int_0^\infty (1-\beta)r^2 dr; \quad \beta = \exp[-w(r)/k_B T]. \qquad (4.28)$$

To rationalize (4.28), note that the Boltzmann exponent β measures the probability of finding a colloid centre in a shell $4\pi r^2 dr$; therefore $1 - \beta$ is proportional to the probability that the shell volume is empty, so the volume integral of $1-\beta$ represents the total excluded volume. Repulsive colloids have a positive excluded volume (Exercise 4.7) and the pressure increase can be understood as the result of isothermal compression of N ideal particles from volume V to a smaller volume $V-(1/2)NV_{ex}$. For attractive colloids, excluded volume and second virial coefficient are negative; the ensuing pressure decrease (studied in Exercise 9.11 for the case of magnetic colloids) can be qualitatively explained by the formation of temporary dimers that reduces the number of particles.

Molecular mass determination

Osmotic pressures, at least for ideal particles, increase linearly with the number of particles in a dispersion. Hence, measuring osmotic pressure comes down to counting of particles—and for a known total colloid mass in the dispersion, the particle mass then follows. In terms of mass concentration $c = \rho m$ of a number density ρ of solutes with mass m, we can rewrite (4.27) to:

$$\frac{\pi}{cRT} = \frac{1}{M}\left(1 + B_2 \frac{c}{m}\right). \tag{4.29}$$

Here $R = k_B N_{av}$ is the gas constant and $M = mN_{av}$ the solute's molecular mass. So from measurements (from a membrane osmometer) of π/c as function of colloid mass concentration c, we find the colloid mass by (linear) extrapolation to zero concentration. The sensitivity of this determination decreases with increasing molar mass: above molar weights of order $M \sim 10^5$ g mol^{-1} osmotic pressures are too low for their accurate determination, and another method should be applied such as ultracentrifugation (discussed in Chapter 7, Section 7.2).

4.5 The Donnan equilibrium

The osmotic pressure exerted by uncharged colloids is generally low because of their large molecular weight and correspondingly low number densities for a given weight concentration. However, for *charged* colloids pressures may increase considerably due to the counter-ions they release in solution: each ion is a 'kinetic unit' carrying on average the same kinetic energy as the colloid and, hence, contributes just as much to the pressure as the colloid does. For a first estimate of osmotic pressure from charged colloids or polyelectrolytes, consider particles with number density ρ_c that each discharge z counter-ions into solution. Application of Van 't Hoff's law to colloids and ions leads to the total osmotic pressure:

$$\frac{\pi_{tot}}{k_B T} = \rho_c + \rho_c z = (1+z)\rho_c. \tag{4.30}$$

As the valency z of a typical polyelectrolyte easily amounts to a few hundred, the pressure increase may indeed be huge. The pressure in (4.30), as we will demonstrate later, is actually the *maximal* pressure that charged colloids can exert, a maximum achieved when the colloids diffuse in a salt-free solution—a solution where only counter-ions are present, which is why (4.30) is also

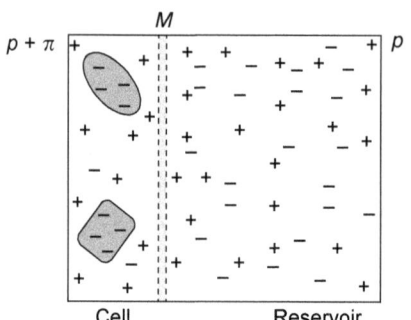
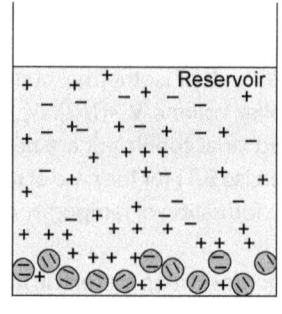

Figure 4.5 Left: An osmotic cell in contact via semi-permeable membrane M with a large volume of salt solution; the latter constitutes a reservoir of constant salt concentration. The charged colloids that cannot pass the membrane are of arbitrary size and shape. **Right**: A Donnan equilibrium in the form of a sediment of colloids in contact with a supernatant solution which acts as the salt solution reservoir. Left it is the membrane that confines the colloids; right it is their own weight.

referred to as the pressure in the **counter-ions-only limit**. When salt is added to the solution osmotic pressures decrease, and at sufficiently high salinity the whole effect of colloidal charge will be undone, resulting in the low pressure $\pi_{\text{tot}} = \rho_c k_B T$ from uncharged colloids. The effect of salt on osmotic pressure can be modelled via the Donnan equilibrium, which we will now address.

The Donnan cell

The Donnan equilibrium (DE) is the thermodynamic salt equilibrium between a dispersion of charged particles and a salt solution. The charged particles are prevented from diffusing into this solution by a semi-permeable membrane or an external force such as gravity (Figure 4.5). The region where colloids are imprisoned is called a **Donnan cell** and the solution outside the cell is named the reservoir. Compared to the cell the reservoir is very large, such that its number concentration ρ_s of (1:1) salt molecules remains constant, which fixes the reservoir's osmotic pressure to

$$\pi_s = 2\rho_s k_B T. \tag{4.31}$$

We note here in passing that when the reservoir is small in comparison to the cell—as in the case of the set-up depicted in Figure 4.6—the reservoir salt concentration increases upon addition of charged colloids to the cell. From this increase the colloid charge number z can be deduced—a topic that you are invited to examine further in Exercise 4.13.

Zero-field approximation

Charged colloids are surrounded by **electrical potential** profiles that decay from surface potential Ψ_0 to $\Psi = 0$ far away from the colloids—we will examine these profiles in Chapter 5. Thus, a dispersion of diffusing, charged colloids

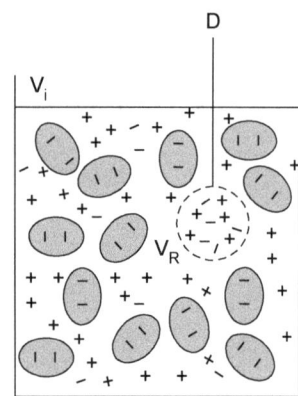

Figure 4.6 Donnan equilibrium between a small salt solution volume V_R separated from a large suspension volume $V_s \gg V_R$ by a semi-permeable membrane (dashed line). The volume V_R adsorbs salt from the suspension in response to mobile counter-ions released by the charge carriers; the relation between salt adsorption and colloid charge is the subject of Exercise 4.13.

harbours an electrical potential landscape that fluctuates in space and time around its average value that is referred to as the **Donnan potential**. The key simplification in the DE is to ignore these fluctuations, and assume that colloids and ions all diffuse in the same spatially constant Donnan potential. In other words: in a Donnan cell as in Figure 4.5 the electric field is taken to be zero. One example of a Donnan cell that we will encounter later is the Debye cube in Chapter 5, Appendix 5A. Other practical instances of charged colloids that can be modelled via a DE are the RBCs that contain charged haemoglobin proteins in equilibrium with blood serum, sediments of charged colloids (Figure 4.5) equilibrating with a reservoir in the form of a supernatant salt solution, and two charged surfaces that closely approach each other—an instance of the DE that we will meet again in Chapter 5, Section 5.5.

Donnan pressure

The total osmotic pressure of a Donnan cell, in excess to the reservoir pressure in (4.31), is

$$\pi_{tot} = \pi_c + \pi_{ion}. \tag{4.32}$$

Here π_c is the osmotic pressure exerted by the colloids in their *uncharged* state and

$$\frac{\pi_{ion}}{k_B T} = \rho_+ + \rho_- - 2\rho_s \tag{4.33}$$

is the excess pressure, also referred to as the **Donnan pressure**, exerted by ions; ρ_+ and ρ_- are the average number densities of, respectively, cations and anions in the cell. Note that (4.33) presupposes ideal ions that obey Van 't Hoff's law; the uncharged colloids in (4.32), however, need not be ideal and pressure π_c could be exerted by a concentrated dispersion of, say, uncharged spheres or neutral polymers. Incidentally, the simple summation in (4.32) of uncharged-colloid pressure π_c and ion pressure π_{ion} is, within the Donnan zero-field model, an exact result (see Philipse and Vrij, 2011). To find the two unknowns ρ_+ and ρ_- in (4.33) we need them in two equations. The first is the condition for thermodynamic equilibrium between ions in the cell and ions in the reservoir, namely the equality of products of ion concentrations—the solubility products:

$$\rho_+ \rho_- = \rho_s^2, \tag{4.34}$$

keeping in mind that the reservoir anion and cation concentration both equal ρ_s. The second equation for ρ_+ and ρ_- follows from the condition that the Donnan cell as a whole is electrically neutral—hence, total numbers of positive and negative charges in the cell are equal:

$$\rho_+ = z\rho_c + \rho_-, \tag{4.35}$$

assuming negative colloids in the cell. From (4.34) and (4.35) we obtain a quadratic equation for the average anion density ρ_- in the cell (Exercise 4.8) with the positive root:

$$\frac{\rho_-}{\rho_s} = -y + \sqrt{1+y^2}; \quad y = \frac{z\rho_c}{2\rho_s}. \tag{4.36}$$

Here y is the ratio of the number density $z\rho_c$ of counter-ions released by the colloids, to the constant ion concentration $2\rho_s$ in the reservoir solution. The choice of y is convenient because its magnitude measures the influence of the colloid charge on the Donnan pressure: a large y signifies that this influence is huge; a small y implies little effect. From (4.36) and (4.35) we find for the cell's cation concentration:

$$\frac{\rho_+}{\rho_s} = y + \sqrt{1+y^2}. \tag{4.37}$$

Substitution of (4.36) and (4.37) in (4.33) leads to the following expression for the Donnan pressure, the excess pressure exerted by ions in the cell:

$$\frac{\pi_{ion}}{\pi_s} = \sqrt{1+y^2} - 1; \quad y = \frac{z\rho_c}{2\rho_s}, \tag{4.38}$$

where $\pi_s = 2\rho_s k_B T$ is the osmotic pressure of the reservoir solution. We see that due to the presence of charged colloids, since $\sqrt{1+y^2} > 1$, the Donnan pressure in the cell always exceeds the osmotic pressure π_s exerted by ions in the reservoir solution.

It is at first sight, perhaps, a bit surprising that (4.38) reveals a *non-linear* increase of ionic pressure with counter-ion concentration $z\rho_c$—how is this, if all ions obey the *linear* Van 't Hoff's law? Indeed, if one could add cations and anions independently of each other to the cell, the Donnan pressure would increase linearly, being simply the sum of pressures from Van 't Hoff's law for the ions. But ionic concentrations *cannot* be varied separately as they are coupled via the equilibrium condition (4.34) and the neutrality requirement (4.35), and it is this coupling that brings about the non-linearity in Donnan pressure (4.38).

Next we will explore what happens to the Donnan osmotic pressure at limiting values of the reservoir salt concentration, or at limiting values of the ion density ratio y.

Low-salt limit

When the reservoir salt concentration is lowered to values such that the ion density ratio $y \gg 1$, the Donnan pressure in (4.38) approaches its maximal value, that is equal to:

$$\frac{\pi_{ion}}{\pi_s} = y \Rightarrow \pi_{ion} = \rho_c z k_B T, \quad \text{for } y = \frac{z\rho_c}{2\rho_s} \gg 1. \tag{4.39}$$

Inserting this ionic Donnan pressure in (4.32) we find for the total pressure a Donnan cell exerts in excess to the reservoir pressure:

$$\pi_{tot} = \pi_c + \rho_c z k_B T = \rho_c (1+z) k_B T, \qquad (4.40)$$

for the case the colloids in their uncharged state are ideal particles—the pressure in the counter-ions-only limit that we already anticipated in (4.30).

High-salt limit

On inspection of (4.38) it will become clear that the Donnan pressure π_{ion} will vanish when the salinity of the reservoir is so high that y is virtually zero. Then the osmotic pressure of the cell becomes equal to the reservoir osmotic pressure $\pi_s = 2\rho_s k_B T$. In other words, as already noted, a sufficiently high salinity nullifies the effect of colloidal charge on a dispersion's osmotic pressure. Upon lowering the salt concentration, the first non-zero contribution to the Donnan pressure follows from (4.38) as:

$$\frac{\pi_{ion}}{\pi_s} = 1 + \frac{1}{2} y^2 - 1 \Rightarrow \pi_{ion} = \left(\frac{z^2}{4\rho_s}\right) \rho_c^2 k_B T, \quad \text{for } 0 < y \ll 1. \qquad (4.41)$$

Inserting this ionic Donnan pressure in (4.32) we find for the cell's total pressure, at small but non-zero ion density ratio y:

$$\frac{\pi_{tot}}{k_B T} = \frac{\pi_c}{k_B T} + \left(\frac{z^2}{4\rho_s}\right) \rho_c^2 = \rho_c + \left(\frac{z^2}{4\rho_s}\right) \rho_c^2, \qquad (4.42)$$

again assuming that colloids in their uncharged state obey Van 't Hoff's law. So here the pressure is a quadratic in the colloid concentration ρ_c. A comparison with the quadratic in (4.27) suggests that the term $z^2/4\rho_s$ is actually a second virial coefficient—that accounts for the pressure increase induced by ions when uncharged colloids are charged up. For the proof that $z^2/4\rho_s$—within the zero-field approximation—indeed follows from the excluded volume as defined in equation (4.28), see Philipse and Vrij (2011).

The Donnan potential

In the Donnan equilibrium there exist not only a difference π_{ion} in ionic pressure between cell and reservoir, but also a difference in electrical potential known as the **Donnan potential**. The jumps in pressure and potential are one-to-one connected, as can be seen as follows. In a DE ions are distributed between cell and reservoir according to the Boltzmann distributions:

$$\rho_\pm = \rho_s \exp[\mp \Phi_D]; \quad \Phi_D = \frac{e \Psi_D}{k_B T}. \qquad (4.43)$$

Here Φ_D is the dimensionless Donnan potential. On insertion of the Boltzmann exponents in (4.33) we obtain for the Donnan pressure:

$$\frac{\pi_{ion}}{k_B T} = \rho_s \exp[-\Phi_D] + \rho_s \exp[+\Phi_D] - 2\rho_s, \qquad (4.44)$$

which can be rewritten to[1]

$$\frac{\pi_{ion}}{\pi_s} = \cosh[\Phi_D] - 1; \quad \cosh x = \frac{e^x + e^{-x}}{2}. \qquad (4.45)$$

Here we see that the Donnan pressure is a single-valued function of the Donnan potential. To give a numerical example, for a Donnan potential of $\Psi_D = 50\,\text{mV}$ at room temperature the dimensionless potential equals $\Phi_D = 2$, which implies a substantial Donnan pressure of $\pi_{ion} = 2.63 \times \pi_s$. The correspondence in (4.45) between electric potential and excess ion pressure is an important one—and a key ingredient in the evaluation in Chapter 5 of the repulsion between two overlapping electrical double layers (see the discussion of equation (5.44)).

The Donnan pressure π_{ion} is expressed in terms of the ion density ratio y in equation (4.38) and in terms of the Donnan potential Φ_D in (4.45). Since these two expressions for π_{ion} must be the same, it follows that

$$\cosh \Phi_D = \sqrt{1 + y^2}; \quad y = \frac{z\rho_c}{2\rho_s}, \qquad (4.46)$$

which can be inverted (Exercise 4.10) to obtain for the magnitude of the Donnan potential:

$$\Phi_D = \cosh^{-1}\left[\sqrt{1 + y^2}\right] = -\sinh^{-1} y; \quad \Phi_D < 1. \qquad (4.47)$$

So, from the magnitude of the ion density ratio y we directly obtain the Donnan pressure via (4.38) and the Donnan potential through equation (4.47).

Salt depletion

In addition to pressure and potential, there is a third quantity that makes a jump going from cell to reservoir, and that is the salt concentration. Due to the negative Donnan potential ($\Phi_D < 0$) in the cell, the Boltzmann distribution $\rho_- = \rho_s \exp[+\Phi_D]$ favours displacement of anions to the reservoir. To preserve the cell's electro-neutrality, every anion must migrate in the company of a cation; that is to say, the cell expels neutral salt molecules to the reservoir—a phenomenon also referred to as **salt depletion**. Hence, the number L_s of

---

[1] Properties of hyperbolic functions like $\cosh x$ and $\sinh x$ are summarized in Appendix 4B.

depleted salt molecules equals the number L_- of anions that has diffused into the reservoir:

$$L_s = L_- = \rho_- - \rho_s. \quad (4.48)$$

We can rewrite (4.48) to obtain the salt depletion as a function of the Donnan potential by making use of the Boltzmann distribution of the anions:

$$\frac{L_s}{\rho_s} = \frac{\rho_-}{\rho_s} - 1 = \exp(\Phi_D) - 1; \quad \Phi_D < 0. \quad (4.49)$$

Thus, making the Donnan potential more negative gradually empties the cell of anions, until only counter-ions remain: the counter-ions-only limit that is reached for a strongly negative potential:

$$\frac{L_s}{\rho_s} = -1. \quad (4.50)$$

So, prior to the addition of charged colloids to the cell, the cell's anion concentration equals the salt number density ρ_s in the reservoir. That number density constitutes the maximal number of anions that the charged colloids can banish from the cell.

Salt depletion in the high-salt limit

One way to evaluate the salt depletion in the high-salt limit is as follows. At high salt the ion density ratio y is small, such that we can approximate the correspondingly small Donnan potential in (4.47) as

$$\Phi_D = -\sinh^{-1} y \approx y, \text{ for } y \ll 1. \quad (4.51)$$

On substitution of this small potential in (4.49) we obtain

$$\frac{L_s}{\rho_s} = \exp(\Phi_D) - 1 \approx \Phi_D = -y, \quad (4.52)$$

which leads to the salt depletion:

$$L_s = -y\rho_s = \frac{1}{2} z \rho_c. \quad (4.53)$$

This interesting result informs us that at high ionic strength, charged colloids with valency z expel a number of anions (as part of a salt molecule) to the reservoir, that equals half of the total number of surface charges on the colloids.

4.6 Depletion forces

On various occasions we have employed Van 't Hoff's law to evaluate an osmotic pressure gradient that entails a net osmotic force, per unit volume. For example, an essential ingredient in the derivation of the diffusion coefficient ('Einstein I') in Chapter 3 is the pressure gradient $d\pi/dx$ that appears in the force balance in equation (3.31). And in Chapter 5 we will meet, in equation (5.37), the ionic osmotic pressure gradient $d\pi_{ion}/dx$, utilized for the computation in Section 5.5 of the repulsion between electrical double layers. In the present chapter we encountered pressure gradients at work in the osmotic pressure difference between solution and pure solvent that in Figure 4.1C puts weight G into motion, until the osmotic equilibrium in Figure 4.1D is reached.

Here, we will address an instance of osmotic pressure gradients that occurs in dispersions comprising not only colloids but also neutral polymers—gradients that, as we shall see, have a peculiar effect on the behaviour of colloids in the 'colloid–polymer mixture'. In Section 4.4 we demonstrated that solute particles exert on a wall that is impermeable to them a force that per unit area equals their osmotic pressure. Thus, polymers exert osmotic pressure on a colloid which they cannot penetrate. Now a single colloid, far away from its neighbours, experiences no net effect of this polymer pressure because any osmotic, normal force on a unit colloid surface area is balanced by the force on an opposing area. However, if two colloids are within a range close enough for polymers not to fit in, an osmotic pressure difference develops between the surrounding polymer solution and the empty solvent region between the colloids. This pressure difference drives colloids together; we speak of a **depletion force** because its origin is the depletion of polymers from the region between adjacent colloids; the polymers, or any other small particles that cause a depletion force, are referred to as **depletants**.

The depletion effect is abundant and is, among the many other cases (discussed in Lekkerkerker and Tuinier, 2011), operative in the clustering of RBCs by serum proteins, the creaming of latex colloids caused by polysaccharides, aggregation of emulsion droplets by surfactants, and polymer-induced phase transitions in concentrated colloidal dispersions. In what follows we will restrict ourselves to the simple case of a dilute collection of colloids in a dispersion of depletants such that only simultaneous interaction between two colloids occurs.

Suppose these two colloids are in contact at a surface-to-surface distance $h=0$. What is the work needed to pull the colloids apart? For depletant particles with diameter d, the space between colloids remains depleted of particles as long as $h \leq d$. Hence if the colloids are brought to a distance $h \leq d$, a volume is created—call it V_{ov}—that is empty of depletants. The work w required to form a void V_{ov} in a solution volume V equals the reversible work for (osmotically) compressing the depletant dispersion from volume V to $V - V_{ov}$:

$$w(h) = -\int_{V}^{V-V_{ov}} \pi_d dV = \pi_d V_{ov}, \quad \text{for } V_{ov} \ll V. \tag{4.54}$$

Here π_d is the osmotic pressure of the depletant dispersion, which remains constant when emptying a small volume $V_{ov} \ll V$ of depletants. Pulling colloids apart is a forced process and, hence, the approach of the colloids is a spontaneous one, which may deliver $-\pi_d V_{ov}$ of reversible work. At distances $h > d$, depletants fill the gap between the two colloids such that the colloids experience uniform osmotic pressure from all sides—from which no net interaction can result.

The OAV potential

To sum up, the interaction (depletion) potential between two colloids is given by:

$$w(h) \begin{cases} = \infty & h < 0 \\ = -\pi_d V_{ov}(h) & 0 \leq h < d. \\ = 0 & h \geq d \end{cases} \qquad (4.55)$$

This potential, known as the **Asakura–Oosawa–Vrij potential**, or AOV potential, also includes the steep repulsion that forbids interpenetration of colloids below the contact distance $h = 0$. The excluded volume $V_{p,ex}$ is the volume surrounding a colloid that is inaccessible for centres of depletant particles. The volume V_{ov} in the AOV potential is that which results from the intersection of two such excluded volumes (see Figures 4.7 and 4.8).

For the two parallel plates at distance x in Figure 4.7, with surface area A (which is very large such that end effects can be ignored), the overlap volume equals $V_{ov} = (d-x)A$. Hence, the attractive part in the AOV potential (energy per unit area) reads:

$$w(x) = -\pi_d(d-x), \quad \text{for } 0 \leq x < d. \qquad (4.56)$$

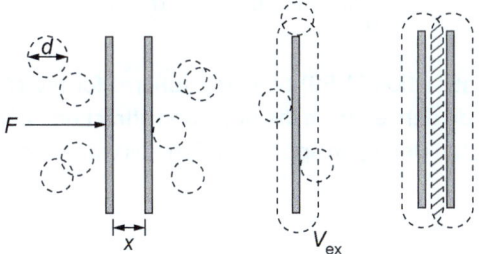

Figure 4.7 Left: Polymer modelled by penetrable spheres (with radius d) that exert osmotic pressure on plates they cannot penetrate. At inter-plate distances $x < d$ the inter-plate region is depleted of polymers such that plates are driven together by uncompensated polymer pressure from the surroundings. **Middle**: A plate excludes polymer sphere centres from a certain volume V_{ex}. **Right**: If excluded volumes from plates overlap (striped region) the volume accessible for polymers increases; hence their entropy rises so adherence of plates is a spontaneous process.

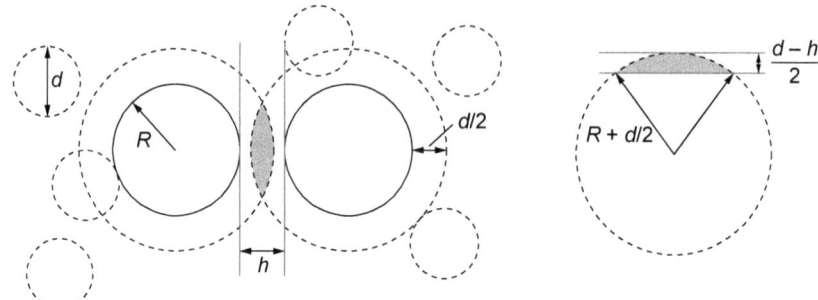

Figure 4.8 Left: The overlap volume (shaded area) for two spheres is a lens composed of two sphere caps. **Right**: Each cap is a slice of height $(d-h)/2$ cut from a sphere with radius $R+d/2$. Here h is the surface-to-surface distance of two spheres with radius R in a solution of depletants with diameter d.

The overlap volume for the two spheres in Figure 4.8 consists of two sphere caps; the same caps, incidentally, that we employed in Chapter 2 to model heterogeneous nucleation (see Figure 2.8 and Exercise 2.6). The volume of one cap with height z cut from a sphere with radius r is $V_{cap} = (\pi/3)z^2(3r-z)$. Since as indicated in Figure 4.8 $z = (d-h)/2$ and $r = R+d/2$, the overlap volume turns out to be:

$$V_{ov} = 2V_{cap} = \frac{\pi}{6}(d-h)^2\left[3R+d+h/2\right]. \tag{4.57}$$

Thus, the depletion attraction between two spheres is

$$\frac{w(h)}{\pi_d} = -\frac{\pi R}{2}(d-h)^2\left[3R+d+h/2\right], \quad \text{for } 0 \leq h < d. \tag{4.58}$$

For depletants that are small in comparison to sphere radius R, (4.58) simplifies to:

$$-\frac{\pi}{6}(d-h)^2 \quad \text{by:} -\frac{\pi R}{2}(d-h)^2 \tag{4.59}$$

The depletion attraction (4.59) between spheres for the case of small depletants can also be derived from the depletion attraction (4.56) between flat plates employing the Derjaguin approximation—as you are requested to verify in Exercise 4.19.

Ghost depletants

For the depletant osmotic pressure π_d to obey Van 't Hoff's law, depletants should be non-interacting. At the same time, however, they should also have a finite volume, otherwise there is no depletion effect in the first place. Vrij's depletant model (Vrij, 1976) resolves the issue by modelling depletant polymer chains as hard spheres (of diameter d) that can penetrate each other, but *not* a colloid. These ghost spheres on one hand obey Van 't Hoff's law, and on the other hand provide a depletion thickness d, in dilute polymer solutions about equal to $d \approx 2R_g$, where R_g is the polymer's radius of gyration.

Contact attraction

The depletion attraction is maximal at contact between two colloids. For two spheres in an ideal depletant solution with pressure $\pi_d = \rho_d k_B T$, the contact attraction follows from (4.58) as

$$\frac{w(h=0)}{k_B T} = -\varphi_d \left(1 + \frac{3R}{d}\right), \qquad (4.60)$$

where $\varphi_d = \rho_d (\pi/6) d^3$ is the depletant volume fraction. Consider, for example, small nanoparticles ($d = 2$ nm) as depletants mixed with colloidal spheres of radius $R = 100$ nm. According to (4.60) the depletant volume fraction required to induce a contact attraction of $-k_B T$ is only about $\varphi_d = d/3R \approx 6.7 \times 10^{-3}$. This example illustrates the effectiveness of small depletants to destabilize a dispersion; that is, to induce gelation or aggregation of colloids.

Reversibility

This destabilization by depletion, it should be noted, is reversible: when depletants are removed from an aggregated dispersion, a stable dispersion of free colloids re-emerges. This reversibility only occurs if depletants do not adsorb on the colloids; if they do, colloids stick together by so-called bridging flocculation—which is an irreversible process. The tunable and reversible depletion attraction gives rise to a rich phase behaviour of colloid–polymer mixtures; this is a topic for which we have no space in this primer, but that is extensively covered in Lekkerkerker and Tuinier (2011).

Summary

- Pressure exerted by ideal particles, either in gases or solutions, is the product of their number density ρ and thermal energy $k_B T$.
- The threefold way to Van 't Hoff's law involves osmotic equilibrium between solvent and solution, forces from ballistic colloidal steps, and colloid distributions near a repulsive wall.
- Ionic pressures caused by charged colloids, relative to a reservoir pressure, follow from the Donnan equilibrium (DE)—that also yields the potential difference between cell and reservoir.
- The DE also models the expulsion of salt to the reservoir—the salt depletion that is the inevitable consequence of the presence of charged particles in solution.
- Osmotic pressure from depleted polymers induces an attraction between colloids, with magnitude and range tuned by depletant size and concentration.

References

H. N. W. Lekkerkerker and R. Tuinier (2011), 'Colloids and the Depletion Interaction', *Lecture Notes in Physics* **833**, Springer.

A. Philipse and A. Vrij (2011), 'On the Thermo-Dynamic Foundation of the Donnan Equation of State', *Phys. Condens. Matter* **23**, 194–206.

J. van Rijssel et al. (2016), 'Thermodynamic Charge-to-Mass Sensor for Colloids, Proteins and Polyelectrolytes', *ACS Sensors* **1**, 1344–1350.

A. Vrij (1976), 'Polymers at Interfaces and the Interactions in Colloidal Dispersions', *Pure Appl. Chem.* **48**, 471.

Appendix 4A: The Derjaguin approximation

To convert an interaction energy $V(x)$ between two flat plates at distance x to an interaction energy $V(h)$ between two spheres at surface-to-surface distance h, we employ the so-called Derjaguin approximation (DA). In the DA, the interaction energy between the two spheres is taken to be the total sum of interactions between parallel rings (Figure 4.9). The distance x between two rings is related to the ring radius y via:

$$\frac{x-h}{2} = R - \sqrt{R^2 - y^2}. \tag{4.61}$$

For given R and h we can differentiate (4.61) to obtain:

$$\frac{1}{2}dx = -d(R^2 - y^2)^{1/2} = \frac{ydy}{R\left[1-(y/R)^2\right]^{1/2}}. \tag{4.62}$$

For big spheres that are at small surface-to-surface distances such that $y/R \ll 1$ we get from (4.62):

$$2ydy = Rdx, \quad \text{for } \frac{y}{R} \ll 1. \tag{4.63}$$

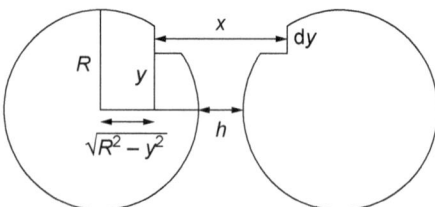

Figure 4.9 The interaction between two parallel flat surfaces and the interaction between two spheres. Within the Derjaguin approximation this conversion comprises summation of all interactions between parallel rings with area $2\pi ydy$.

If $w(x)$ is the interaction energy per unit area between two rings (each with area $2\pi y \, dy$) the total interaction $w(h)$ between the two spheres is:

$$w(h) = 2\pi \int_0^\infty w(x) y \, dy = \pi R \int_h^\infty w(x) \, dx. \qquad (4.64)$$

As the DA is limited to short distances, the effect of layers at large y and large x is negligible such that the upper limit of the integration can be put to infinity. Substitution of the attractive part (4.56) of the AOV potential in (4.64) and carrying out the integration (Exercise 4.19) yields:

$$\frac{w(h)}{\pi_d} = \begin{cases} -\dfrac{1}{2}\pi R(d-h)^2, & \text{for } 0 \le h < d \text{ and } d/R \ll 1 \\ 0, & \text{for } h \ge d \end{cases} \qquad (4.65)$$

For the DA to apply accurately, the sphere radius R must be much larger than the interaction range, here equal to the size d of the depletant; hence (4.54) is accurate only when $d/R \ll 1$.

Appendix 4B: Hyperbolic functions

In the calculation of properties of charged colloids, so-called hyperbolic functions regularly make their appearance. Examples are the hyperbolic cosine in the Donnan pressure (4.45) and the inverse hyperbolic sine in expression (4.47) for the Donnan potential. You will also frequently encounter hyperbolic functions in the treatment of electrical double layers in Chapter 5. These functions, plotted in Figure 4.10, are specified below, together with some of their properties.

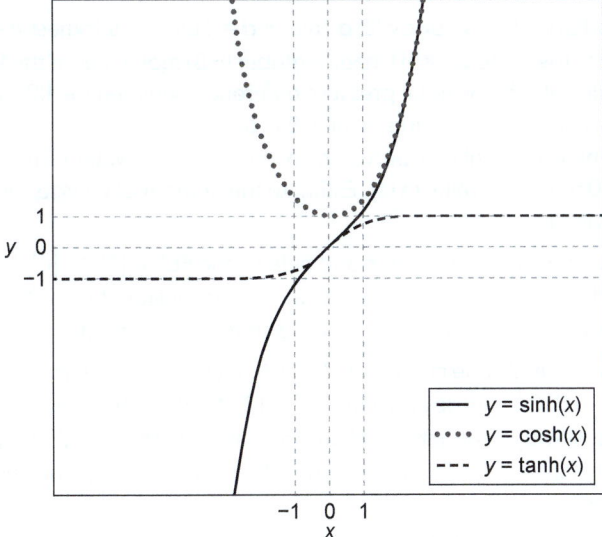

Figure 4.10 Hyperbolic sine, cosine, and tangent. Note: $\cosh(x) = \cosh(-x)$; $\sinh(x) = -\sinh(-x)$.

$$\sinh(x) = \frac{e^x - e^{-x}}{2}; \quad \cosh(x) = \frac{e^x + e^{-x}}{2}; \quad \tanh(x) = \frac{e^x - e^{-x}}{e^x + e^{-x}} = \frac{e^{2x} - 1}{e^{2x} + 1}.$$

Taylor expansions:

$$\sinh(x) = x + \frac{x^3}{3!} + \frac{x^3}{5!} + \ldots; \quad \cosh(x) = 1 + \frac{x^2}{2!} + \frac{x^4}{4!} + \ldots; \quad \tanh(x) = x - \frac{x^3}{3} + \ldots$$

Inverse hyperbolic functions:

$$z = \sinh(x) \to x = \sinh^{-1}(z); \quad z = \cosh(x) \to x = \cosh^{-1}(z);$$
$$z = \tanh(x) \to x = \tanh^{-1}(z).$$

Logarithmic representations:

$$\sinh^{-1}(z) = \ln\left[z + \sqrt{z^2 + 1}\right]; \quad \cosh^{-1}(z) = \ln\left[z + \sqrt{z^2 - 1}\right]; \quad \tanh^{-1}(z) = \frac{1}{2}\ln\left(\frac{1+z}{1-z}\right).$$

Exercises

4.1 **(a)** Calculate the osmotic pressure (in bar) at 27 °C exerted by a salt solution containing 10 g NaCl/L.
(b) What is the height of a water column that exerts the same pressure?
(c) How many grams of glucose have to be dissolved in water for the sugar solution to be isotonic with 0.9 wt% NaCl? M(NaCl) = 58 g/mol; M(glucose) = 180 g/mol.

4.2 Calculate the Coulombic potential energy between 1 mole of protons and 1 mole of electrons at a distance of 1 metre, in water with a dielectric constant $\varepsilon = 78.4$.

4.3 We model an RBC as a disc with diameter $d = 8$ μm and thickness $h = 2$ μm. Suppose there are 25×10^7 haemoglobin (HB) molecules in the RBC.
(a) Calculate the osmotic pressure difference between the RBC and plasma, if π is only due to the HB proteins.
(b) A measurement of π between an RBC and pure water turns out to be 100 times the value in (a). Estimate the number of charges on an HB molecule.

4.4 The ratio between [Cl⁻] ions in and outside the RBC is $[Cl^-]_i / [Cl^-]_u = 0.69$. The pH outside the cell is 7.40. What is the pH inside the cell? Which assumption(s) do you have to make to solve this problem?

4.5 A dialysis membrane contains HB ($c = 6$ g/L). In equilibrium [Na⁺] = 0.018 M inside and [Na⁺] = 0.012 M outside the membrane. Is HB positively or negatively charged? Calculate the so-called equivalent weight of the HB ion, defined as the ratio of HB weight to the HB charge number.

4.6 Suppose inside an RBC the Donnan potential is −9 mV. Calculate the pH in the RBC, if the pH in the surrounding plasma is 7.40.

4.7 Calculate the second virial coefficient B_2 in (4.27) for hard spheres of radius R, employing the hard-sphere potential: $w(r) = \infty$ for $0 \leq r < 2R$ and $w(r) = 0$ for $r \geq 2R$.

4.8 Formulate and solve the quadratic equation referred to in the text above (4.36). Why must we discard the negative root of the solution of this quadratic?

4.9 (a) Find from equation (4.38) the first-order correction term to the pressure π_c exerted by uncharged colloids.
(b) Formulate the total osmotic pressure, assuming that the uncharged colloids obey Van 't Hoff's law.

4.10 Verify equation (4.47).

4.11 A dispersion contains a weight concentration c_{tot} of colloids that are polydisperse in size. Show that the measured osmotic pressure of the dispersion yields the *number* averaged molecular mass M_n.

4.12 If the reservoir in the Donnan equilibrium from Figure 4.5 contains 0.1 M NaCl and the cell 0.1 M of a monovalent colloidal anion, then what is the Na^+ concentration in the cell?

4.13 A sensor for charge-to-mass ratio: A volume V_s with a dispersion of charged colloids or polyelectrolytes is connected to a small salt solution volume V_R (the sensor) by a semi-permeable membrane (Figure 4.6). The salt concentration ρ_s^R in the sensor is measured as a function of colloid weight concentration c in the dispersion. Assuming that for each colloid concentration a Donnan equilibrium between dispersion and sensor is achieved, show that

$$\left(\frac{\rho_s^R}{\rho_0}\right)^2 = 1 + \left(\frac{z}{m}\right)\frac{c}{\rho_0}, \quad \text{for } V_R \ll V_S. \qquad (4.66)$$

Here z is the number of counter-ions injected by a colloid of mass m into solution; ρ_0 is the (constant) number density of fully dissociated (1:1) salt in the dispersion. For charge-to-mass (z/m) determination employing equation (4.66) see van Rijssel et al. (2016).

4.14 Explain why the reservoir in Exercise 4.13 must be small.

4.15 The osmotic pressure exerted by North Sea water is about 27 bar. Calculate the salt molarity of this water. Assume T = 20 °C and that the sea only contains 1:1 electrolyte.

4.16 Human blood exerts an osmotic pressure of 7.8 bar. A solution with a higher salt concentration than blood will, when mixed with blood, lead to shrinkage of blood cells. An isotonic NaCl solution will cause neither shrinkage nor swelling of blood cells; what is the NaCl concentration in an aqueous isotonic solution? Assume that blood has a temperature of 37 °C.

4.17 One-dimensional velocity distributions: For the distribution of velocity components v_x of the thermally moving particles, the relevant energy in the Boltzmann factor is the particles' kinetic energy. Hence the probability distribution for v_x reads:

$$P(v_x) = C\exp[-E_{kin,x}/k_BT] = C\exp[-mv_x^2/2k_BT].$$

Here $E_{kin,x}$ is the contribution of the x-component of the velocity to the kinetic energy of a particle with mass m.

(a) Determine the constant C that follows from the normalization requirement

$$C\int_{-\infty}^{+\infty} \exp\left[-mv_x^2/2k_BT\right]dv_x = 1. \quad \text{Hint: employ the Gaussian integral:}$$

$$\int_{-\infty}^{+\infty} e^{-ay^2} dy = (\pi/a)^{1/2}.$$

(b) Evaluate $<v_x^2>$ and the average squared speed $<u^2>$.
(c) Calculate the rms speed $u_{rms} = <u^2>^{1/2}$ for N_2 and CO_2 molecules at $T = 298$ K.
(d) How does average kinetic energy depend on temperature?

4.18 Solve integral (4.23); employ boundary conditions (4.20).

4.19 Verify how via the Derjaguin approximation the depletion attraction between two plates can be converted to the depletion attraction between two spheres. Which assumption do you have to make for this conversion to be accurate?

4.20 Show how the diffusion coefficient $D = k_BT/f$ follows from the force K_{MF} on a colloid in (4.21) and the Boltzmann distribution (4.19). Hint: evaluate the stationary particle flux j_C this force would generate, and require that in equilibrium the sum of j_C and the diffusion flux j_D from Fick's first law must be zero.

Find the solutions to these exercises at the end of the book.

5 Charged Colloids and Electrical Double Layers

5.1 Introduction

In Chapter 2 we have studied the preparation of colloids via precipitation in a supersaturated solution, a preparation route that relies on the poor solubility of the material of which the colloids are composed. Such poorly soluble colloids are also labelled as **lyophobic** colloids—which literally means that they 'greatly fear their dissolution'. In contrast, colloids also exist that love to disperse themselves in the solvent, and they are classified as **lyophilic**. Examples of the latter are polyelectrolytes such as gelatine and a variety of proteins that spontaneously dissolve in water. Also, association colloids and micelles (the subject of Chapter 10) are lyophilic, as they form thermodynamically stable dispersions.

The class of lyophobic colloids is large and diverse; a few illustrative members are: the clay platelets in soil; alumina and zirconia colloids (dispersions of which are employed in ceramic shaping methods); the carbon-black particles in rubber tyres; titania and iron-oxide particles that are the inorganic pigments in, respectively, white and red paints; and the magnetite colloids in the ferrofluids that will be investigated in Chapter 9. The defining characteristic of all these lyophobic particles is their inherent thermodynamic instability—that is to say, their natural, spontaneous tendency to flocculate and stick together to form **aggregates**. This instability frequently manifests itself at an early stage in the preparation of these colloids, if that occurs via the precipitation process examined in Chapter 2. What happens is that newly formed, nucleated particles flocculate almost instantaneously to form visible precipitates, various examples of which are shown in Figure 5.1. The colloidal instability is a consequence of the colloids' thermal energy being overruled by the strong Van der Waals attraction between contacting particles (Figure 5.2). In other words, Brownian motion is unable to maintain lyophobic colloids in a dispersed state as single particles.

The rate at which particles flocculate, the flocculation kinetics, can be retarded considerably by diffusive ion clouds, **electrical double layers** (EDLs), that surround electrically charged colloids. Overlap of EDLs of two approaching particles brings about an EDL repulsion that, in contrast to the Van der

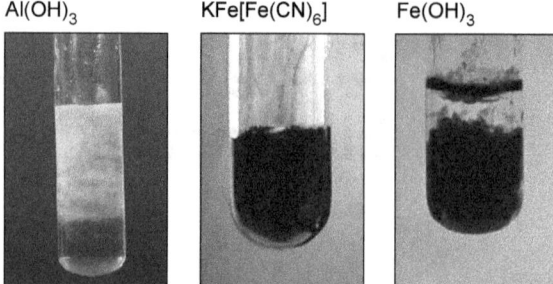

Figure 5.1 Precipitates of aluminium hydroxide (**left**), Prussian blue (**middle**), and iron hydroxide (**right**). These precipitates are actually agglomerates of small, unstable colloids that almost immediately flocculate after coming into existence by the nucleation process discussed in Chapter 2.

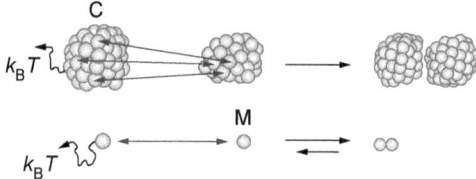

Figure 5.2 Why lyophobic (poorly soluble) colloids are inherently thermodynamically unstable. The Van der Waals attraction between molecules M is comparable to their thermal energy $k_B T$ so dimerization is reversible, to an extent set by temperature T: increasing T shifts the dimerization equilibrium to the higher-entropy side of two monomers. Colloids have at given T the same thermal energy $k_B T$ as molecules, but now this energy is dwarfed by the strong inter-colloid attraction—the sum of very many molecular Van der Waals attractions. Hence colloid aggregation is a spontaneous, irreversible process.

Waals attraction that joins colloids, causes them to *dis*join—which is why the EDL repulsion is also called a **disjoining force**. When the disjoining force is sufficiently strong such that no flocculation occurs on a timescale of, say, months, we speak of a dispersion with **colloidal stability**. This stability, it should be noted, is not a thermodynamic but merely a kinetic one. A famous example of truly long-term colloidal stability is exhibited by the gold sols prepared by Michael Faraday in 1856; in the Faraday museum in London the stable, reddish gold dispersions are still on display.

The electrical surface charges that underlie an EDL repulsion are formed when the (in most cases inorganic) surfaces of lyophobic colloids are immersed in water or polar organic solvents, usually by dissociation of surface groups or ion adsorption from solution. Consider, as an illustrative example, the dissociation of the weakly acidic silanol groups from glass or silica surfaces. Silica becomes negatively charged by releasing positive **counter-ions** (here protons) into solution; thermal energy of the protons opposes the electrostatic attraction from the surface such that the counter-ions diffuse only

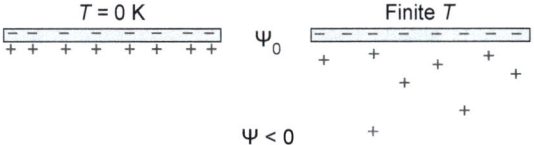

Figure 5.3 Charged surfaces with electrical surface potential $\psi_0 < 0$ are neutralized by positive counter-ions. The surface is immersed in a salt-free solution: co-ions are deficient. In the absence of thermal energy (**left**) coulombic attraction permanently condenses counter-ions onto the surface. At finite temperatures (**right**), however, thermal ionic pressure counteracts ion accumulation. When the effect of coulombic attraction balances ionic osmotic pressure, an equilibrium ion density profile results, given by the Boltzmann distribution for counter-ions in (5.3).

a limited distance into the solution (Figure 5.3). Oppositely charged **co-ions** (e.g. chloride anions) migrate further away into the solution. The mixture of diffuse counter-ions and co-ions is called the diffuse part of the EDL: the EDL's static part is the charges that are fixed on the surface—the $\equiv SiO^{-1}$ ions in the case of a glass or silica surface.

The equilibrium distribution of ideal ions in an EDL near a charged surface will be further examined in Section 5.2, with special attention to the characteristic feature that EDLs comprise fewer salt molecules but more ions (per volume) than the bulk solution. These latter excess ions, as will be discussed in Section 5.5, generate the repulsion between overlapping EDLs. The potential profile in an EDL can be found via the **Poisson–Boltzmann** (PB) equation, which is explained in Section 5.3. The potential profile in a single, free EDL is derived in Section 5.4, and commentary is provided on important features such as the potential's exponential decay, and its contraction by salt addition. Two overlapping EDLs are the subject of Section 5.5; here the disjoining force π_{dis} is evaluated, together with the corresponding repulsive interaction energy. The latter is combined in Section 5.6 with the Van der Waals attraction to arrive at an approximate expression for the Derjaguin–Landau–Verwey–Overbeek (**DLVO**) interaction potential between two colloidal spheres. On the basis of the DLVO-potential curve, flocculation electrolyte concentrations can be estimated, with magnitudes that, as we shall see, strongly depend on ion valency z.

5.2 Ion distributions near a charged surface

In the diffuse part of an EDL ions adopt certain concentration profiles that will be examined here for a charged planar surface immersed in a bulk solution of fully dissociated $z:z$ electrolyte. In the x-direction perpendicular to the surface an electrical potential profile ψ develops that, as illustrated in Figure 5.4, extends from a surface potential ψ_0 to zero potential $\psi = 0$ in the

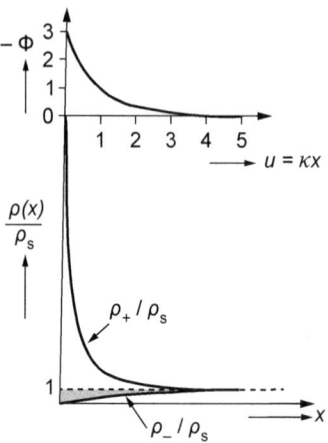

Figure 5.4 Electrical potential profile $\Phi(u)$ and local ion concentrations $\rho_+(x)$ and $\rho_-(x)$ near a negatively charged surface, with surface potential $\psi_0 = -(75/z)$ millivolts; $\Phi_0 = -3$. Ion number densities are scaled on their value ρ_s in the salt bulk solution. The hatched area represents the amount of depleted salt L_s as given by equation (5.4).

Source: Figure adapted from Verwey and Overbeek (1948).

bulk far away from the surface. The chemical potential $\mu_\pm$ of the z-valent ions in solution is given by:

$$\mu_\pm = \mu_s + k_B T \ln(\rho_\pm/\rho_s) \pm ze\psi. \tag{5.1}$$

Here ρ_+ and ρ_- are number densities of, respectively, cations and anions; $e = +1.6022 \times 10^{-19}$ C is the proton charge, and μ_s is the chemical potential of ions in the bulk solution, far away from the surface, which contains a number density ρ_s of salt molecules. The term $ze\psi$ in (5.1) is the reversible electrical work for transferring a charge of ze coulombs from the bulk solution where $\psi = 0$ to a location at potential ψ. The logarithmic, ideal-gas term in (5.1) represents the change in chemical potential when ideal ions migrate from number density ρ_s in the bulk to a volume element with ion number density $\rho_\pm$.

Ideal ions

The assumption of ideal ions in (5.1) is, incidentally, a very convenient one, which greatly simplifies the analysis of charged interfaces that follows in this chapter. For ideal ions, for example, the ionic osmotic pressures required in Section 5.5 to evaluate the EDL repulsion directly follow from Van 't Hoff's law—taking non-ideality of ions into account here would considerably complicate the investigation. How do we know when ions are behaving ideally? They do so when their thermal energy far exceeds their coulombic energies of interaction. This will be the case when the average distance between ions is much smaller than the **Debye screening length**—the length scale that will be introduced in Section 5.3.

This criterion for ion ideality is further examined in Appendix 5A. One numerical example from this examination is that in an aqueous 10^{-3} M NaCl solution (at $T = 298$ K) the joint thermal energy of ions is about 180 times larger than their coulombic interaction energy, implying ideal behaviour.

Boltzmann distributions

We return to the chemical potentials in (5.1), with the assessment that in thermodynamic equilibrium they are independent of distance x from the charged surface at $x = 0$. That is to say, in an equilibrium ion distribution no work is involved in the displacement of ions. Thus the equilibrium condition is:

$$\frac{d\mu_{\pm}}{dx} = k_B T \frac{d\ln \rho_{\pm}}{dx} \pm ze\frac{d\psi}{dx} = 0, \tag{5.2}$$

from which we obtain the Boltzmann distributions for the equilibrium number densities $\rho_{\pm}$ of ideal ions:

$$\rho_{\pm} = \rho_s \exp[\mp \Phi] \; ; \; \Phi = ze\psi/k_B T, \tag{5.3}$$

where Φ is the dimensionless electrical potential. To evaluate the equilibrium ion densities $\rho_{\pm}(x)$ at distance x from the surface, the potential profile $\Phi = \Phi(x)$ is needed—which for a single EDL we will derive in Section 5.4.

The Boltzmann distributions (5.3) entail that in the diffuse part of an EDL the co-ion concentration ρ_- at location x is much smaller than the counter-ion concentration ρ_+ at that location, as can also be seen in Figure 5.4. Thus, due to the charged surface, an amount of co-ions has been transferred to the bulk solution. To preserve electro-neutrality that transfer must take place via neutral salt molecules: on average, each co-ion is chaperoned by a counter-ion. We see here the salt depletion at work that we met earlier in Section 4.5 for the Donnan equilibrium. In the latter case, the electrical potential is a spatial constant; hence ρ_- and the salt depletion $L_s = \rho_- - \rho_s$ in equation (4.48) are constant. For the potential profile in an EDL, the salt depletion follows from the integration:

$$L_s = \int_0^{\infty}(\rho_- - \rho_s)dx = \rho_s \int_0^{\infty}(e^{\Phi} - 1)dx. \tag{5.4}$$

In Section 5.4 we will evaluate this integrated salt depletion for the (Gouy–Chapman) potential profile in a single EDL. Here we continue with the assessment that, where the salt concentration in the EDL falls below the bulk salt concentration, the ion density in the EDL is *higher* than in the bulk solution. The ion number density in excess of the bulk ion density $2\rho_s$ equals

$$\rho_+ + \rho_- - 2\rho_s = 2\rho_s(\cosh(\Phi) - 1), \tag{5.5}$$

where we have substituted the Boltzmann distributions from equation (5.3). The ion excess results from the Boltzmann factor for counter-ions in (5.3) being so much larger than for co-ions. These excess ions, as we will see in Section 5.5, are ultimately responsible for the net repulsion that exists between two overlapping EDLs.

5.3 The Poisson–Boltzmann equation

We have just seen that, in a volume element of the solution in an EDL, positive counter-ions numerically outweigh the negative co-ions (see also Figure 5.4) such that everywhere in the EDL $\rho_+ > \rho_-$. As a consequence a volume element carries a net (here positive) electrical charge. The net space charge density ρ^* (with dimension C m^{-3}) in the volume element is for z-valent ions given by:

$$\rho^* = \rho_+(ze) + \rho_-(-ze) = ze(\rho_+ - \rho_-) = -2\rho_s ze \sinh \Phi, \quad (5.6)$$

where we have substituted for ρ_+ and ρ_- the Boltzmann distributions from equation (5.3). Any net coulombic charge density entails divergence of the electrical field $\vec{E}$ (Figure 5.5, left). For a field changing only in the x-direction that divergence is the gradient in E:

$$\frac{dE}{dx} = \frac{\text{net charge density}}{\varepsilon \varepsilon_0}, \quad (5.7)$$

where $\varepsilon \varepsilon_0$ is the di-electrical constant. Equation (5.7), known as **Gauss's flux theorem**, expresses that a positive unit charge is the source of an electric field line whereas a negative charge is a field line's end-point (Figure 5.5, left). So when a volume element comprises equal numbers of cations and anions,

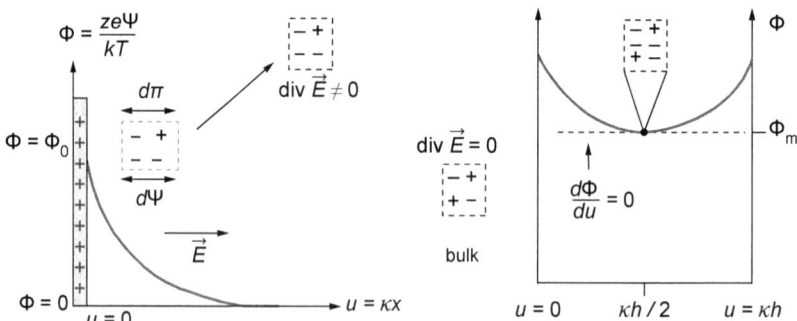

Figure 5.5 Left: Electrical potential profile Φ near a charged surface; Φ_0 is the surface potential and u is the dimensionless distance $u = \kappa x$ from the surface at $u = 0$. An EDL volume element experiences gradients in ion osmotic pressure π and potential ψ. Bulk electrolyte is electrically neutral, but an EDL volume element carries net charge that produces the electric field divergence given by Gauss's theorem (5.7). **Right**: Symmetric potential profile Φ in two overlapping EDLs from two charged surfaces. In the midplane the electric field is zero; the pressure exerted by ions in the midplane region equals, as shown in Section 5.5, the disjoining pressure π_{dis} that pushes the surfaces apart.

the numbers of field lines that enter and leave the element are the same and dE/dx is zero. When the volume element contains a surplus of either cations or anions, there is a net flux of field lines either leaving or entering the element, measured by the field gradient in (5.7). The magnitude of the electric field is, by definition, equal to minus the gradient in the electric potential ψ:

$$E = -\frac{d\psi}{dx}. \quad (5.8)$$

Inserting this expression for the electric field in (5.7) we obtain the so-called **Poisson equation**:

$$\frac{d^2\psi}{dx^2} = -\frac{\text{net charge density}}{\varepsilon\varepsilon_0}. \quad (5.9)$$

To solve the Poisson equation to obtain the potential profile ψ we need to specify how the charge density is related to ψ. Now the charge in Gauss's theorem (5.7) may stem from either static or mobile charge distributions; in an EDL, of course, charges are carried by ions that are in thermal motion. So we can equate the net charge density in the Poisson equation to the diffusive space charge density ρ^* that depends on the electrical potential according to equation (5.6). We thus obtain:

$$\frac{d^2\psi}{dx^2} = \frac{2\rho_s ze}{\varepsilon\varepsilon_0}\sinh\Phi, \quad (5.10)$$

a result that is known as the **Poisson–Boltzmann (PB) equation**, as it combines the Poisson equation and the Boltzmann distributions that are present in the net space charge density (5.6). Replacing the electrical potential ψ in the left-hand side of (5.10) with its dimensionless counterpart $\Phi = ze\psi/k_B T$, we find the PB equation in its compact, dimensionless form:

$$\frac{d^2\Phi}{du^2} = \sinh\Phi; \quad u = \kappa x. \quad (5.11)$$

Here u is the perpendicular distance x from the charged surface scaled on the Debye screening length $1/\kappa$, defined by:

$$\kappa^2 = \frac{2\rho_s(ze)^2}{\varepsilon\varepsilon_0 k_B T}. \quad (5.12)$$

The physical significance of the Debye length κ^{-1} is that it provides a measure for the spatial extent of an EDL, for which reason κ^{-1} is also referred to as the 'thickness' of a diffuse EDL. The Debye length is relevant for the EDL as a whole—and we seize here the opportunity to introduce another length scale, namely the **Bjerrum length** r_B that is pertinent to individual ions only. The distance r_B is the separation between two ions when their coulombic interaction energy equals their thermal energy $k_B T$. Now the interaction potential $V(r)$ between two z-valent ions at distance r is, according to Coulomb's law,

$$V(r) = \frac{(ze)^2}{4\pi\varepsilon\varepsilon_0 r}. \quad (5.13)$$

Hence by putting $V(r=r_B) = k_B T$ we find for the Bjerrum length

$$r_B = \frac{(ze)^2}{4\pi\varepsilon\varepsilon_0 k_B T}. \tag{5.14}$$

On substitution of (5.14) in (5.12) we can rewrite the Debye length in terms of the Bjerrum length

$$\kappa^{-1} = \frac{1}{\sqrt{8\pi\rho_s r_B}}. \tag{5.15}$$

Since the Debye length decreases with increasing salt concentration as $\kappa^{-1} \propto \rho_s^{-1/2}$, it follows that EDLs shrink upon increase of the solution's salinity—which decreases the stability of charged colloids, as we will see in Section 5.6.

5.4 A single, flat EDL

The electrical potential profile $\Phi = \Phi(u)$ in a single, flat EDL, also referred to as the Gouy–Chapman (GC) profile, is a solution of the Poisson–Boltzmann (PB) equation that was developed in the previous Section 5.3. Below we will derive the GC profile, followed by a discussion of its important characteristics, including the contraction of the profile due to salt addition, and the significant amount of salt that the single EDL expels to the solution reservoir. We start with rewriting the PB equation (5.11) to:

$$d\left(\frac{d\Phi}{du}\right) = \frac{\sinh(\Phi)}{d\Phi/du} d\Phi, \tag{5.16}$$

which can be integrated to obtain the square of the electric field $d\Phi/du$ present at location u in the EDL:

$$\left(\frac{d\Phi}{du}\right)^2 = 2\cosh(\Phi) + C. \tag{5.17}$$

The value of the integration constant C is determined by a physical boundary condition, namely that at infinite distance $u \to \infty$ from the surface located at $u = 0$, both the potential Φ and the electric field $d\Phi/du$ have fully decayed to zero. Hence $0 = 2 + C$ so the integration constant equals $C = -2$ such that the square of the electric field equals

$$\left(\frac{d\Phi}{du}\right)^2 = 2\cosh(\Phi) - 2 = 4\sinh^2(\Phi/2), \tag{5.18}$$

where we have employed the identity $\cosh(\Phi) - 1 = 2\sinh^2(\Phi/2)$, which you are requested to verify in Exercise 5.23. Equation (5.18) has two solutions:

$$\frac{d\Phi}{du} = +2\sinh\left(\frac{\Phi}{2}\right); \frac{d\Phi}{du} = -2\sinh\left(\frac{\Phi}{2}\right). \tag{5.19}$$

Now for a negatively charged surface both Φ and $\sinh(\Phi/2)$ are negative but the gradient in the potential is the positive slope $d\Phi/du > 0$. Hence, we must disregard the positive root in (5.19) to find for the electric field strength:

$$\frac{d\Phi}{du} = -2\sinh(\Phi/2). \tag{5.20}$$

In Exercise 5.20 you are asked to show that integration of (5.20) yields

$$\tanh(\Phi/4) = \tanh(\Phi_0/4)e^{-u}; \; u = \kappa x, \tag{5.21}$$

which is the GC potential profile that we set out to determine. Examples of GC profiles are shown in Figure 5.6 for various values of the surface potential. Let us examine some features and consequences of this GC potential profile—the potential landscape in the vicinity of a flat, charged surface immersed in a bulk solution with a number density ρ_s of salt molecules.

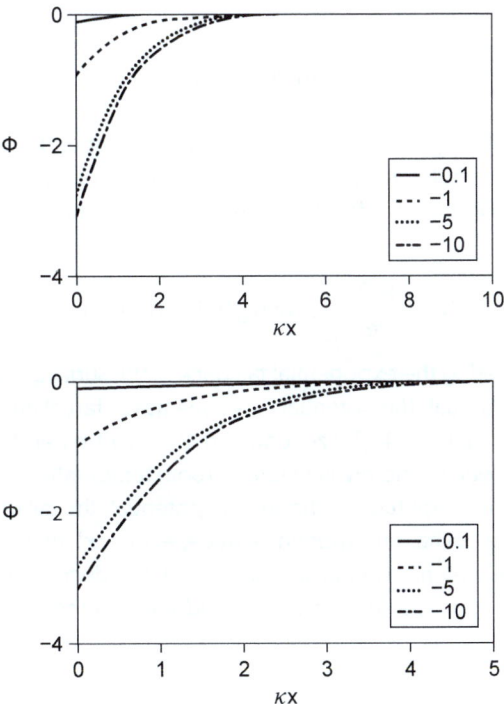

Figure 5.6 Top: Gouy–Chapman potential profiles from equation (5.21) for various values of the dimensionless (negative) surface potential Φ_0. **Bottom**: A zoomed-in version. Note that high (absolute) values of the surface potential, the tails of the curves at large κx fall on top of each other, in accordance with equation (5.24). In other words, the magnitude of potentials in the tail are virtually independent of a high surface potential.

Source: Figure prepared by Bonny Kuipers.

Exponential decay

First we will look at the profile's spatial decay. Sufficiently close to the surface at $u = 0$ there is an exponential decay of the quantity $\tanh(\Phi/4)$. Sufficiently far away from the surface, however, there is an exponential decay of the potential Φ itself; for the small potentials in the tail we have $\tanh(\Phi/4) \sim \Phi/4$, which on substitution in (5.21) yields

$$\Phi = 4\tanh(\Phi_0/4)e^{-u}, \text{ for } \Phi \ll 1. \quad (5.22)$$

The **Debye–Hückel (DH) approximation** entails that surface potentials are small ($\Phi_0 \ll 1$) such that $\tanh(\Phi_0/4) \sim \Phi_0/4$ and (5.22) reduces to the DH potential (i.e. a potential profile that decays exponentially all the way from surface to bulk solution):

$$\Phi = \Phi_0 e^{-u}, \text{ for } \Phi_0 \ll 1. \quad (5.23)$$

When, on the other hand, the surface potential is high, we have $\tanh(\Phi_0/4) \sim 1$ and, consequently, expression (5.22) for the tail of the potential profile simplifies to

$$\Phi = 4e^{-u}, \text{ for } \Phi_0 \gg 1 \text{ and } \Phi \ll 1. \quad (5.24)$$

So for high surface potentials the profile's tail is independent of the magnitude of the surface potential, a trend that can also be observed in the GC profiles depicted in Figure 5.6. The absolute magnitude of the electrical potential in the tail is:

$$\Psi = \frac{4k_B T}{ze} e^{-u}, \text{ for } \Phi_0 \gg 1 \text{ and } \Phi \ll 1. \quad (5.25)$$

Thus, extrapolating the exponential tail back to the surface at $u = 0$, according to the DH potential, the potential curve seems to start at an 'effective' DH surface potential of $\Psi_0 = 4k_B T/ze$, equal to $\Psi_0 = 100$ mV and 50 mV for, respectively, monovalent and bivalent ions at room temperature.

So whatever the magnitude of the surface potential, the tail of the potential in a single EDL always has an exponential decay $\exp(-\kappa x)$, with a decay length set by the Debye screening length κ^{-1}. Since the latter decreases with salt concentration as $\kappa^{-1} \propto \rho_s^{-1/2}$, it follows that salt addition to the solution reservoir makes an EDL shrink.

Surface potential

Salt addition not only makes an EDL contract, but also lowers the electrical surface potential Φ_0—at least for surfaces with constant charge number density σ. The relation between Φ_0 and salt concentration ρ_s is a consequence of the condition of overall electro-neutrality, as can be seen as follows. A surface with a constant number density σ of negative charges carries $-\sigma e$ coulombs per unit area. For the whole system to be electro-neutral this net surface charge must add up with the integrated net positive charge in solution to zero:

$$-\sigma e + \int_0^\infty \rho^* dx = 0; \quad \rho^* > 0. \tag{5.26}$$

Upon insertion of the Poisson equation (5.9) we obtain for the surface charge number density:

$$\sigma e = -\varepsilon\varepsilon_0 \int_0^\infty \frac{d^2\Psi}{dx^2} dx = \varepsilon\varepsilon_0 \left(\frac{d\Psi}{dx}\right)_{x=0}. \tag{5.27}$$

Here we have employed the boundary condition that the electric field $d\Psi/dx$ is zero at infinity. What (5.27) expresses is that the initial slope of the electrical potential in the diffuse EDL at the surface (i.e. the electric field at $x = 0$) is proportional to the surface charge density σ. The dimensionless electrical field at the surface follows from (5.20) as

$$\frac{d\Phi_0}{du} = -2\sinh(\Phi_0/2). \tag{5.28}$$

On combination of (5.27) and (5.28) we find for the connection between surface charge density and surface potential Φ_0:

$$\sigma e = \frac{\varepsilon\varepsilon_0 \kappa k_B T}{ze}\left(\frac{d\Phi}{du}\right)_{u=0} = \frac{-2\varepsilon\varepsilon_0 \kappa k_B T}{ze}\sinh(\Phi_0/2). \tag{5.29}$$

The relation between surface potential and salt concentration ρ_s now follows from inversion of the hyperbolic sine in (5.29):

$$\Phi_0 = -2\sinh^{-1}\left[\frac{ze^2\sigma}{2\varepsilon\varepsilon_0 k_B T}\kappa^{-1}\right]. \tag{5.30}$$

This equation reveals that the surface potential adapts to salt concentration ρ_s as follows. The proportionality $\kappa^{-1} \propto \rho_s^{-1/2}$ entails that an increase in ρ_s decreases the argument of the inverse hyperbolic sine function in (5.30). In other words, salt addition to a bulk solution lowers the electrical potential on a constant-charge density surface immersed in it.

Diffuse condenser

A surprisingly simple condenser model for a GC EDL emerges on inspection of the DH limit of low surface potentials $\Phi_0 \ll 1$. In that limit we can put $\sinh^{-1}(\Phi_0/2) \sim \Phi_0/2$ in (5.29) to obtain:

$$-\sigma e = \frac{\varepsilon\varepsilon_0 \kappa k_B T}{ze}\Phi_0 = \frac{\varepsilon\varepsilon_0}{\kappa^{-1}}\Psi_0. \tag{5.31}$$

This result is reminiscent of an electrical condenser composed of two parallel plates carrying opposite charges $\pm e\sigma$, separated by a distance d and a potential difference between the plates equal to Ψ_0:

$$-\sigma e = C_{ap}\Psi_0; \quad C_{ap} = \frac{\varepsilon\varepsilon_0}{d}. \tag{5.32}$$

Here C_{ap} is the capacitance of the condenser, which measures how much charge the plates can accommodate per unit of applied potential. Comparing (5.31) and (5.32) we see that a low-potential EDL mimics a condenser composed of a charged surface with an oppositely charged layer of counter-ions at a distance $d = \kappa^{-1}$. Since, in the DH limit, the average distance between counter-ions and the surface is equal to κ^{-1} (Exercise 5.9), the condenser model of an EDL amounts to placing all counter-ions at that average distance.

We will return to the condenser model of an EDL in Chapter 11, Section 11.4, where we will employ this model to assess the effect of an EDL on the speed of electrophoresis of a charged colloid.

Salt depletion

Knowing the electric field in the GC EDL, we can now compute the amount of salt depleted by a single EDL, represented by the hatched area in Figure 5.4. First we rewrite the salt depletion integral (5.4) to

$$\frac{L_s}{\rho_s \kappa^{-1}} = \int_{\Phi_0}^{0} \frac{e^{\Phi} - 1}{d\Phi/du} d\Phi. \tag{5.33}$$

Substituting the electric field from (5.20), and performing the integration (Exercise 5.10), the following result for the salt depletion by an EDL near a negatively charged surface is obtained:

$$\frac{L_s}{2\rho_s \kappa^{-1}} = e^{\Phi_0/2} - 1; \quad \Phi_0 < 0. \tag{5.34}$$

It is evident that an increase in (absolute value of) the surface potential enhances the amount of expelled salt. For high surface potentials the salt depletion approaches its maximal value:

$$L_s = -2\rho_s \kappa^{-1}, \quad \text{for} - \Phi_0 \gg 1, \tag{5.35}$$

which manifests a situation in which the large majority of co-ions has been expelled from the EDL as salt molecules to the bulk solution.

5.5 Two interacting EDLs

Having examined the single, free EDL in the previous section, our task is now to consider the case of overlap between two EDLs, in particular to evaluate the disjoining pressure π_{dis} that makes the charged surfaces spontaneously drift apart (Chu, 1976). Consider two negatively charged, parallel flat surfaces with constant charge number density σ, immersed at constant temperature T in a large electrolyte reservoir with an invariable salt number density ρ_s and constant osmotic pressure $\pi_s = 2\rho_s k_B T$. The disjoining force can be found by summing up the two forces that ions experience in the vicinity of a charged surface: ions are subjected to an electrical force F_E due to a gradient in electrical potential, and they experience an osmotic force F_π due to a gradient in the total ion number density. The electric field strength E equals the force per proton charge; there is a surplus ρ^* of positive charges per volume, so the x-directed electrical force per volume equals

$$F_E = \rho^* E = -\rho^* \frac{d\Psi}{dx} \quad (Nm^{-3}), \quad (5.36)$$

where ρ^* is the net charge density from (5.6). The total osmotic force (again per volume) experienced by the ions is

$$F_\pi = -\frac{d\pi_{ion}}{dx} \quad (Nm^{-3}). \quad (5.37)$$

Here π_{ion} is the total osmotic pressure jointly exerted by cations and anions. In equilibrium the forces in (5.36) and (5.37) add up to zero: $F_E + F_\pi = 0$, from which it follows that

$$\frac{d\pi_{ion}}{dx} + \rho^* \frac{d\Psi}{dx} = 0. \quad (5.38)$$

On substitution in (5.38) of the net charge density ρ^* from the Poisson equation (5.9) we obtain

$$\frac{d}{dx}\left[d\pi_{ion} - \varepsilon\varepsilon_0 \frac{d^2\Psi}{dx^2} d\Psi \right] = \frac{d}{dx}\left[d\pi_{ion} - \frac{\varepsilon\varepsilon_0}{2} d\left(\frac{d\Psi}{dx}\right)^2 \right] = 0, \quad (5.39)$$

which upon integration yields:

$$\pi_{ion} - \frac{\varepsilon\varepsilon_0}{2}\left(\frac{d\Psi}{dx}\right)^2 = \text{constant}. \quad (5.40)$$

Let us inspect the physical meaning of the various terms in this equation (which are also explained in Figure 5.7). The total osmotic pressure π_{ion} in (5.40) that is jointly exerted by cations and anions inclines to expand an EDL. The negative term in the left-hand side of (5.40)—also known as the **Maxwell stress**—accounts for the net electrostatic attraction between ions and the charged surface, which tends to shrink an EDL. We see in (5.40) that the difference between ionic osmotic pressure and Maxwell stress has the same constant value everywhere between the plates in Figure 5.5, right. That constant is nothing but the net ionic pressure, the disjoining pressure π_{dis}, that drives the surfaces apart. Now, midway between the plates at $x = h/2$ the electric field $d\Psi/dx$ is zero, for reasons of symmetry (see also Figure 5.5, right). Hence, the integration constant in (5.40), the disjoining pressure π_{dis}, equals the ionic pressure $\pi_{ion,m}$ in the midplane region—in excess of the pressure π_s exerted by the salt solution reservoir:

$$\pi_{dis} = \pi_{ion,m} - \pi_s. \quad (5.41)$$

The excess ion number density in the midplane region follows from (5.5) as

$$\rho_{+,m} + \rho_{-,m} - 2\rho_s = 2\rho_s(\cosh(\Phi_m) - 1), \quad (5.42)$$

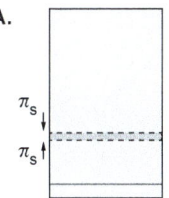

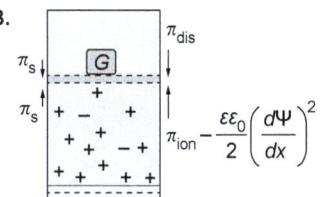

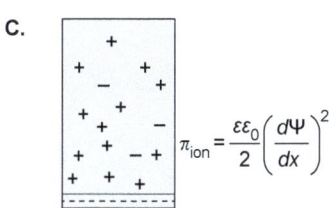

Figure 5.7 **(A)** A salt solution exerts uniform osmotic pressure π_s on both sides of a membrane permeable only to solvent and not to ions. **(B)** An EDL exerts a disjoining pressure π_{dis} equal to ionic pressure π_{ion} minus Maxwell stress $(d\Psi/dx)^2$. An external pressure π_{dis} caused by weight G prevents EDL expansion. **(C)** In the absence of external pressure an EDL freely expands until the GC-equilibrium profile is reached where at any location ionic pressure balances the Maxwell stress.

where Φ_m is the (dimensionless) electrical potential midway between the plates, at $x = \kappa h/2$. The excess ionic pressure in the midplane zone is the thermal energy $k_B T$ times the ion number density in (5.42):

$$\pi_{ion,m} - \pi_s = 2\rho_s k_B T(\cosh(\Phi_m) - 1). \tag{5.43}$$

From (5.41) and (5.43) we obtain what is known as the **Langmuir equation** for the disjoining pressure:

$$\frac{\pi_{dis}}{\pi_s} = \cosh(\Phi_m) - 1. \tag{5.44}$$

Note that for uncharged surfaces $\cosh(\Phi_m = 0) = 1$ such that the disjoining pressure is of course zero. For charged surfaces, whatever the sign of the midplane potential, $\cosh(\Phi_m) > 1$, which entails a disjoining pressure that is always positive—in contrast to the Maxwell stress in (5.40) which represents a pressure that is always negative.

Before we move on to further evaluation of the disjoining pressure in (5.44), we reflect for a moment on the similarity between the Langmuir equation (5.44) and the result $\pi_D/\pi_s = \cosh\Phi_D - 1$ for the Donnan osmotic pressure that we found in Chapter 4; see equation (4.45). More than being similar, the two results are actually the same! Recall from Chapter 4 that a Donnan equilibrium involves ions diffusing in a region of zero-electric field, a region that is in thermodynamic salt equilibrium with a reservoir solution. The midplane zone is precisely such a region as it harbours ideal ions diffusing in a constant Donnan potential $\Phi_D = \Phi_m$.

In what follows we will evaluate the magnitude of the pressure π_{dis} in (5.44) for the limiting cases of strong overlap (SO) and weak overlap (WO) of the two involved EDLs. These limits can be discerned in the potential profiles from Figure 5.8—which are numerical solutions of the PB equation (5.11). When the interplate distance h is much larger than the Debye screening length κ^{-1}, the EDLs are in the WO regime, characterized by a small potential midway between the plates. When distance h decreases, the electrical potential profile rises and flattens, signifying a declining electric field. For distances $\kappa h < 1$ the two EDLs enter the SO regime characterized by high potentials but a vanishingly small electric field.

Strong-overlap (SO) regime

When two negatively charged surfaces are in the SO regime, the electrical potential is strongly negative. As a result all co-ions have been expelled (via neutral salt molecules) to the bulk. Only counter-ions are left between the surfaces, with number density ρ_+ (see also Figure 5.9). For a strongly negative midplane potential, the right-hand term in (5.44) equals $\cosh(\Phi_M) - 1 \approx (1/2)\exp(-\Phi_m)$ such that

$$\frac{\pi_{dis}}{k_B T} = 2\rho_s \left[\frac{1}{2}\exp(-\Phi_m)\right] = \rho_+, \text{ for } \Phi_m \ll 0. \tag{5.45}$$

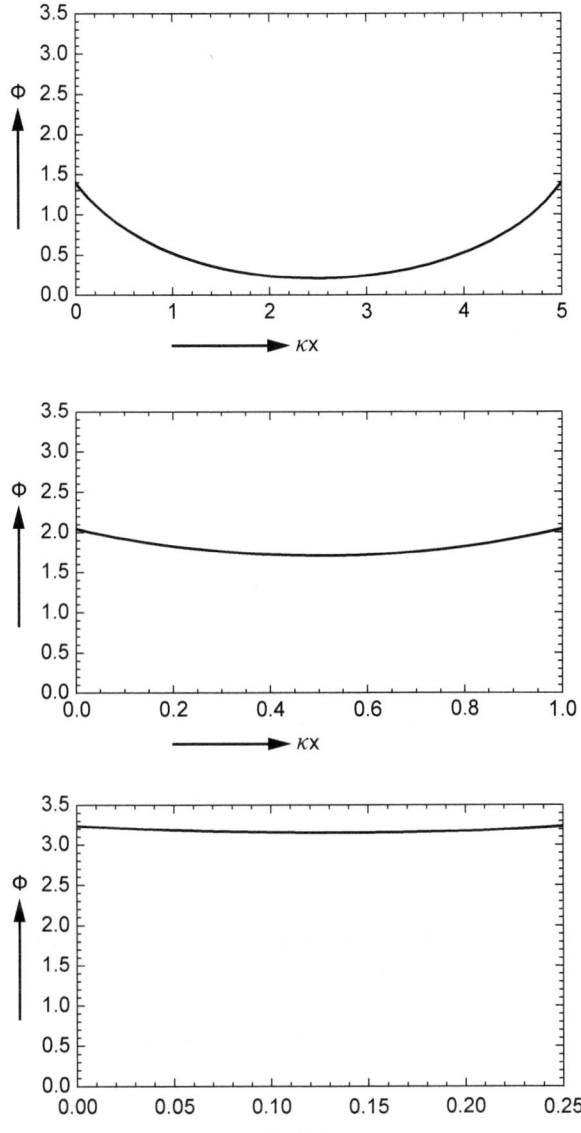

Figure 5.8 Two parallel plates with constant charge density at positions $\kappa x = 0$ and $\kappa x = \kappa h$; h is the plate–plate distance and κ^{-1} the Debye length. Drawn lines are numerical electrical potential profiles Φ from the PB equation (5.11). **Top**: Two EDLs in the weak-overlap (WO) regime with small midplane potential leading to the exponentially decaying disjoining pressure (5.50). **Middle**: Leaving the WO regime, electrical potentials rise and flatten, corresponding to a weakening electric field. **Bottom**: For $\kappa h < 1$ the two EDLs enter the strong-overlap (SO) regime where $\Phi_m \gg 1$.

Source: Figure prepared by Bonny Kuipers.

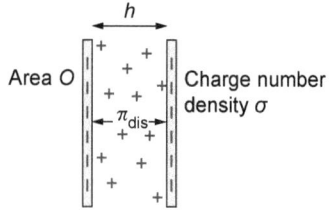

Figure 5.9 Illustration of the strong-overlap (SO) limit: upon very close approach of two constant-charge surfaces, all co-ions are expelled to the environment. Van 't Hoff's osmotic pressure law for the counter-ions entails a disjoining pressure $\pi_{dis} = 2\sigma/h$.

We recognize here Van 't Hoff's osmotic pressure law for a solution of ideal counter-ions. For two surfaces at distance h, with area O and constant charge number density σ, there are $2\sigma O$ counter-ions in a volume Oh, so the pressure in the SO regime can also be written as

$$\frac{\pi_{dis}}{k_B T} = \frac{2\sigma O}{Oh} = \frac{2\sigma}{h}. \tag{5.46}$$

We see that this pressure diverges in the limit $h \to 0$—a consequence of compressing a constant number of counter-ions in a vanishingly small volume.[1]

The weak-overlap (WO) regime

In a dilute dispersion of charged colloids, each surrounded by an EDL, only interaction between two colloids occurs, because the probability of more than two colloids simultaneously interacting is negligible. When the two colloids approach each other by Brownian motion, their EDLs start to overlap; in most cases a modest overlap will suffice to produce a repulsive energy of several $k_B T$ that will make the colloids diffuse apart. We will now address the EDL repulsion in the weak-overlap (WO) regime for two flat surfaces, and then convert the result to spheres via the Derjaguin approximation from Appendix 4A.

In the WO regime, the distance h between two charged surfaces is at least several Debye lengths, such that only the tails of their EDLs overlap (Figure 5.10). The electrical potential Φ_m in the midplane region at $x = h/2$ is then small, and can be approximated by the sum of the potential $\Phi(x = h/2)$ from two unperturbed, single double layers. In Section 5.5 we found that, for the tail of the potential profile for a single EDL,

$$\Phi = 4\tanh(\Phi_0/4)e^{-u}, \text{ for } \Phi \ll 1, \tag{5.47}$$

a result that places no restrictions on the magnitude of the surface potential Φ. Hence in the WO regime the midplane potential is:

$$\Phi_m = 2\Phi(x = \kappa h/2) = 8\tanh(\Phi_0/4)e^{-\kappa h/2}; \quad \Phi_m \ll 1. \tag{5.48}$$

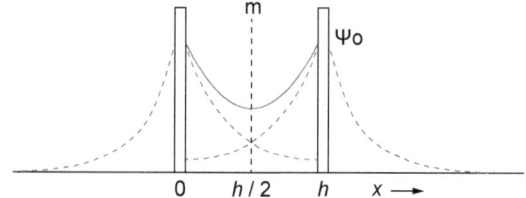

Figure 5.10 Sketch of the weak-overlap (WO) regime: two charged surfaces are at a large distance h such that only the tails of their EDLs interpenetrate and, consequently, the midplane potential Ψ_m approximately equals the sum of the potentials at $h/2$ of two single EDLs (these are represented by dashed lines).

---

[1] The $1/h$ distance dependence of the disjoining pressure in the counter-ions-only limit has been observed for silica surfaces—see B. Stojimirovic et al. (2020).

For small Φ_m we retain only the first term in the Taylor expansion of (5.44):

$$\frac{\pi_{dis}}{\pi_s} = \cosh(\Phi_m) - 1 = \frac{1}{2}\Phi_m^2, \quad \text{for } \Phi_m \ll 1, \quad (5.49)$$

which on substitution of (5.48) yields for the disjoining pressure

$$\frac{\pi_{dis}}{\pi_s} = 32\tanh^2(\Phi_0/4)e^{-\kappa h/2}, \quad \text{for } \Phi_m \ll 1. \quad (5.50)$$

The essential finding here is that in the WO regime, where two charged plates are sufficiently far apart, the disjoining pressure exhibits an exponential distance decay, with a decay length set by the (salt-dependent) Debye screening length.

Repulsive interaction energy

The disjoining pressure π_{dis} pushes surfaces apart and to fix the surfaces at a distance h an external pressure that equals π_{dis} has to be applied to the plates. A very slight increase of that external pressure brings the surfaces an infinitesimal distance dh closer. The associated small amount dw of reversible work performed on the plates is $dw = -\pi_{dis}dh$. The interaction energy A_{rep} is defined as the reversible work needed to bring the two surfaces from infinity to distance h, and hence follows from the integration:

$$A_{rep} = \int dw = -\int_\infty^h \pi_{dis}\, dh. \quad (5.51)$$

Note that A_{rep} is an interaction energy per unit area. On substitution of (5.50) in (5.51) we find for the repulsive interaction energy (Exercise 5.14):

$$\frac{A_{rep}}{k_B T} = 64\rho_s\kappa^{-1}\tanh^2(\Phi_0/4)e^{-\kappa h}. \quad (5.52)$$

We also see that in the WO regime the double-layer repulsive energy has an exponential decay set by the Debye screening length.

Debye–Hückel approximation

When the surface potential is small, we can substitute in (5.52): $\tanh(\Phi_0/4) \sim \Phi_0/4$, to obtain for the repulsive interaction energy:

$$\frac{A_{rep}}{k_B T} = 64\rho_s\kappa^{-1}(\Phi_0/4)^2 e^{-\kappa h} = 4\rho_s\kappa^{-1}\Phi_0^2 e^{-\kappa h}, \quad \text{for } \Phi_0 \ll 1. \quad (5.53)$$

One way to reach in practice the Debye–Hückel (DH) limit of small surface potentials is to bring the pH of a dispersion close to the isoelectric point, though not *too* close, otherwise the colloids might aggregate—as happens to the casein colloids in acidified milk (see Figure 5.11). The repulsion in (5.53) can be rewritten to (Exercise 5.21) the repulsive energy per unit area

$$A_{\text{rep}} = 2\varepsilon\varepsilon_0 \kappa \, \Psi_0^2 e^{-\kappa h}. \tag{5.54}$$

To evaluate the pre-exponential factor in (5.54) (i.e. the amplitude of the EDL repulsion in the DH limit), we need to know the surface potential Ψ_0. An estimate for Ψ_0 can be made by equating it to the so-called **zeta potential** ζ derived from electrophoresis (a characterization method that is further explained in Chapter 11). The treatment of electrophoresis in Section 11.4 will make clear that the zeta potential ζ is lower than Ψ_0, but by precisely how much is uncertain. Thus electrophoresis provides at best a lower bound for surface potential Ψ_0 and hence a lower bound for an EDL repulsion in the WO regime.

Figure 5.11 Left: The glass of milk has pH = 6.7 such that casein colloids with an isoelectric point (IEP) of pH ≈ 4.5 are sufficiently negatively charged and colloidally stable. **Right**: When the milk from the left glass is mixed with an equal volume of vinegar, the pH lowers to the IEP, where protons discharge the casein micelles, which lose their colloidal stability and almost immediately flocculate, in what must be an instance of the fast flocculation process discussed in Chapter 6.

5.6 The DLVO potential

The occurrence of aggregation in a dispersion of charged colloids demonstrates that, in addition to the double-layer repulsion, attractive forces between the colloids must also be present. These attractions are the Van der Waals attractions, caused by the interaction between fluctuating and permanent electrical molecular dipoles. The potential energy of attraction between two molecules at distance r is of the form:

$$V_{att} = -\frac{c}{r^6}, \quad (5.55)$$

where c is a positive constant. The Van der Waals attraction between two colloids is the combined effect of the attractions between all atoms of one colloid and all atoms of the other one. For the case of two parallel, flat surfaces at distance x the outcome of this summation is (Appendix 5B):

$$A_{att} = -\frac{A_H}{12\pi x^2} \quad (J/m^2). \quad (5.56)$$

Here A_{att} is the attractive energy per unit plate area; the constant A_H is known as the Hamaker constant. Note that the decay of Van der Waals attraction between two colloids is much slower than for two molecules: $1/x^2$ versus $1/r^6$, owing to the cumulative effect of many molecular Van der Waals attractions between the colloids. Together with the WO repulsion (5.52) the total interaction energy between two flat plates becomes

$$A_{tot}(x) = 64k_B T \rho_s \kappa^{-1} \tanh^2(\Phi_0/4)e^{-\kappa x} - \frac{A_H}{12\pi x^2} \quad (J/m^2). \quad (5.57)$$

The combination of a Van der Waals force and an exponentially decaying double-layer repulsion is usually called a DLVO potential.[2]

We should keep in mind that the repulsive part of the DLVO potential in (5.57) only applies to weakly overlapping EDLs. Using the **Derjaguin approximation (DA)** (which is explained in Appendix 4A) we can convert the interaction (5.57) for two flat plates into a DLVO potential $A_{tot}(h)$ for two spheres at surface-to-surface distance h:

$$A_{tot}(h) = \pi R \int_h^\infty A_{tot}(x)dx. \quad (5.58)$$

Substitution of (5.57) in (5.58) ultimately yields the DLVO potential for two spheres, with weakly overlapping EDLs:

$$\frac{A_{tot}(h)}{k_B T} = 64\pi \rho_s R \kappa^{-2} \tanh^2(\Phi_0/4)e^{-\kappa h} - \frac{A_H}{k_B T}\frac{R}{12h}, \text{ for } \kappa^{-1} \ll R. \quad (5.59)$$

The DA is only accurate when the colloid radius R is much larger than the interaction range, here set by the Debye screening length κ^{-1}. Hence the DLVO potential of the form (5.59) applies for a WO repulsion and a small Debye length $\kappa^{-1} \ll R$.

---

[2] 'DLVO' refers to the names of the Russian physicists Derjaguin and Landau, and the Dutch colloid scientists Verwey and Overbeek, who developed the theory of the stability of lyophobic colloids; see Verwey and Overbeek (1948).

One barrier and two minima

Due to the algebraic decay of the Van der Waals attraction, it dominates the total interaction at small interplate distances h, forming a deep dip in the potential near contact, usually referred to as the primary minimum. Also, at sufficiently large h the net interaction is an attraction as the EDL repulsion decays faster than Van der Waals attraction, resulting in a wide and very shallow secondary minimum. This minimum, if low enough, will lead to coagulation of colloids, though it is of a different character to the irreversible aggregation between particles in immediate contact in the deep primary minimum. The 'secondary coagulation' will be more reversible due to the shallowness of the minimum and absence of a potential barrier that prevents colloids from diffusing apart.

At intermediate distances the double-layer repulsion may surpass the Van der Waals attraction to produce an energy barrier (Figure 5.12) that prevents the surfaces from flocculating in the primary minimum. The height A_{max} of the DLVO-potential barrier decreases upon addition of salt, and when the barrier is below about one $k_B T$ the surfaces will stick.

Flocculation concentration

To estimate the flocculation salt concentration, we will employ as the flocculation criterion that the energy barrier A_{max} in the DLVO potential is (much) smaller than the thermal energy $k_B T$ such that colloids stick unhindered together by Van der Waals attractions. Hence $A_{max} \approx 0$, which implies that:

$$A_{tot} \approx 0 \text{ and } \frac{dA_{tot}}{dh} \approx 0. \tag{5.60}$$

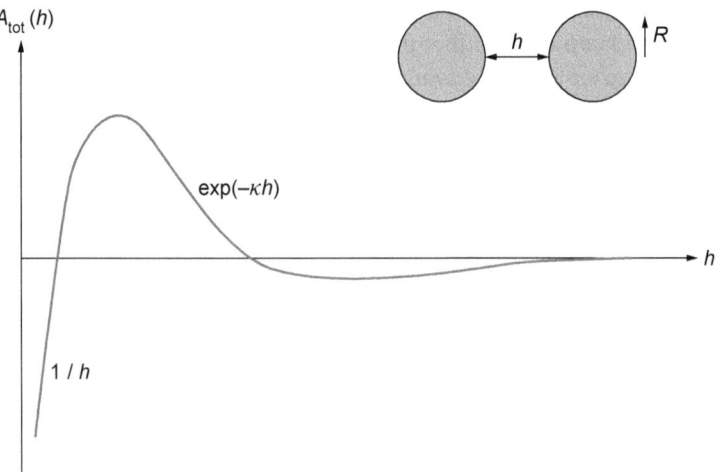

Figure 5.12 Sketch of the DLVO potential $A_{tot}(h)$ from equation (5.59) between two spheres at surface-to-surface distance h. $A_{tot}(h)$ is the sum of an EDL repulsion with an exponentially decaying tail, and a Van der Waals attraction with an h^{-1} distance dependence. The attraction causes a deep primary minimum to the left and a wide, very shallow secondary minimum to the right of the repulsive energy barrier.

On substitution of the potential (5.59) in criterion (5.60) we find, for a symmetrical $z:z$ electrolyte, the following proportionality for the electrolyte concentration ρ_{floc} (Exercise 5.15):

$$\rho_{\text{floc}} \propto A_H^{-2} z^{-6} \tanh^4(\Phi_0/4); \quad \Phi_0 = ze\Psi_0/k_B T. \tag{5.61}$$

For the low surface potentials in the DH approximation this proportionality modifies to

$$\rho_{\text{floc}} \propto A_H^{-2} \Psi_0^4 z^{-2}, \text{ for } \Phi_0 \ll 1. \tag{5.62}$$

Noteworthy is the strong effect of surface potential in the DH limit, which explains why, near their isoelectric point, colloids already aggregate at low ionic strength. Important too is the decrease in flocculation concentration with increasing ion valencies according to a dependence in the range $1/z^2 - 1/z^6$; the strong decrease of $\rho_{\text{s,floc}}$ with ion valency is known as the Schulze–Hardy rule.

The marked effect of ion valency z can be qualitatively understood from its appearance in the counter-ion-density Boltzmann distribution:

$$\rho_+ = \rho_s \exp\left[-ze\Psi/k_B T\right]. \tag{5.63}$$

Hence, an increase in valency shrinks the profile in an exponential fashion, leading to 'thinner' EDLs of two charged surfaces that can come closer together. So, an increase in valency of the ions and an increase in their concentration have a similar, contracting effect on EDLs that may lead to loss of colloidal stability.

A striking, practical example of the influence of ion valence is the colloidal instability of clay colloids in soil due to salts containing bivalent Ca^{2+} ions; when, in the aftermath of a seawater inundation, these salts are gradually replaced by NaCl, the clay colloids become stable, with disastrous consequences for plant growth (as further elucidated in Figure 5.13).

Repeptization phenomena

Once electrolyte concentrations are brought to the flocculation value discussed above, the kinetic scenario is that colloids encounter each other by Brownian motion followed by irreversible sticking in the primary minimum. This irreversibility will also be a key assumption underlying the kinetics of fast flocculation treated in Chapter 6. Sometimes, however, flocculation can be forced to reverse, in a process referred to as **repeptization**, a process in which a flocculated dispersion returns to a stable one by washing out the salt that induced the aggregation. However, here it is not sufficient to decrease salt concentrations to just below the flocculation values. Instead, the salinity of dispersions must be brought to a very low value for repeptization to take place.

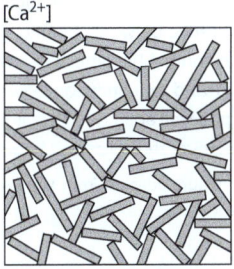
[Ca^{2+}]

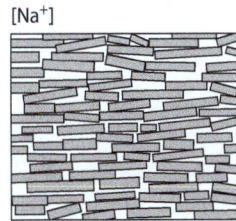
[Na^+]

Figure 5.13 Illustration of clay platelets in soil kept in an open-structured flocculated state (**top**) by salts comprising bivalent Ca^{2+} ions. Inundation by seawater replaces these salts by NaCl; after reclaiming flooded soil, sodium chloride is washed out by rain and the clay colloids become stable, leading to dense sediments (**bottom**) with a permeability for water and air much too low to sustain growth of plants and trees. The restoration period of soil thus ruined by colloidal stability can be shortened by addition of Ca^{2+} ions in the form of gypsum.

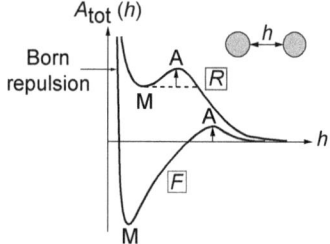

Figure 5.14 Sketch providing a qualitative explanation of the repeptization (R) process. Colloids flocculate (F) by diffusing across a repulsive barrier of insignificant height. By desalination of the dispersion, the double-layer repulsion strongly increases such that the difference between minimum M and maximum A becomes less than a few $k_B T$ and repeptization occurs—aggregates fall apart due to colloids diffusing over the collapsed barrier.

Source: Figure adapted from Verwey and Overbeek, 1948.

All this can be understood once we realize that the primary minimum will certainly not have the infinite depth implicated by taking the limit $h \to 0$ in equation (5.59), for at least two reasons. First, a steeply rising repulsive force appears, the so-called **Born repulsion**, when electronic shells of atoms from two surfaces near contact start to interpenetrate each other. As a result, the infinitely deep primary minimum changes into a minimum with a finite depth. The second reason that dampens the Van der Waals contact attraction is a more practical one, namely surface roughness of the involved colloids. Colloids never have atomically flat surfaces, and consequently particles make contact at some protruding points. It is at these points that the Van der Waals interaction is strongly negative, but the joint effect of these sticky protrusions may be significantly smaller than the attraction between two perfectly smooth surfaces.

Consider in Figure 5.14 the sketches of the interaction-potential curves for the limits of flocculation (F) and repeptization (R). The flocculation salt concentration is set by the maximum A just above the h-axis; for repeptization the distance between minimum M and maximum A has to be less than a few $k_B T$. For a colloid to be able to follow route R—and to break free from an aggregate—a much higher repulsion is needed than the small repulsive barrier a colloid has to pass to aggregate along route F. Consequently, for repeptization to occur, electrolyte must be removed to a very large extent. In practice this can be achieved by dialysis of dispersions against pure water for extensive periods of time, or by repetitive sedimentation of flocculated particles followed by repeated replacement of the supernatant by pure solvent.

Summary

- Lyophobic colloids are thermodynamically unstable and require, for charged colloids, a surrounding electrical double layer (EDL) for kinetic stabilization.
- An EDL harbours a balance between electrostatic forces on ions that contract the EDL and ion osmotic pressures that expand it. This force balance entails the EDL electrical potential profile, which also follows from the PB equation.
- Overlap between the EDLs generates a repulsion, equal to the excess pressure exerted by ions midway between the two charged surfaces. The tail of this repulsion decays exponentially, over a distance set by the Debye screening length κ^{-1}.
- The DLVO interaction potential comprises an EDL-repulsion and a Van der Waals attraction. Colloidal stability entails a repulsive barrier that shields colloids from the primary Van der Waals contact attraction. A high ionic strength eliminates the repulsive barrier and induces flocculation.
- The flocculation electrolyte concentration strongly decreases with increasing ion valency. Washing out most electrolyte may help a flocculated dispersion to regain stability.

References

B. Chu (1976), *Molecular Forces—based on the Baker Lectures of Peter J. W. Debye*. Interscience Publishers, New York.

B. Stojimirovic et al. (2020), 'Experimental evidence for algebraic double-layer forces', *Langmuir* **36**, 47–54.

E. J. W. Verwey and J. T. G. Overbeek (1948), *Theory of the Stability of Lyophobic Colloids*. Elsevier, New York. Reprint (1991): Dover Publications, Mineola, NY.

Appendix 5A: The Debye cube

In this chapter, and in the others, whenever ions appear in discussion they are assumed to be **ideal particles**. Here we examine a way to check the validity of this assumption. Ideal behaviour of ions implies that their coulombic interaction energies are negligible in comparison to their kinetic energy which is proportional to the thermal energy $k_B T$. One way to weigh ionic interaction energies against kinetic energy is to count the number N_D of ions in a Debye cube (Figure 5.15). A Debye cube is a cube of salt solution with an edge length that equals the Debye screening length κ^{-1}. Recall that the Debye length is the decay distance for the electrical potential in a salt solution; hence ions located inside a Debye cube volume experience a (nearly) flat potential and therefore a (virtually) zero electric field. In a solution with a number density ρ_s of $z:z$ electrolyte, the ion population N_D in a Debye cube with volume κ^{-3} is:

$$N_D = \rho_s \kappa^{-3} = \rho_s^{-1/2} \left[8\pi r_B\right]^{-3/2}; \quad \kappa^2 = 8\pi \rho_s r_B; \quad r_B = \frac{(ze)^2}{4\pi \varepsilon \varepsilon_0 k_B T}, \quad (5.64)$$

where r_B is the Bjerrum length for the z-valent ions. Note a curious feature of the Debye cube: its ion population *increases* when the salt number density ρ_s is lowered (see also Figure 5.15). The inequality $N_D \gg 1$ entails that kinetic energy $E_{kin} = (3/2)k_B T$ of ions outweighs their coulombic interaction energy. The electrical interaction energy for two nearby z-valent ions is approximately:

$$V_{pot} \approx \frac{(ze)^2}{4\pi \varepsilon \varepsilon_0 \langle r \rangle} = \frac{r_B}{\langle r \rangle} k_B T. \quad (5.65)$$

Here $<r>$ is the average separation distance between neighbouring ions which approximately equals $\langle r \rangle \sim \rho_s^{-1/3} = 8\pi r_B N_D^{2/3}$. Hence, the interaction energy scales with the population of the Debye cube as

$$V_{pot} \sim \frac{k_B T}{8\pi} N_D^{-2/3}, \quad (5.66)$$

confirming that the ionic electrostatic potential energy vanishes for a growing number of Debye cube inhabitants. The ratio of thermal energy $(3/2)k_B T$ to potential energy scales with N_D as

$$\frac{E_{kin}}{V_{pot}} \sim 12\pi N_D^{2/3}. \quad (5.67)$$

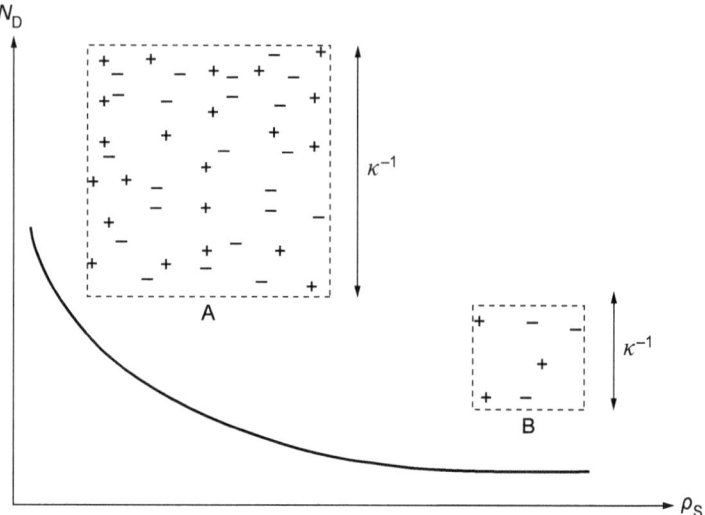

Figure 5.15 A Debye cube is a cube of salt solution with an edge length that equals the Debye screening length κ^{-1}. On decrease of salt concentration ρ_s the number N_D of ions in a Debye cube increases, in accordance with equation (5.64). When $N_D \gg 1$, as in cube **A**, ions are ideal in the sense that their joint coulombic interaction energy is negligible in comparison with their thermal energy.

So the kinetic energy already dominates for a modest number of residents in the Debye cube. For example, a 10^{-3} M solution of NaCl in water at $T = 298$ K has about $N_D \approx 11$ ions in a Debye cube (here with an edge of $\kappa^{-1} = 9.7$ nm) such that $E_{kin}/V_{pot} \approx 183$.

Appendix 5B: Van der Waals attraction between two parallel plates

The inherent instability of lyophobic colloids (as explained in Figure 5.1) results from the many intermolecular Van der Waals attractions that pull colloids together. Here we will evaluate the sum of all these molecular attractions to obtain the total Van der Waals force between two parallel plates (Chu, 1976). Using the Derjaguin approximation (dealt with in Appendix 4A), this force between flat surfaces is then converted in Section 5.6 to the Van der Waals attraction between two spheres—the attraction that appears in the DLVO potential for two spheres in equation (5.59).

The two parallel plates L and R (Figure 5.16) are at a surface-to-surface distance h; the thickness of the plates is very much larger than h, and plate areas O are large such that $O \gg h^2$ so any end effects can be disregarded. To evaluate the Van der Waals attraction between these semi-infinite plates,[3] we first compute the attraction between plate L and one atom α at a distance z from its surface.

---

[3] The assumptions here are that molecular forces are pairwise additive and a discrete sum over pair attractions can be replaced by an integration—the integrals in equations (5.70), (5.71), and (5.72).

COLLOID SCIENCE

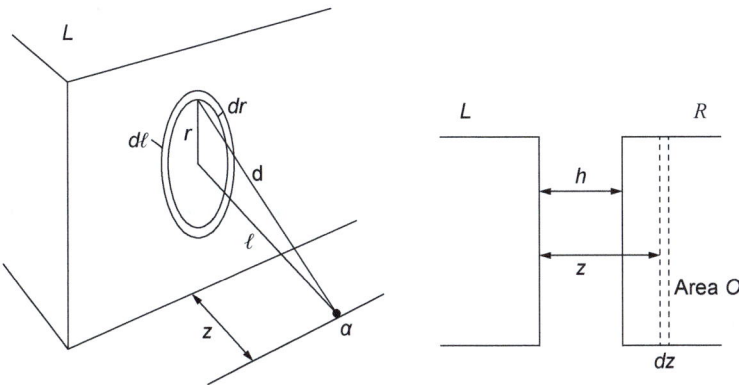

Figure 5.16 Left: Sketch accompanying the computation in Appendix 5B of the Van der Waals attraction V_a between atom α and the atoms in an annulus with radius r located inside a large plate on a slab with thickness $d\ell$. Subsequent integration over r and ℓ yields the attraction $V_a(z)$ between plate L and the atom at a distance z from its surface. **Right**: Integration of $V_a(z)$ from $z = h$ to infinity provides the attraction $V(h)$ between two large plates L and R at surface-to-surface distance h.

The atom is at distance ℓ from the centre of an annulus with radius r and area $2\pi r dr$, which is located on a slab (Figure 5.16, left) with thickness $d\ell$. The number of atoms in the ring composed of two annuli separated by distance $d\ell$ is:

$$\rho 2\pi r \, drd\ell, \tag{5.68}$$

where ρ is the (constant) atom number density in the plates. The attraction between α and an atom in the ring equals:

$$V_\alpha = -\frac{c}{d^6} = -\frac{c}{(r^2 + \ell^2)^3}; \quad c > 0. \tag{5.69}$$

Here d is the distance between α and a ring atom. The attraction between atom α and the whole slab with thickness $d\ell$ is:

$$V_\alpha(\ell) = -2\pi\rho c d\ell \int_0^\infty \frac{r}{(r^2+\ell^2)^3} dr = -\frac{\pi\rho c}{2} \frac{d\ell}{\ell^4}. \tag{5.70}$$

The attraction between atom α and the whole plate L follows from integration over ℓ:

$$V_\alpha(z) = -\frac{\pi\rho c}{2}\int_z^\infty \frac{d\ell}{\ell^4} = -\frac{\pi\rho c}{6z^3}. \tag{5.71}$$

In plate R a slab with thickness dz contains $\rho O dz$ atoms. Hence, the total attraction per unit area between L and R is

$$\frac{V(h)}{O} = -\rho \int_h^\infty V_\alpha(z) dz = -\frac{\pi \rho^2 c}{12h^2}. \tag{5.72}$$

The Hamaker constant A_H is defined as $A_H = \pi^2 \rho^2 c$, and for the attraction per unit area we will use the symbol $A_{att}(h)$. Hence:

$$A_{att}(h) = -\frac{A_H}{12\pi h^2}. \qquad (5.73)$$

This expression for the Van der Waals attraction is not applicable for very short distances where the Van der Waals attraction is a discrete sum over pairwise attractions.

Exercises

5.1 Explain why charged surfaces expel salt to the surrounding bulk salt solution.

5.2 Calculate the Debye screening length for water ($\varepsilon = 80.3; T = 298\,K$) containing $10^{-5}, 10^{-3}, 0.1$, and $1\,M$ NaCl.

5.3 The self-ionization of ethanol is the equilibrium:

$$C_2H_5OH + C_2H_5OH \rightleftharpoons C_2H_5OH_2^+ + C_2H_5O^-.$$

Other examples of self-ionizing solvents are ethylene diamine and water. Fill in the twelve missing numbers in Table 5.1.

5.4 A student, with a keen eye for applications, proposes to exploit the electrical potential difference between a charged surface and the reservoir solution to light a battery, arguing that 'any electrical potential difference induces motion of electrical charge and, hence, can deliver electrical work'. Do you agree?

5.5 How does the Debye screening length change when the surface potential is doubled?

5.6 For which value of the electrical surface potential Ψ_0 at room temperature is the dimensionless surface potential $\Phi_0 = e\Psi_0/kT = 1$?

Table 5.1 Self-ionizing solvents

$T = 298\,K$	Water	Ethylene diamine	Ethanol
ε_s	78.3	12.9	24.55
ρ_s /M Average ion concentration from self-ionization of the pure solvent	10^{-7}	$10^{-7.6}$	$10^{-9.4}$
r_B/nm Bjerrum and Debye length, see equation (5.64)	–	–	–
κ^{-1}/μm	–	–	–
N_D Number of ions in Debye cube defined in equation (5.64)	–	–	–
E_{kin}/V_{pot} Energy ratio estimated from equation (5.67)	–	–	–

5.7 A colloidal silica sphere (radius $R = 100$ nm) in ethanol ($\varepsilon = 24.55$) has a zeta potential of $\zeta = 50$ mV.
  (a) Estimate the number σ of charges on the sphere, assuming the surface potential Ψ_0 equals $\zeta = 50$ mV.
  (b) How much surface area (in nm^2) is available for each charge?
  (c) What is the average distance between two surface charges?

5.8 Show how the force balance (5.40) on ions near a charged surface leads to the Gouy–Chapman (GC) potential profile (5.21) for a single EDL. Hint: what will the value be of the constant in (5.40) for a single, freely expanding double layer?

5.9 **(a)** Make a sketch of the counter-ion density $\rho_+(x)$ near a charged surface, in excess to the counter-ion number density ρ_s at infinity.
  (b) Evaluate the average distance $<x>$ between excess counter-ions and the surface at $x = 0$, assuming that the surface potential is small: $\Phi_0 \ll 1$. Hint: what would be the probability for finding a counter-ion at x, in excess of the counter-ion density in the bulk far away from the surface?

5.10 Solve the salt-depletion integral (5.33).

5.11 Rivers discharging in seas tend to form shorter deltas than those liberating water into a lake. Explain why.

5.12 The half-distance $x_{1/2}$ is the distance at which the (dimensionless) potential in an EDL has dropped to $\Phi = \Phi_0/2$. Calculate $x_{1/2}$ for the GC potential profile (5.21) and show that in the DH limit $x_{1/2} = (\ln 2)\kappa^{-1}$.

5.13 The silanol (SiOH) groups on a silica surface are weak acids. Discuss what might happen when two (negatively) charged silica surfaces, submerged in water, closely approach each other.

5.14 Verify the evaluation of the repulsive energy in equation (5.52).

5.15 **(a)** Give the expression for the DLVO potential $A_{tot}(\kappa h)$—in terms of the distance h normalized by the Debye length κ^{-1}—between two flat, charged surfaces, with the repulsive interaction energy given by equation (5.53).
  (b) Evaluate how the flocculation concentration ρ_{floc} (that induces aggregation of the surfaces) depends on ion valency z, Hamaker constant A_H, and surface potential Ψ_0. Hint: define ρ_{floc} as the salt concentration at which the repulsive energy barrier disappears (i.e. $dA_{tot}(h)/dh = 0$ and $A_{tot}(h) = 0$).
  (c) What does the dependence from (b) look like in the limits of high and low surface potentials?

5.16 Explain why, when the two electrical double layers from two parallel plates start to overlap, the net force between the plates is repulsive.

5.17 Why does the DLVO potential have a secondary minimum?

5.18 Argue whether the DH approximation works better for a charged sphere or for a charged flat surface.

5.19 Prove the identity $e^x - 1/e^x + 1$ by $(e^x - 1)/(e^x + 1)$ (an identity that you will need in Exercise 5.20).

5.20 Solve the differential equation in (5.20) to obtain the GC potential profile in the form (5.21). Hint: employ the substitutions $e^{\Phi/2} = x$ and $d\Phi = (2/x)\,dx$ and simplify the solution that you get for $\Phi(u)$ using the identity from Exercise 5.19.

5.21 Verify that (5.54) follows from (5.53).

5.22 Show that the solution of the PB equation (5.11) for low potentials is the DH potential profile from (5.23).

5.23 Verify the identity $\cosh(\phi) - 1 = \sinh^2(\phi/2)$ that is employed in equation (5.18).

Find the solutions to these exercises at the end of the book.

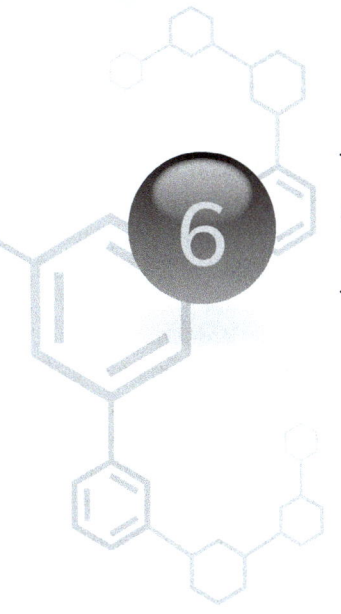

6 Colloidal Kinetics

6.1 Introduction

The kinetic stability of charged colloids, as we saw in Chapter 5, stems from a repulsive barrier in the DLVO interaction potential between two colloids. This barrier, which results from overlap between two electrical double layers (EDLs), perishes upon addition of sufficient electrolytes, which contract EDLs such that their overlap vanishes. Consequently, colloids can approach each other sufficiently close to stick together by the Van der Waals attraction. When the primary minimum of the latter (see also Figure 5.12) at particle contact is much deeper than the thermal energy $k_B T$, we speak of **irreversible flocculation** because the doublet formation cannot be undone by Brownian motion.

What we are interested in here is the rate at which this flocculation occurs. How much time, for example, does it take for two single colloids to aggregate, in what is called the initial stage of flocculation? And what can we say about the kinetics in the later stage of flocculation in which ever-growing clusters are formed? One might think that the initial flocculation rate will be determined by the average microscopic particle speed that is set by the colloid's average kinetic energy, assuming that this speed determines how long it takes for a colloid to contact its nearest neighbour. This kinematic picture, however, is wholly inadequate: microscopic speeds are only at play when colloids make their tiny ballistic steps—the random moves of atomic size on the momentum relaxation timescale that we discussed in Chapter 3. But the distance colloids have to migrate to encounter each other is very much larger than that: it is one that colloids have to cover by Brownian motion: flocculation occurs in the diffusive time regime and, hence, is a diffusion problem: how frequently do colloids meet each other by Brownian motion—upon which they permanently stick together?

Section 6.2 starts the analysis of flocculation kinetics with the evaluation of the frequency at which colloids meet *one* target sphere by diffusion. Next this encounter frequency is employed in Section 6.3 to assess rate constants for early-phase flocculation—the initial stage in which only a 'reaction' between two singlet colloids occurs. Section 6.4 addresses retardation of flocculation by an energy barrier—a retardation that, as it turns out, can be modelled via a reduced effective diffusion coefficient. Late-stage coagulation, treated in Section 6.5, involves the diffusional formation of particle triplets, quadruplets, etc. which, in turn, also meet by Brownian motion to form increasingly larger clusters. The kinetics of this late-stage process can be computed under the assumption that all rate constants equal the rate constant for reaction between two colloids in the early-phase stage. Section 6.6 tackles a special instance of a (late-stage) coagulation process in the form of large spherical clusters that grow by the diffusional uptake from monomeric particles from their environment. Stirring or shaking a dispersion may strongly accelerate flocculation, for reasons explained in Section 6.7.

6.2 Diffusion to a target sphere

We begin our analysis of the flocculation kinetics of mutually attracting colloids by evaluating the frequency by which colloids meet one another via Brownian motion. Within the so-called Smoluchowski diffusion model, this encounter frequency is computed via the irreversible absorption of freely diffusing particles onto a target sphere. This model works as follows.

Consider a collection of Brownian spheres with radius R_j diffusing in the vicinity of a target sphere with radius R_i centred at the origin (Figure 6.1). When a j-sphere encounters the target sphere, it irreversibly sticks due to a contact attraction that is much larger than the thermal energy k_BT. The target acts as a 'sink' for particles that touch it; hence, a concentration profile $c_j(r)$ of j-spheres will develop around the target. In the stationary state this concentration profile remains unchanged in time—this state is reached in about R^2/D seconds, which is the typical time needed for j-particles to diffuse a distance roughly equal to a sphere diameter. The stationary diffusion flux $J(j \to i)$ follows from Fick's first diffusion law as:

$$J(j \to i) = 4\pi r^2 D_{ij} \frac{dc_j}{dr} = \text{constant}. \tag{6.1}$$

Here D_{ij} is the **mutual diffusion coefficient** for the Brownian motion of j-particles relative to the target i-sphere—which itself is also in diffusive motion. The collision radius R_{ij} is the distance between centres of i and j when particles make contact (Figure 6.1). The concentration of free j-spheres at the contacting distance is zero, which leads to the boundary condition:

$$c_j = 0 \text{ at } r = R_{ij}. \tag{6.2}$$

On integration of (6.1) and applying boundary condition (6.2) we obtain the stationary concentration profile of j-spheres around the target:

$$c_j = \frac{J(j \to i)}{4\pi D_{ij}} \left[R_{ij}^{-1} - r^{-1} \right]. \tag{6.3}$$

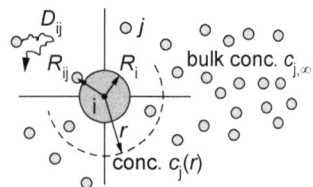

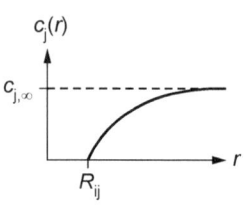

Figure 6.1 Top: j-spheres diffuse from a bulk (with number concentration $c_{j,\infty}$) towards a diffusing tracer sphere i with radius R_i which acts as a sink from which no j-sphere can escape. **Bottom**: The stationary concentration profile of j-particles according to equation (6.3).

Brownian encounter frequencies

Suppose $c_{j,\infty}$ is the constant bulk concentration of j-spheres, and that this bulk concentration is reached at a distance $r = R_i + \delta$ from the origin in Figure 6.1:

$$c_j(r = R_i + \delta) = c_{j,\infty} \qquad (6.4)$$

From (6.3) and boundary condition (6.4) the encounter frequency of j-spheres on one i-sphere follows as:

$$J(j \to i) = 4\pi D_{ij} R_{ij} c_{j,\infty} \left[1 - \frac{R_{ij}}{R_i + \delta}\right]^{-1}. \qquad (6.5)$$

The question is now which value should we take for the 'diffusion-zone thickness' δ—in other words, where in Figure 6.1 does the bulk begin? Fortunately, for a single target sphere in a large dispersion volume of j-particles, we do not need to specify δ any further than that it is much larger than the target radius R_i. So, for the steady-state diffusion flux we can take the limit

$$\lim_{\delta \to \infty} J(j \to i) = 4\pi D_{ij} R_{ij} c_{j,\infty}. \qquad (6.6)$$

The simple, asymptotic result for Brownian encounter frequencies is a fortuitous consequence of Brownian motion in three-dimensional space: diffusion on a two-dimensional plane or one-dimensional Brownian motion on a line involves an undetermined distance δ (Exercise 6.1).

Mutual Brownian motion

To further work out the encounter frequency (6.6), we need to evaluate the mutual diffusion coefficient D_{ij} that accounts for the diffusion of j-particles relative to the (also diffusing) origin of Figure 6.1. In Chapter 3 we studied Einstein's equation II which relates the diffusion coefficient D of a colloid to its mean-squared displacement $<x^2>$ in a given direction in time t:

$$D = \frac{<x^2>}{2t}. \qquad (6.7)$$

Suppose two diffusing particles i and j have, at some instance in time, their centres at position vectors $\vec{x}_i$ and $\vec{x}_j$. The mean-squared value of the magnitude of the vector $\vec{x}_i - \vec{x}_j$ connecting the two particle centres (Figure 6.2) defines the mutual diffusion coefficient via:

$$D_{ij} = \frac{\langle (\vec{x}_i - \vec{x}_j).(\vec{x}_i - \vec{x}_j) \rangle}{2t} = \frac{\langle (x_i - x_j)^2 \rangle}{2t}. \qquad (6.8)$$

If the two particles are diffusing independently from each other, we can work out (6.8) to:

$$D_{ij} = \frac{\langle x_i^2 \rangle}{2t} - 2\frac{\langle x_i x_j \rangle}{2t} + \frac{\langle x_j^2 \rangle}{2t} = D_i + D_j. \qquad (6.9)$$

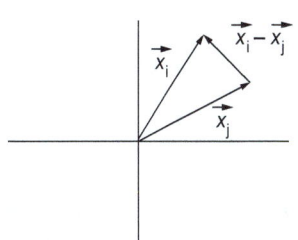

Figure 6.2 Position vectors $\vec{x}_i$ and $\vec{x}_j$ of two diffusing particles; the mutual diffusion coefficient D_{ij} accounts for the mean-squared value of the displacement $x_i - x_j$ separating the two particles.

Since x_i and x_j are statistically uncorrelated, it follows that in (6.9) $<x_i x_j> = <x_i><x_j> = 0$, bearing in mind that for random Brownian motion average displacements $<x>$ are zero. Hence, the mutual diffusion coefficient for spheres i and j equals:

$$D_{ij} = D_i + D_j = \frac{k_B T}{6\pi\eta}\left(\frac{1}{R_i} + \frac{1}{R_j}\right). \qquad (6.10)$$

Encounters between dissimilar spheres

For solid spheres the collision radius equals $R_{ij} = R_i + R_j$ (see also Figure 6.1). On substitution of R_{ij} and D_{ij} in (6.6) we find how the diffusion flux depends on the radii R_i and R_j of the particles involved:

$$J(j \to i) = \frac{2k_B T}{3\eta}\left(2 + \frac{R_i}{R_j} + \frac{R_j}{R_i}\right) c_{j,\infty}. \qquad (6.11)$$

A remarkable feature of this diffusion flux $J(j \to i)$ is its minimum value for spheres of identical size (see Exercise 6.3):

$$J(i=j) = \frac{8k_B T}{3\eta} c_{j,\infty}, \text{ for } R_i = R_j. \qquad (6.12)$$

In other words, for a given volume fraction, polydispersity always accelerates Brownian encounter frequencies in comparison to monodisperse spheres. To rationalize that (6.12) is the minimal flux, consider a monodisperse system in which we shrink all spheres except the target sphere; then the collision frequency increases due to the enhanced diffusion of the shrunk spheres. If, on the other hand, in the monodisperse dispersion only the target sphere is reduced in size, it will diffuse faster around in the collection of j-spheres, which also increases the diffusion flux $J(j \to i)$.

To get an idea of the collision frequencies involved, we monitor a target sphere with radius $R_i = 1\,\mu m$ diffusing in an aqueous host dispersion of spheres with radius R_j at a volume fraction $\phi_j = 0.01$. Equation (6.11) yields the collision frequencies on the target sphere shown in Table 6.1.

Thus, the micron-sized target is bombarded quite frantically in a host dispersion of nanoparticles, whereas for an equal volume fraction of micron-sized hosts, the target must wait for more than half a minute for the next Brownian encounter to occur.

6.3 Rapid early-phase flocculation

The early phase of a flocculation or **coagulation** process is defined as the initial stage in which only encounters between two sticky singlet particles occur, leading to the irreversible formation of a dimer. On the basis of the diffusion flux $J(j \to i)$ in (6.6), from colloids onto one target particle, we will compute the decrease of singlet concentration in time in this initial stage. This decrease, which can be called a 'reaction rate', obeys a differential equation that can be constructed as follows.

Table 6.1 Brownian collision frequencies on a target sphere with radius $R_i = 1\,\mu m$

R_j		$J(j \to i)$
10	nm	68×10^2 sec^{-1}
100	nm	80×10^{-2} sec^{-1}
1,000	nm	26×10^{-3} sec^{-1}

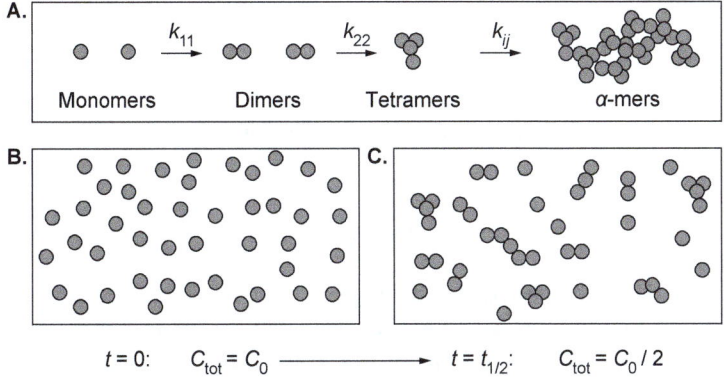

Figure 6.3 (A) In the initial flocculation stage singlet colloids irreversibly form dimers; in the later stage (discussed in Section 6.5) i-mers and j-mers coagulate to a-mers, where $a = i + j$. **(B)** Initially the total number density of particles equals the concentration c_0 of free monomers. **(C)** The half time for the late-stage flocculation is defined as it the time it takes for the *total* number c_{tot} of particles and clusters to drop to $c_{tot} = c_0/2$.

Suppose in a solution, solute particles labelled i and j irreversibly form dumbbells or dimers (Figure 6.3A); c_i and c_j denote bulk number concentrations of the solutes; the subscript ∞ previously denoting bulk values has been dropped. The probability to find an i-particle in some small volume element is proportional to concentration c_i, and the probability that a j-particle is in that same volume element (where it contacts particle i) is proportional to c_j. Since every contact removes an i- and a j-particle, concentrations decrease in time according to:

$$\frac{dc_i}{dt} = \frac{dc_j}{dt} \propto -c_i c_j. \tag{6.13}$$

The reaction rates dc/dt are called 'second order in concentration' because they decrease—note the minus sign in (6.13)—in proportion to the product of two concentrations. The constant of proportionality in (6.13) is the so-called **rate constant** k_{ij}:

$$\frac{dc_i}{dt} = \frac{dc_j}{dt} = -k_{ij} c_i c_j. \tag{6.14}$$

Rate constants and colloidal size

To compute changes in concentration in time t according to (6.14) we need first to evaluate the magnitude of rate constant k_{ij}. For the diffusion-controlled kinetics modelled in the previous section, the rate constant directly follows from the flux in equation (6.6). The total encounter frequency between particles i and j equals $J(j \rightarrow i)c_i$, and since every encounter removes a free i and j particle we have:

$$\frac{dc_i}{dt} = \frac{dc_j}{dt} = -J(j \rightarrow i)c_i. \tag{6.15}$$

Since equations (6.14) and (6.15) must be identical, it follows that:

$$k_{ij}c_ic_j = J(j \to i)c_i, \quad (6.16)$$

which implies that the rate constant for the diffusion-controlled flocculation is equal to

$$k_{ij} = J(j \to i)/c_j = 4\pi D_{ij}R_{ij}, \quad (6.17)$$

where we have substituted the diffusive flux $J(j \to i)$ from equation (6.6). Remarkably, for monodisperse particles this rate constant is independent of particle size—for if we substitute in (6.17) $R_{ij} = 2R_1$ and $D_{ij} = 2D_1$ it turns out that

$$k_{11} = 16\pi D_1 R_1 = \frac{8k_BT}{3\eta}. \quad (6.18)$$

This size independence suggests that k_{11} should also account for diffusion-controlled reactions between small molecules or ions. For example, for

$$OH^- + NH_4^+ \xrightarrow{k_r} NH_3 + H_2O,$$

the experimental rate constant is $k_r = 5.6 \times 10^{-17} \, m^3 s^{-1}$; from (6.18) we obtain for water ($\eta = 0.84$ mPa s) at $T = 298$ K: $k_{11} = 1.2 \times 10^{-17} \, m^3 s^{-1}$, which is indeed correct in order of magnitude.

Rate constants and colloidal shape

In addition to its weak *size* independence, the rate constant in (6.17), derived for spheres, is also fairly insensitive to the *shape* of the flocculating Brownian particles. We will illustrate this for the very non-spherical case of thin rods with diameter D and length $L \gg D$. The orientationally averaged diffusion coefficient of a thin rod is (Dhont, 1996):

$$D = \frac{k_BT}{3\pi\eta L}\ln\left(\frac{L}{D}\right), \quad \text{for } \frac{L}{D} \gg 1. \quad (6.19)$$

A freely rotating rod sweeps by rotational diffusion an approximately spherical volume with diameter L. Brownian encounters between two rods may occur when their 'rotation volumes' overlap. Thus the collision distance for the rods is about $R_{11} \approx L$, which on substitution together with (6.19) in (6.17) yields:

$$k_{11} = 8\pi D R_{11} = \frac{8k_BT}{3\eta}\ln\left(\frac{L}{D}\right). \quad (6.20)$$

So even for thin Brownian rods or fibres, (6.18) provides a reasonable estimate of the rate constant; the rod's aspect ratio only slightly enhances the rate constant k_{11} via its logarithm. The implication is that kinetic results derived in what follows for spheres also approximately hold for non-spheres.

The underlying reason is that non-sphericity increases the average collision distance R_{ij} between colloids, but this enhancing effect on collision frequency is compensated by the slower diffusion of non-spheres.

Flocculation half-life time

Even though the rate constant is independent of particle size, the particle number density due to flocculation decreases in time at a rate that strongly depends on the sphere radius R, as can be seen as follows. Consider the initial stage of flocculation in which only doublets of spheres are formed. Equation (6.14) reads for identical spheres:

$$\frac{dc_1}{dt} = -k_{11}(c_1)^2. \tag{6.21}$$

The solution of (6.21) shows (Exercise 6.2) that in the initial stage the concentration $c(t)$ of free singlet spheres decreases as

$$c(t) = \frac{c_0}{1+(t/t_{1/2})}; \quad t_{1/2} = \frac{1}{k_{11}c_0}. \tag{6.22}$$

Here $t_{1/2}$ is the **half-life time** of the singlet spheres and c_0 the initial singlet number density at $t=0$; see also Figure 6.3B. For a starting volume fraction $\phi_0 = c_0(4/3)\pi R^3$ of singlet particles the half-life time equals:

$$t_{1/2} = \frac{4\pi R^3}{3k_{11}\phi_0}. \tag{6.23}$$

It is evident that—due to the R^3 dependence of the half-life time in (6.23)—colloids in the micron size range flocculate relatively slowly, whereas for nanoparticles flocculation occurs very much faster. At a given volume fraction, according to equation (6.23), $t_{1/2}$ for nanoparticles with radius $R=1\,nm$ is a staggering factor of 10^9 smaller than the half-life time of colloids with radius $R=1\,\mu m$.

6.4 Slow flocculation

When colloids repel each other by, for example, the electric double-layer (EDL) repulsion treated in Chapter 5, the rate of flocculation will obviously go down as the inter-colloid repulsion decreases the colloid's Brownian encounter frequencies. The process of retarded coagulation can be visualized as colloids that slowly diffuse up a potential energy curve in the direction of a neighbour colloid, that pass the maximum of the EDL repulsion, and that subsequently diffuse relatively rapidly downhill to meet the irreversibly sticky neighbour. To investigate slow flocculation we will restrict ourselves to colloids that diffuse in one linear direction only, towards a sink that irretrievably removes free particles from solution; this simplest instance of a diffusion scheme suffices to comprehend and model the retardation of early-phase flocculation rates (Overbeek and Kruyt, 1956).

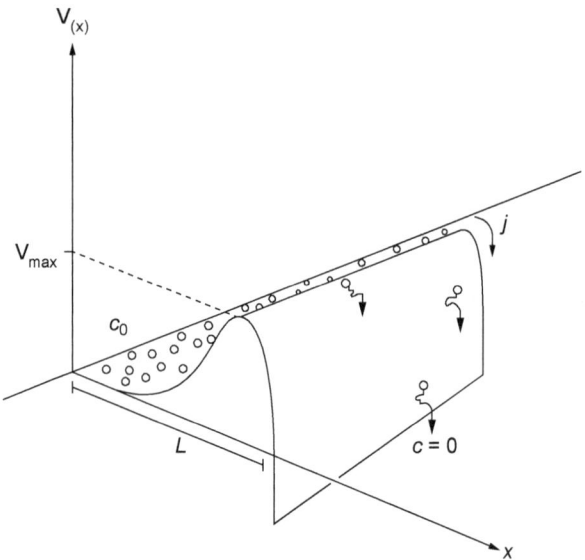

Figure 6.4 Particles diffuse in the x-direction from a source with constant concentration c_0, over a distance L across a potential barrier with height V_{max} into an absorbing sink where the concentration of free particles remains zero.

Consider, then, colloids in Figure 6.4 that diffuse out of a large reservoir with constant colloid number density c_0 in the x-direction and that, after crossing an energy barrier with height V_{max}, end up in a sink where the concentration of *free* colloids is zero. Our point of departure is the constant, steady-state diffusion flux j_d in the absence of any repulsion (so $V_{max} = 0$), which according to Fick's first law reads

$$j_d = -D\frac{dc}{dx}. \tag{6.24}$$

The boundary conditions for source and sink in Figure 6.4 are, respectively, $c = c_0$ at $x = 0$ and $c = 0$ at $x = L$. So we can integrate (6.24)

$$j_d \int_0^L dx = -D \int_{c_0}^0 dc, \tag{6.25}$$

to obtain for the constant diffusion flux of colloids with diffusion coefficient D:

$$j_d = D\frac{c_0}{L}. \tag{6.26}$$

Diffusion in a force field

We now switch on a potential of interaction, denoted by V, between two colloids; necessarily associated with this potential is a force K exerted on each colloid which is defined by:

$$K = -\frac{dV}{dx}; \quad V = V(x). \tag{6.27}$$

Here we have indicated explicitly that the interaction potential depends on distance x—and on x only. Force K imparts on colloids a stationary speed u that, according to **Stokes' law**, entails a frictional force fu (f is the Stokes friction factor) that balances K:

$$fu = K. \tag{6.28}$$

The directed, convective flux j_c of particles set up by force K equals the concentration $c = c(x)$ of particles times their speed:

$$j_c = c\,u. \tag{6.29}$$

Combining equations (6.27) to (6.29) we obtain for the convective, force-induced flux of colloids:

$$j_c = -\frac{c}{f}\frac{dV}{dx}. \tag{6.30}$$

The *total* steady-state flux j is the sum of diffusion flux in (6.24) and convective flux in (6.30)

$$j = j_d + j_c = -D\left[\frac{dc}{dx} + \frac{c}{k_B T}\frac{dV}{dx}\right]; \quad D = \frac{k_B T}{f}. \tag{6.31}$$

To solve this differential equation, we need to make a convenient substitution for the variable c, the colloid concentration profile in the x-direction in Figure 6.4. The choice for this substitution will become clear after first solving (6.31) for the case $j = 0$.

Equilibrium colloid profiles

The flux in Figure 6.4 can be switched off by closing the sink such that colloids no longer disappear upon their diffusional arrival at $x = L$. What then happens is that colloids adopt the equilibrium concentration profile $c_{eq}(x)$ along the x-direction in Figure 6.4. On putting $j = 0$ in (6.31) we obtain:

$$\frac{dc}{c} = -\frac{dV}{k_B T}, \tag{6.32}$$

the solution of which is the Boltzmann distribution for colloids in the potential profile $V = V(x)$:

$$c_{eq}(x) = c_0 \exp\left[-V/k_B T\right]. \tag{6.33}$$

Non-equilibrium profiles

If we now open again the sink in Figure 6.4, the constant steady-state flux $j \neq 0$ will reappear, and colloid concentrations take on values $c = c(x)$ that deviate from the equilibrium values $c_{eq}(x)$ in (6.33). Supposing these deviations are accounted for by a factor γ, then for the non-equilibrium concentrations we have:

$$c = \gamma c_{eq}(x) = \gamma c_0 \exp\left[-V/k_B T\right], \tag{6.34}$$

where $\gamma = \gamma(x)$ is an x-dependent function that in equilibrium has the value $\gamma = 1$; (6.34) is the 'convenient substitution' referred to above, as it leads to the solution of (6.31) as follows. On substitution of (6.34) in (6.31) we obtain (Exercise 6.12) the following equation for γ:

$$\frac{d\gamma}{dx} = -\frac{j}{Dc_0}\exp[V/k_BT], \qquad (6.35)$$

which has the solution

$$\gamma = 1 - \frac{j}{Dc_0}\int_0^x \exp[V/k_BT]dx. \qquad (6.36)$$

Note that this solution satisfies the boundary condition that we imposed on γ, namely that its equilibrium value (for $j = 0$) equals $\gamma = 1$. The second boundary condition on γ is $\gamma(x = L) = 0$; that is to say, the colloid concentration is zero at the entrance at $x = L$ of the sink in Figure 6.4.

Steady-state colloid flux

On substitution of $\gamma = 0$ and $x = L$ in (6.36) we then finally obtain the solution of (6.31), namely the constant, steady-state colloid flux:

$$j = \frac{Dc_0}{\int_0^L \exp[V/k_BT]dx}. \qquad (6.37)$$

Note that for zero interaction potential ($V = 0$) the integral in the denominator equals L such that the flux in (6.37) reduces to the purely diffusional flux in (6.26); for a repulsive interaction ($V > 0$) the integral exceeds L and the flux is reduced with respect to (6.26). For an estimate of this reduction we note that the value of the integral in (6.37) is dominated by the exponent of the maximum V_{max} of the potential curve in Figure 6.4—a dominance that justifies the approximation:

$$\int_0^L \exp[V/k_BT]dx \approx \frac{1}{2}L\exp[V_{max}/k_BT]. \qquad (6.38)$$

Hence, the magnitude of the particle flux is about:

$$j \approx 2D\exp[-V_{max}/k_BT]\frac{c_0}{L} = D_{eff}\frac{c_0}{L}. \qquad (6.39)$$

Effective diffusion coefficients

What we see here is that the retardation of early-phase flocculation rates by inter-colloid repulsions can be approximately accounted for by replacement of the free-particle diffusion coefficient D in Fick's first law (6.26) by the effective diffusion coefficient

$$D_{eff} = 2\exp[-V_{max}/k_BT]D. \qquad (6.40)$$

Due to the exponential dependence of D_{eff} on $-V_{max}$, inter-colloid repulsions may strongly reduce flocculation rates. To give a numerical example: for a barrier of $V_{max} = 10\,k_B T$, we have in order of magnitude $D_{eff} \sim 10^{-4} D$; since the half-life time for early-phase flocculation depends, according to equations (6.22) and (6.17), on the inverse of the diffusion coefficient, $t_{1/2} \propto 1/D$, it follows that a repulsive barrier of $10\,k_B T$ increases the half-life time by a factor of order 10^4. For barriers larger than, say, $V_{max} \approx 20 k_B T$ flocculation will be an immeasurably slow process—the defining characteristic of colloidal stability of a dispersion.

6.5 Kinetics of late-stage coagulation

Beyond the initial stage of dimer formation that we discussed in Section 6.3, further aggregation of attractive particles by Brownian motion produces triplets, quadruplets, etc. (see Figure 6.3A and C) which, in turn, also encounter each other by diffusion to form larger clusters. Smoluchowski showed that this—at first sight hopelessly complicated—kinetic problem can be approximately solved as follows (Overbeek and Kruyt, 1956).

Consider the number density c_α of aggregates that contain α spheres. Such α-mers are formed by the encounters of smaller aggregates and disappear by the uptake of any other particle or aggregate. The change in α-mer concentration in time (see also Exercise 6.10) is:

$$\frac{dc_\alpha}{dt} = \frac{1}{2}\sum_{i=1}^{\alpha-1} k_{i,\alpha-i} c_i c_{\alpha-i} - \sum_{i=1}^{\infty} k_{i,\alpha} c_i c_\alpha. \qquad (6.41)$$

This equation can be solved easily if we ignore any difference between reaction rate constants k_{ij} and, consequently, put all rate constants equal to $k_{ij} = k_{11}$:

$$\frac{dc_\alpha}{dt} = \frac{1}{2} k_{11} \sum_{i=1}^{\alpha-1} c_i c_{\alpha-i} - k_{11} c_\alpha \sum_{i=1}^{\infty} c_i. \qquad (6.42)$$

Before we move on, let us first see how we can justify this (at first sight rather drastic) assumption that all rate constants equal k_{11}; it implies, for example, that the rate constant for irregular aggregates of, say, 10 particles equals the rate constant for single spheres. Note, however, that the diffusion coefficient D of a cluster is inversely proportional to the typical cluster size R_c, which entails that the rate constant $k \propto D R_c$ is indeed fairly insensitive to the shape and size of the aggregates that form in the flocculation process. In other words, growth of a cluster makes it a larger target, but also a slower diffuser. Let us return to differential equation (6.42) and observe that in terms of the total number density

$$c_{tot} = \sum_{i=1}^{\infty} c_i, \qquad (6.43)$$

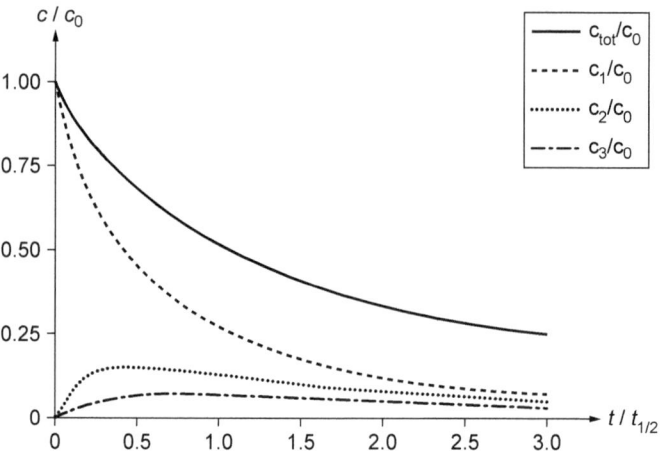

Figure 6.5 Change in species concentration in time according to equation (6.46). Time t is scaled here on the half-life time $t_{1/2}$ of the total concentration c_{tot}; c_0 is the initial number density of spherical colloids at $t=0$.

it can be rewritten (see also Exercise 6.11) to the differential equation of a simpler form

$$\frac{dc_{tot}}{dt} = -\frac{1}{2} k_{11} c_{tot}^2, \qquad (6.44)$$

which has the solution

$$c_{tot}(t) = \frac{c_0}{1 + t/t_{1/2}} \; ; \quad t_{1/2} = \frac{2}{k_{11} c_0}. \qquad (6.45)$$

Here the half-life time for the *total* number of particles (monomers, dimers, trimers, etc.) turns out to equal twice the half-life time of singlet spheres in equation (6.22). Apart from the total particle number density, we can also evaluate the concentration of all the various particle species (α-mers) in time. From equation (6.42):

$$\frac{dc_1}{dt} = -k_{11} \sum_{i=1}^{\infty} c_i c_1 ; \quad \frac{dc_2}{dt} = \frac{1}{2} k_{11} c_1^2 - k_{11} \sum_{i=1}^{\infty} c_i c_2, \; etc. \qquad (6.46)$$

Again, all rate constants are taken to be equal: $k_{ij} = k_{11}$, which leads to the time-dependencies of concentrations of the various species as sketched in Figure 6.5.

6.6 Diffusional growth

Here we will address a special case of a late-stage coagulation process that comprises large clusters growing by the diffusional uptake of monomeric particles from their environment (Philipse, 2018). So, the monomers do not aggregate with each other, but only with the cluster. To this process, which models the growth of colloids in a molecular solution, we can directly apply the Smoluchowski diffusion model from Section 6.2. For a colloidal target sphere i that is much larger than monomers j, the collision frequency of monomers on the target follows from (6.6) as:

$$J(j \to i) \approx 4\pi D_j R_i c_{j,\infty}. \tag{6.47}$$

Here we have put the mutual diffusion coefficient D_{ij} in (6.6) equal to the diffusion coefficient D_j of the rapidly diffusing monomers, and replaced the collision diameter R_{ij} by the radius R_i of the big cluster. Considering that each j-particle contributes a volume v_j to the volume V_i of the growing target cluster, we can write for V_i the differential equation:

$$\frac{dV_i}{dt} = J(j \to i) v_j; \quad V_i = (4/3)\pi R_i^3. \tag{6.48}$$

Substitution of (6.47) and integration yields for the colloid radius R_i at time t:

$$R_i^2(t) - R_i^2(t_0) = 2D_j \phi_j (t - t_0); \quad \phi_j \approx c_{j,\infty} v_j. \tag{6.49}$$

Here $R_i(t_0)$ is the target radius at time t_0; the volume fraction ϕ_j actually surpasses the true volume fraction of j-particles because the volume contribution v_j to the growing sphere volume exceeds the volume of j-spheres. In equation (6.49), incidentally, we can discern the scaling $R \sim t^{1/2}$ —the colloid radius growth in proportion to the square-root of time, which is characteristic for diffusion-controlled growth. The growth equation (6.49) is indeed an instance of Einstein's law for quadratic displacement (given in equation (6.7)) here in the form of a particle radius squared that grows linearly in time. The Stokes–Einstein single-particle diffusion coefficient D is replaced here by an effective diffusion coefficient $D_j \phi_j$.

To compute, in order of magnitude, the colloid growth rate from equation (6.49) we employ the rule of thumb that diffusion coefficients of small molecules or ions in water are of order $D_j \sim 10^{-5}$ cm^{2}s^{-1}. So, for a volume fraction $\phi_j = 0.01$, equation (6.49) predicts a sphere growth rate of about $dR_i^2/dt \approx 20$ μm^{2}s^{-1}. This is very fast: for a colloid radius of $R_i = 1$ μm this would entail a radius growth rate of $dR_i/dt \approx 10$ μm s^{-1}. Growth of colloids in a supersaturated solution is often very much slower than expected from equation (6.49). Retarding factors include exhaustion of the bulk (decrease of ϕ_j in time) or a chemical process that only slowly generates the particles j. Also, a repulsion between colloid and monomers will reduce the diffusive monomer flux towards the colloids. This diminished flux is an instance of the slow flocculation that is treated in Section 6.4.

Here we have looked at diffusional growth of a homogeneous sphere. An analysis of the growth of irregular, fractal-like objects is presented in Appendix 2A of Chapter 2.

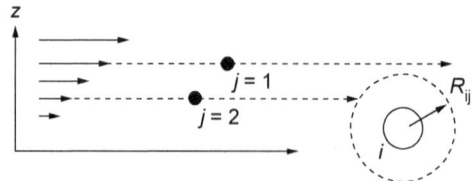

Figure 6.6 Centres of particles j move in a shear flow towards target particle i. Shear-induced flocculation is auto-catalytic: the larger the particle aggregate, the more rapidly it will catch other particles in a stirred solution.

6.7 Flow-induced flocculation

The EDL repulsion between charged colloids, as we saw in Chapter 5, decreases when the salt concentration of a dispersion is increased, or when the pH is brought closer to the colloid's isoelectric point. When the colloids get 'sticky' due to Van der Waals attractions but, nevertheless, still aggregate only quite slowly, the colloidal dispersion is said to be 'marginally stable'. Surprisingly, shaking or stirring such a marginally stable dispersion may strongly enhance the flocculation rate. In a quiescent dispersion, colloids only encounter each other via undirected Brownian motions. In a stirred dispersion, however, colloids move ('convect') at different speeds along different flow lines which enhances the colloids' encounter frequency and, hence, accelerates the aggregation rate (Figure 6.6). This so-called **flow-induced flocculation** is auto-catalytic in nature: the larger a convecting aggregate, the faster it grows by uptake of other particles.

To model flow-induced flocculation (Overbeek and Kruyt, 1956), and its auto-catalytic character, we consider colloids that are propelled by solvent flowing at speed $v(z)$ at height z (see Figure 6.7). The change in solvent speed in the z-direction is called the velocity gradient $\dot{\gamma} = dv/dz$. Imagine a sphere labelled i that has its centre at position $z = 0$. The j-particles at height z move at speed $v(z) = \dot{\gamma}z$; on inspection of Figure 6.7 we infer that the flux of j-particles in the direction of the i-sphere is:

$$c_j v(z) 2\left(R_{ij}^2 - z^2\right)^{1/2} dz; v(z) = \dot{\gamma}z. \tag{6.50}$$

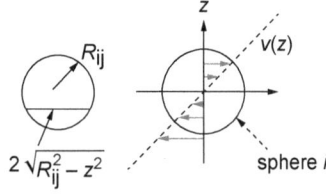

Figure 6.7 Drawings accompanying the derivation of the rate constant in equation (6.52) for flow-induced flocculation of colloids propelled by flowing solvent.

Here R_{ij} is the collision diameter and c_j is the number density of j-particles. Note that j-spheres with their centres at altitudes $z > R_{ij}$ do not collide with the i-sphere. Hence integration of (6.50) from $z = 0$ to $z = R_{ij}$ gives the flux of j-particles on the upper hemisphere of particle i. The absolute magnitude of the flux on the total surface of the i-sphere is therefore:

$$J(j \to i) = 2 \times c_j \dot{\gamma} 2 \int_0^{R_{ij}} z\left(R_{ij}^2 - z^2\right)^{1/2} dz = \frac{4}{3}\dot{\gamma} R_{ij}^3 c_j. \tag{6.51}$$

The corresponding rate constant is:

$$(k_{ij})_{flow} = \frac{4}{3}\dot{\gamma} R_{ij}^3. \tag{6.52}$$

We compare this flow-induced rate constant to the purely diffusional rate constant derived in Section 6.2:

$$\frac{(k_{ij})_{flow}}{(k_{ij})_{diff}} = \frac{4\dot{\gamma}R_{ij}^3}{3 \times 4\pi D_{ij}R_{ij}} \approx \frac{4R^3\eta\dot{\gamma}}{k_B T}. \quad (6.53)$$

Here we ignore any size difference between i- and j-particles and employ the Stokes–Einstein diffusion coefficient $D = k_B T/6\pi\eta R$. From (6.53) it follows that the larger the colloids, the more important flow-induced flocculation becomes in comparison to pure diffusion-induced coagulation. The R^3-dependence in (6.53) also explains the already-mentioned auto-catalytic character of flow-induced aggregation: once flocculation has started (by either diffusion or stirring), further stirring strongly accelerates the process: the larger the aggregates or flocs, the more rapidly they will catch other aggregates in the agitated suspension.

Summary

- The kinetics of irreversible flocculation of attractive colloids is determined by the frequency at which colloids encounter by Brownian motion.
- The rate constant k_{11} for early-phase flocculation at temperature T in solution with viscosity η is independent of particle size and equals $k_{11} = 8k_B T/3\eta$
- The half-time for initial flocculation is inversely proportional to k_{11} and increases, for given initial particle volume fraction, strongly with particle size.
- Retardation of flocculation kinetics by an inter-colloid repulsive barrier can be modelled via effective diffusion coefficients that decay exponentially with barrier height.
- Later-stage coagulation kinetics is computable when all rate constants nearly equal k_{11}, signifying that the product of cluster size and diffusion coefficient is about constant.
- Stirring dispersions may strongly, and auto-catalytically, enhance flocculation rates: the larger the convecting aggregates, the faster they grow by catching other particles.

References

J. K. G. Dhont (1996), *An Introduction to the Dynamics of Colloids*. Elsevier, Amsterdam.

J. T. G. Overbeek (1956), 'Kinetics of Flocculation', in *Colloid Science*, vol. 1, 'Irreversible Systems', H. R. Kruyt (ed.). Elsevier, Amsterdam, pp. 278–301.

A. P. Philipse (2018), *Brownian Motion: Elements of Colloid Dynamics*. Springer, Cham, Switzerland.

Exercises

6.1 Derive the equivalent of (6.5) for Brownian motion on a flat plane, for discs with radius R_j towards a target disc with radius R_i. Consider the limit $\delta \to \infty$. What is your conclusion?

6.2 Solve (6.21) to find (6.22).

6.3 Show that (6.12) is the minimum of (6.11).

6.4 Calculate the half-life for the flocculation of identical spheres in the initial stage, for $\phi = 0.01$ and sphere radii of $R = 10$ nm and $R = 10$ μm.

6.5 Consider a mixture of spheres with a certain distribution in the sphere radius. Show that by diffusional growth the distribution will sharpen. Hint: Consider two spheres from the distribution with radii R and $R + \Delta R$ ($\Delta R > 0$), and assess how ΔR changes in time.

6.6 Refer to the reaction rate ratio in (6.53). For which sphere radius R does this ratio equal 1, for a reasonable shear rate of $\dot{\gamma} = 1\,\text{s}^{-1}$? The spheres are dispersed in water with a viscosity $\eta = 0.89$ mPa sec at $T = 298$ K.

6.7 Virions are the complete, infective forms of a virus outside a host cell. They hitch-hike through an animal host via convection of air currents or body fluids. Close to a cell's surface, however, extracellular fluid flow is strongly damped such that virions must rely on their Brownian motion to make the final step to the surface. Consider a single host cell with radius $R = 2$ μm in an aqueous solution with a weight concentration c_w (g/L) of virions. The solution is not stirred so the rate at which virions find the cell is diffusion controlled. Assume each encounter leads to irreversible adsorption of a virion on the cell surface. The virions, modelled by spheres of radius a, have a specific volume of $\bar{V} = 0.7\,\text{cm}^3\,\text{g}^{-1}$.
(a) Formulate the expression for the diffusive virion flux J towards the host cell.
(b) Compute the missing values of this flux in Table 6.2.

Table 6.2 Encounter frequencies J (s^{-1}) between virions and host cell.

c_w (g/L)	10^{-3}	10^{-2}	10^{-1}
a (nm)			
10	–	–	–
20	–	–	–
40	–	–	–

Solution viscosity $\eta = 10^{-3}\,\text{kgm}^{-1}\text{s}^{-1}$; solution temperature; $T = 298$ K

6.8 The following numbers were obtained by counting free silica spheres in water ($T = 298$ K, viscosity 0.89 mPa s) during flocculation by excess sodium chloride:

t/min	0	2	4	7	12	20
$c/10^8\,\text{cm}^{-3}$	100	14	8.2	4.6	2.8	1.7

Calculate the second-order rate constant k_{11} and compare it with the prediction based on the assumption that the flocculation is an irreversible, diffusion-controlled process.

6.9 Colloidal spheres with radius $R = 500$ nm and volume fraction of $\phi_1 = 0.01$ are mixed in an aqueous solution (viscosity $\eta = 0.89$ mPa s at $T = 298$ K) with small nanoparticles with radius $r = 10$ nm and volume fraction $\phi_2 = 0.05$. Calculate the Brownian encounter frequency between small spheres and one big sphere.

6.10 Spell out all rate equations for the formation and disappearance of tetramers, and formulate the net change dc_4/dt of tetramers in time. Verify that on substitution of $4 = \alpha$ you arrive at equation (6.41).

6.11 Demonstrate how (6.44) follows from (6.42).

6.12 Verify that equation (6.35) for γ follows from substitution of (6.34) in (6.31).

6.13 Does the rate equation (6.14) also hold for identical particles (with index $i = j = 1$)?

Find the solutions to these exercises at the end of the book.

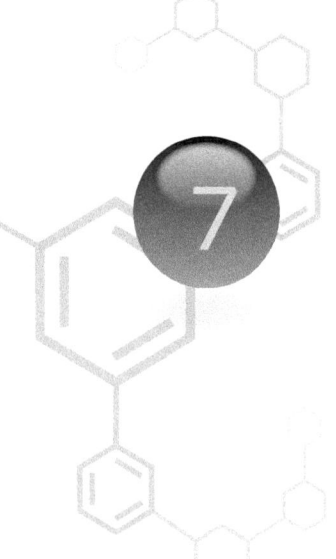

7 Sedimentation, Flow in Porous Media, and Random Particle Packings

7.1 Introduction

The settling of colloids in a dispersion in Earth's gravity field or a centrifuge, a process known as **sedimentation**, is employed for the purpose of separating colloids from the dispersion, or to determine the size of dispersed particles. The crux of the size determination is the balance between the gravitational pull on the particle and the viscous resistance a settling particle experiences—a balance that yields a particle dimension, as explained in Section 7.2. A viscous force is exerted on a sedimenting colloid because it moves relative to the stagnant solvent; this motion gives rise to the Stokes friction factor f—the same factor that, as discussed in Chapter 3 on Brownian motion, also appears in the colloid's diffusion coefficient $D = k_B T/f$. Of course, if we fixed the particle in space and allowed the solvent to move at the same speed relative to the colloid, the viscous friction would be the same. This brings us to the insight that the sedimentation of a group of Brownian particles in a stagnant solvent and liquid filtration through a group of static particles are equivalent processes—or, to be precise, *nearly* equivalent, because the Brownian motion of the sedimenting particles somewhat dampens the frictional force per particle, as we will see in Section 7.4.

Examples of static particle collections are the random particle packings shown in Figure 7.1. Liquid motions through such packings are instances of viscous flow in porous media—materials comprising an open, usually tortuous pore structure with a high specific surface area of the solid phase. Modelling of flow in **porous media**—as will take place in Section 7.4—is a worthwhile undertaking as liquid migration occurs in a wide range of porous materials, including soil, porous rocks, separation membranes, packed-bed reactors, drying paints, and powdered coffee.

Flow through porous media in the form of particle packings also occurs in a **colloidal filtration** process. Here, pressurized liquid migrates through a stacking of particles that is deposited on a porous substrate. Colloidal filtration is employed, for example, in the ceramics industry in the slip-casting of ceramic dispersions; in this casting process particles accumulate on the inside of a porous mould that draws in solvent. Upon drying, the deposited particle packing forms a 'green shape' that is sintered to, for instance, plates, cups, and crucibles. Another large-scale example of colloidal filtration occurs in the paper-making industry, where aqueous suspensions of cellulose fibres (with additives in the form of silica or clay colloids) are filtered on a wire, to form wet paper sheets.

In this chapter we start in Section 7.2 with examining sedimentation speeds and how they yield information on particle size and interactions between particles. Settling of colloids continues until equilibrium is reached in the form of a time-independent **sedimentation–diffusion equilibrium (SDE)** concentration profile. Section 7.3 explains how this SDE profile yields the buoyant colloid mass—and clarifies why for *charged* particles this mass determination only works for high-salinity solutions. Section 7.4 further explores the analogy between sedimentation and liquid permeation, to arrive at Darcy's law for viscous flow in particle packings and porous media. The liquid permeability defined by Darcy's law depends, in a way elucidated in Section 7.5, on volume fraction and specific surface area of a porous medium's solid phase. These solid-volume fractions are a priori known for the random sphere and rod packings discussed in Section 7.6.

7.2 Single-particle sedimentation

Sedimentation mobility

We will start our treatment of sedimentation with the case of a single free colloid that settles under gravity in a large volume of solvent; 'free' refers here to a particle settling far away from other particles and vessel walls. The downward force on the colloid with mass m is its weight mg, g being the acceleration induced by gravity. The upward or buoyant force on the colloid equals—according to Archimedes' principle—the weight of the solvent displaced by the colloid. Thus the apparent buoyant mass Δm of a sedimenting colloid equals m minus the mass m_s of relocated solvent:

$$\Delta m = m - m_s = m\left(1 - \frac{\delta V}{m}\right) = m\left(1 - \delta \bar{V}\right). \tag{7.1}$$

Here δ is the solvent mass density and $\bar{V}$ the specific particle volume—the displaced solvent volume V divided by colloid mass m. The colloid settles in a liquid under gravity at a stationary speed v, achieved when its buoyant weight equals the frictional force fv exerted on it:

$$(\Delta m)g = fv, \tag{7.2}$$

where f is the Stokes friction factor and g the gravitational acceleration. The **sedimentation mobility** μ of the colloid is defined as $\mu = v/g$. Hence:

$$\mu = \frac{v}{g} = \frac{\Delta m}{f}(s). \tag{7.3}$$

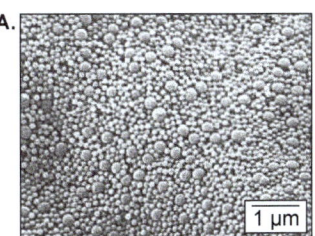

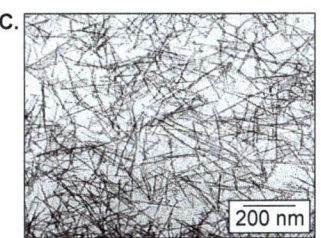
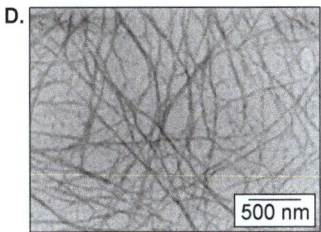

Figure 7.1 Electron micrograph images of particle packings of **(A)** bi-disperse silica (SiO_2) spheres, **(B)** silica ellipsoids, **(C)** silica needles, and **(D)** alumino-silicate (imogolite) fibres. Random particle packing densities decrease with increasing particle aspect ratio (as shown in Section 7.6); so, porosity of the depicted micro-structures increases going from A to D.

Source: Images courtesy of (A) Thies-Weesie and Philipse (1994); (B) Sacanna et al. (2007); (C) Philipse, Nechifor, and Patmamanoharan (1994); and (D) Koenderink, Kluijtmans, and Philipse (1999).

The sedimentation mobility has the unit of time, and equals the time it typically takes for colloids with buoyant mass Δm, starting from a standstill at time $t = 0$, to reach a stationary sedimentation speed (Exercise 7.2a). This time span—that actually matches the momentum relaxation time treated in Section 3.2—is so short (Exercise 7.2b) that in practice one always measures particles moving at constant, stationary speed.

Svedberg relation

For the determination from sedimentation mobilities of (buoyant) masses of colloids or any other solutes such as proteins and biopolymers, we need to specify in equation (7.3) the Stokes friction f. However, for most particle shapes, this factor is theoretically unknown. Fortunately, the friction factor also appears in the diffusion coefficient $D = k_B T / f$ of the solute in question, which we can determine via dynamic light scattering—as detailed in Chapter 11. On substitution of $f = k_B T / D$ in equation (7.3) we obtain the so-called **Svedberg relation**

$$\mu = \frac{D \Delta m}{k_B T} \tag{7.4}$$

which shows how the buoyant particle mass Δm can be determined, without knowledge of particle size or shape, from sedimentation and diffusion data.

Stokes radius

The radius a in a sphere's friction factor $f = 6\pi\eta a$ is also known as the sphere's hydrodynamic radius or the **Stokes radius**. A sphere with mass density δ_p has a buoyant mass $\Delta m = (4\pi/3)(\delta_p - \delta)a^3$ so the sphere's mobility that follows from equation (7.3) is

$$\mu = \frac{(4\pi/3)(\delta_p - \delta)a^3}{6\pi\eta a} = \frac{2a^2}{9\eta}(\delta_p - \delta), \tag{7.5}$$

showing that measured sedimentation speeds yield the square of the Stokes radius. For settling entities of unknown size or shape, sedimentation mobilities can be converted via equation (7.5) to an *effective* radius a_{eff}—the radius of spheres that would settle as fast as the experimental particles. A value for a_{eff} that is much larger than the expected free-particle size very likely demonstrates settling of particle clusters, in which case a_{eff} provides a typical cluster dimension. A too-small a_{eff} presumably manifests retardation of particle settling due to repulsions between the colloids—a retardation that we will briefly address for the case of sedimenting spheres.

The Bachelor coefficient

The mobility μ in equation (7.5) applies to a single sphere that sediments, unaffected by the settling of other spheres, in a very dilute dispersion. When the sphere concentration rises, sedimentation mobilities decrease. For uncharged spheres at low volume fractions sedimentation speeds diminish linearly with concentration (Russel, 1987):

$$\frac{\mu(\varphi)}{\mu} = \frac{f}{f(\varphi)} = 1 - K\varphi, K = 6.55; \varphi < 0.05. \quad (7.6)$$

Here $f(\varphi)$ is the hydrodynamic friction factor of a sphere in a dispersion with sphere volume fraction φ, and K is the so-called Bachelor coefficient, for uncharged hard-spheres equal to $K = 6.55$. The spheres slow down each other's sedimentation mainly because of the retarding effect of solvent backflow induced by sphere settling in a closed vessel, as illustrated in Figure 7.2, left. The EDLs surrounding charged spheres (the subject of Chapter 5) keep the spheres at larger inter-particle distances—which entails more exposure to this backflow, such that sphere settling is slower than expected from (7.6). Attractive spheres, in contrast, form temporary doublets in which they partly shield each other, resulting in faster sedimentation than predicted from the Bachelor relation (7.6). Thus measurement of the concentration-dependent sedimentation of colloids provides information about colloid–colloid interactions—information that is actually quite accurate: even weak colloid interaction energies of less than $k_B T$ can be detected in the concentration dependence of $\mu(\varphi)$ (see Planken et al., 2009).

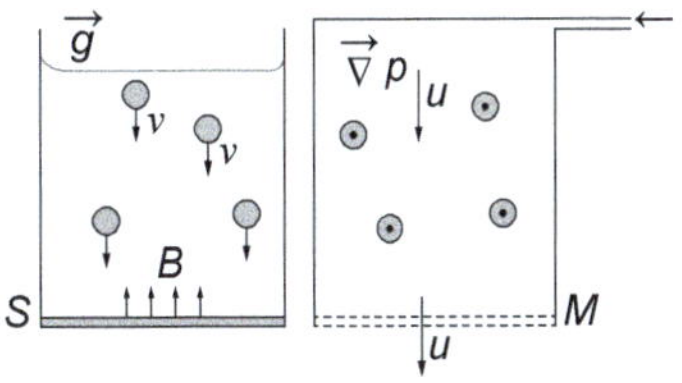

Figure 7.2 **Left**: Colloids (in Brownian motion) sediment downwards under gravity in a dispersion with stationary speed v towards a solid wall S, from which solvent backflow B is going upwards such that net solvent displacement is zero. **Right**: Particles not in Brownian motion but located at fixed positions along which solvent is moving downwards, at speed u driven by pressure gradient $\vec{\nabla}p$, leaving the vessel through a porous membrane M.

Analytical ultracentrifugation

To determine the sedimentation mobility of colloids or macromolecules with masses too small to be probed by gravity, an **analytical ultracentrifuge (AUC)** is needed. In an AUC a spinning rotor exerts a centripetal force on a sedimentation cell, which is directed towards the rotation axis. The corresponding centripetal acceleration of the cell at a distance r from this axis is $a = \omega^2 r$, where ω is the angular rotor velocity in radians per second. The colloids move towards the bottom of the cell (Figure 7.3), experiencing an effective weight increase, equivalent to an enhancement of the gravitational acceleration from g to $\omega^2 r$. Thus, the Svedberg relation (7.3) remains exactly the same, with the sedimentation mobility given by $\mu = v/\omega^2 r$. The determination of μ is as follows. Suppose the boundary between dispersion and supernatant moves at a rate $v = dr_b/dt$. Integration of $\omega^2 r\mu = dr_b/dt$ yields

$$\ln\frac{r_b(t)}{r_b(t_0)} = \omega^2 \mu(t - t_0), \quad (7.7)$$

where $r_b(t)$ is the position of the boundary at time t. The sedimentation mobility μ, therefore, follows from a graph of the logarithmic term in (7.7) versus $(t - t_0)$.

7.3 Sedimentation–diffusion equilibrium

Colloidal particles settle under the influence of gravity until an SDE concentration profile is established (Figure 7.4). This equilibrium arises from the balance between the downward particle flux due to gravity and a back-flux

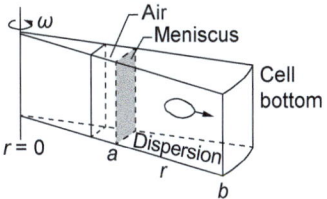

Figure 7.3 Schematic of an ultracentrifugation experiment. Colloids in a dispersion migrate radially to the bottom of a sector-shaped cell, with an apparent weight $(m - m_0)\omega^2 r$ at a distance r from the axis that rotates at angular velocity ω.

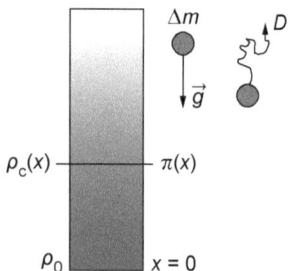

Figure 7.4 When the downward sedimentation flux under gravity of uncharged, ideal colloids with buoyant weight $(\Delta m)g$ is balanced by an upward diffusion flux, the colloid number density $\rho_c(x)$ (Exercise 7.17) obeys the barometric Boltzmann distribution (7.9).

due to diffusion, which opposes the concentration gradient created by gravity. The equilibrium concentration profile $\rho_c(x)$ follows from the isothermal force balance between the gradient in osmotic pressure π and the buoyant particle weight per volume of dispersion:

$$\frac{d\pi}{dx} + \rho_c(x)\Delta mg = 0, \qquad (7.8)$$

where x is the distance to the bottom of the vessel at $x = 0$. For non-interacting, ideal particles, for which Van 't Hoff's law $\pi = \rho_c k_B T$ applies, the solution of differential equation (7.8) is the exponentially decaying SDE profile:

$$\rho_c(x) = \rho_0 \exp(-x/l_g); \quad l_g = \frac{k_B T}{\Delta mg} \qquad (7.9)$$

In equation (7.9), also known as the barometric height distribution, ρ_0 is the particle concentration at $x = 0$, and l_g is the so-called **gravitational length**, which is a measure of the thickness of the profile. The buoyant colloid mass can, in principle, be resolved from an SDE profile $\rho_c(x)$, but in practice this mass determination is a challenge: vessels should be rigorously kept at constant temperature because concentration profiles are easily disturbed, or even washed out, by liquid convections that are caused by even minute temperature gradients. A better option for mass determination are the centrifugal equilibrium profiles discussed next, because an analytical ultracentrifuge is specially designed to avoid disturbance of samples in sedimentation cells by convections or vibrations.

Equation of state

Interactions between colloids lead to deviations from the barometric profile (7.9). The osmotic **equation of state** of the interacting colloids—that is, the osmotic pressure versus colloid concentration—can be determined from the measured non-barometric SDE profile $\rho_c(x)$. To see how this determination works (Piazza, 2014), integrate the force balance (7.8) to obtain:

$$\pi_b = \pi_h + \Delta mg \int_b^h \rho_c(x) dx, \qquad (7.10)$$

where h is an altitude that is sufficiently high, and consequently where the colloid concentration is sufficiently low, for the pressure to obey Van 't Hoff's law. So, by changing the integral's lower boundary b, the pressure π_b as a function of colloid number density ρ_b is recovered. If profiles are much more extended than the gravitational length l_g, osmotic pressures will be much larger than expected from Van 't Hoff's law for ideal, uncharged colloids. This pressure increase will be observed for charged colloids (see below). Attractions between the particles shrink the equilibrium profile, though attraction between colloids may also lead to voluminous non-equilibrium gels.

Centrifugal equilibrium profiles

To determine the colloid mass from an SDE profile in a centrifuge, a lower rotor speed is used than that employed for a velocity experiment—packing of all colloids near the bottom of the cell must be avoided. Instead, it is desirable to achieve a profile which is sufficiently extended for data fitting, in particular of the dilute tail of the profile where colloidal interactions are insignificant. The concentration profile follows from the net centrifugal force $F = (\Delta m)\omega^2 r$ acting on the colloids. The potential energy $U(r)$ of a colloid at radial position r equals the work needed to bring a colloid to r, starting from the meniscus (see Figure 7.3) at radial location a:

$$U(r) = -\int_a^r F dr' = (\Delta m)\omega^2 \frac{1}{2}(a^2 - r^2). \qquad (7.11)$$

The Boltzmann distribution for ideal particles, in the cell from Figure 7.3, is therefore

$$\rho_c(r) = \rho(a)\exp\left[\frac{-U(r)}{k_B T}\right] = \rho(a)\exp\left[\frac{r^2 - a^2}{2\lambda^2}\right]; \quad \lambda^2 = \frac{k_B T}{(\Delta m)\omega^2}. \qquad (7.12)$$

The thickness of the profile, set by the length λ, can be adjusted by changing rotor speed ω. A graph of $\ln \rho_c(r)$ versus r^2 will yield the length λ and, therefore, the buoyant colloid mass.

This—in principle—quite accurate mass determination has frequently been checked for monodisperse biomolecules (proteins, viruses, DNA fragments); molecular masses match the values known from elemental compositions very well (see Harding et al., 1992).

Charged colloids

Electrical charge on colloids may significantly expand the SDE profile, corresponding to a reduction of the effective colloid weight. This weight reduction is the more substantial the lower the ionic strength of the dispersion under study. In Section 4.5 it was shown that at very low salt concentrations, counter-ions exert a pressure $\pi_{ion} = z\rho_c k_B T$, so for uncharged colloids that obey Van 't Hoff's law $\pi_c = \rho_c k_B T$, the total osmotic pressure in the counter-ions-only limit is

$$\pi = (1+z)\rho_c k_B T. \qquad (7.13)$$

Here z is the number of free, mobile counter-ions that each colloid releases into solution. On substitution of (7.13) in the force balance (7.8), we find the concentration profile

$$\rho_c(x) = \rho_0 \exp\left[-x/l_g(1+z)\right], \qquad (7.14)$$

showing a gravitational length which, compared with the uncharged state, has been increased by a factor $(1+z)$—a quite substantial inflation of the SDE profile

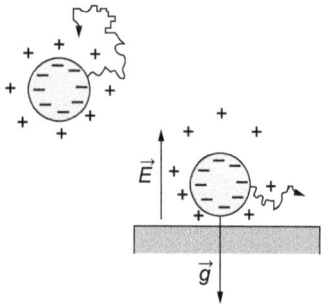

Figure 7.5 **Left**: A charged colloid in free Brownian motion is surrounded by a thermally fluctuating, on average spherically symmetric, counter-ion cloud. **Right**: Gravity biases diffusive displacements of colloids towards the earth but has no effect on the virtually weightless ions that spill into the solution, staying however in the colloid's vicinity to limit separation of opposite charges. Since centres of positive and negative ion distributions do not coincide, an electric field $\vec{E}$ is present that lifts colloids and, hence, reduces their effective weight.

since z may be of order $10^2 – 10^3$. The weight of colloids is reduced because SDE profiles of charged particles harbour an electric field (as explained in the caption to Figure 7.5) that pushes particles upwards against gravity or a centrifugal force. Addition of salt lowers osmotic pressures and, hence, shrinks the SDE profile with respect to (7.14), to reach in the high-salt limit the exponential SDE profile in (7.9) for uncharged colloids (see Raşa and Philipse, 2004).

7.4 Flow in particle packings and Darcy's law

In sedimenting dispersions colloids migrate through the solvent (Figure 7.2) forced by gravity or a centrifuge. The particle velocity relative to the solvent determines the hydrodynamic friction which retards particle sedimentation speeds. One can also fix the particles in a stationary array (Figure 7.2, right), through which liquid is then driven by a pressure gradient. Now it is the liquid flow that is slowed down by the hydrodynamic friction. Such particle arrays, or the particle packings obtained from sedimentation or filtration (Figure 7.1), are instances of porous media—through which viscous liquids permeate at a rate given by **Darcy's law**.

Darcy's law

Suppose liquid flows with speed u through a static array (Figure 7.2, right) comprising a number density ρ of spheres. When the hydrodynamic friction force on each particle in the array equals $f_{sa}(\varphi)u$, then the total frictional force per volume is

$$\rho f_{sa}(\varphi) u, \tag{7.15}$$

where $f_{sa}(\varphi)$ is the hydrodynamic friction of a particle in a sphere array with solid volume fraction φ. If the flow speed u in the array is caused by a pressure gradient, with on average magnitude ∇p, we have the following balance between pressure force and viscous force:

$$\rho f_{sa}(\varphi) \vec{u} = -\vec{\nabla} p. \tag{7.16}$$

Thus, the liquid flow velocity is:

$$\vec{u} = -\frac{1}{\rho f_{sa}(\varphi)} \vec{\nabla} p = -\frac{k(\varphi)}{\eta} \vec{\nabla} p. \tag{7.17}$$

Here

$$k(\varphi) = \frac{\eta}{\rho f_{sa}(\varphi)} \tag{7.18}$$

is the liquid permeability of the particle packing or porous medium through which the viscous liquid is flowing. The linear relation (7.17) between fluid speed and the pressure gradient driving the flow is known as Darcy's law. Note that since the friction factor in the denominator of (7.18) is proportional to solvent viscosity η, permeability k does not depend on η but only on structure and geometry of a porous medium—as is also illustrated by the permeability for a dilute sphere array in (7.19).

The friction factor $f_{sa}(\varphi)$ for spheres fixed in an array exceeds the Stokes factor $f(\varphi)$ for spheres in Brownian motion; the latter motion influences the collective friction, resulting in the $O(\varphi)$ correction in (7.6) to the free-particle friction. For permeation of disordered fixed spheres the concentration dependence is $\sqrt{\varphi}$ (see Saffman, 1973), implying a larger hydrodynamic friction than for Brownian spheres with the same sphere volume fraction. For only a very dilute array of spheres, $f_{sa}(\varphi)$ equals the friction factor $f_{sa}(\varphi \to 0) = f = 6\pi\eta a$ for a single, free sphere; the corresponding permeability is:

$$k_0 = \frac{\eta}{\rho f} = \frac{\eta}{(3\varphi/4\pi a^3)(6\pi\eta a)} = \frac{2}{9}a^2\varphi^{-1}, \quad (7.19)$$

for an array with sphere volume fraction $\varphi \ll 1$. Note that the permeability in (7.19) and sedimentation mobility μ in (7.5) both depend on the sphere radius squared—which underlines the analogy between liquid permeation and sedimentation referred to in the Introduction.

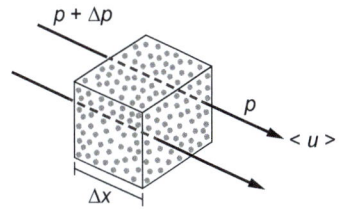

Figure 7.6 Viscous liquid flow with constant, average speed $<u>$ through a small porous cube with edges Δx driven by a pressure difference Δp. The balance between pressure forces and frictional forces on the liquid leads to Darcy's law in the form of equation (7.20).

Unidirectional flow

Darcy's law (7.17) simplifies for one-dimensional viscous flow (Figure 7.6) in the x-direction to:

$$<u> = -\frac{k}{\eta}\frac{dp}{dx}. \quad (7.20)$$

Here $<u>$ is the fluid flow speed, averaged over the volume element of porous medium in Figure 7.6. For a total pressure drop ΔP across a macroscopic medium extending from $x = 0$ to $x = L$ Darcy's law reads in its integral form (Exercise 7.1)

$$<u> = \frac{k}{\eta}\frac{\Delta P}{L}. \quad (7.21)$$

The permeability k in (7.21) is assumed to be independent of x, that is to say, the porous medium is homogeneous and has everywhere the same permeability k. The case of an inhomogeneous medium with a spatially varying k is addressed in Exercise 7.1.

Colloidal filtration

In a colloidal filtration process liquid flows driven by a pressure gradient, through a stacking of particles that is deposited on a porous substrate. The thickness of particle deposits grows in time so percolating liquid experiences an increasing hydrodynamic friction, causing filtration rates to decrease in time. To quantify this rate reduction, we consider the case of the one-directional filtration process sketched in Figure 7.7. A dispersion containing a colloid volume fraction φ forms a particle packing with volume fraction φ_c and time-dependent thickness $L(t)$. Conservation of particle volume entails for the growth rate of the particle 'cake' (Exercise 7.8):

$$\frac{dL(t)}{dt} = \frac{\varphi}{\varphi - \varphi_c}<u>. \quad (7.22)$$

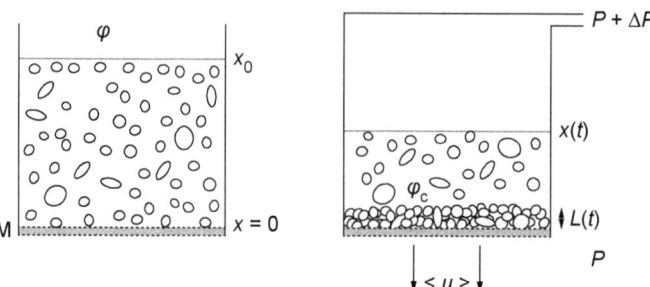

Figure 7.7 Schematic of a colloidal filtration set-up: a pressure difference ΔP drives water, at average flow speed $<u>$; colloids (with volume fraction φ in dispersion) are deposited on a porous substrate M into a particle packing with thickness $L(t)$ and volume fraction φ_c. The liquid meniscus descends from an initial value x_0 at $t=0$ to $x(t)$ at time t.

The average flow rate $<u>$ in Figure 7.7 follows from Darcy's law in its integrated form:

$$<u> = -\frac{k}{\eta}\frac{\Delta P}{L(t)}, \qquad (7.23)$$

where ΔP is the absolute pressure drop across the particle packing. On substitution of (7.23) in the growth rate (7.22) and subsequent integration, we find how the thickness of the particle cake increases in time:

$$L^2(t) = \frac{2\varphi}{\varphi_c - \varphi}\frac{k}{\eta}\Delta P t. \qquad (7.24)$$

Hence, the colloid deposit grows at a rate $dL/dt \propto t^{-1/2}$, implying a fast growth rate in the early stage of the filtration process (which starts at $t = 0$), followed by a significant slowing down of the filtration in a later stage of the process. The square-root time dependence $L \propto t^{1/2}$ of the deposit thickness might take us back to the diffusional growth discussed in Section 6.6—where it was concluded that a cluster radius R increases in proportion to $R \propto t^{1/2}$. However, the increase in deposition thickness is *not* controlled by diffusion but merely by convection of particles, propelled by flowing liquid at a rate that totally dominates their Brownian motion. The square-root time dependence of the layer thickness L stems solely from the increase in time of the viscous force exerted by the growing particle deposit on percolating solvent.

7.5 The Kozeny–Carman scaling relation

Permeabilities of simple geometries like a cylindrical or a rectangular pore can be found by solving the Stokes equation for viscous flow (as we will, in Chapter 8, Section 8.4). The solution of the Stokes equation is the fluid velocity profile in the pore, from which the average flow speed $<u>$ is obtained; dividing $<u>$ by the average pressure gradient driving the flow then provides, in accordance with Darcy's law (7.21), the liquid permeability k of the pore.

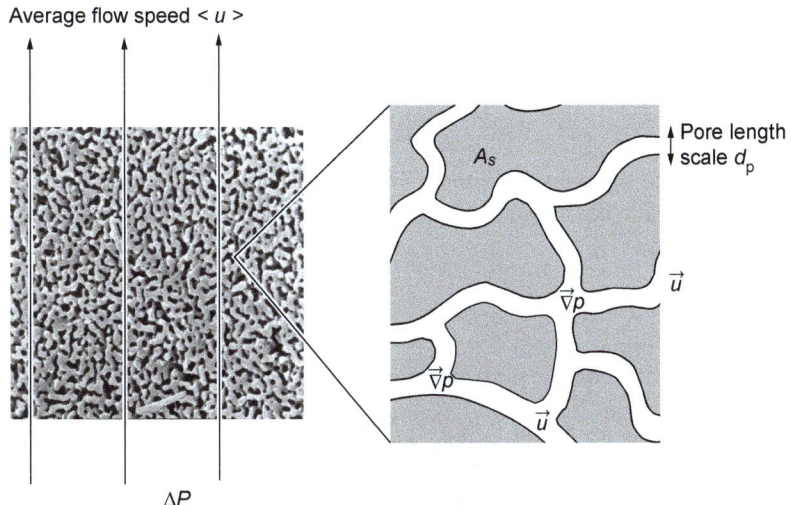

Figure 7.8 Average flow speed $<u>$ (**left**) in a porous medium is, in principle, the volume average of the local flow velocities $\vec{u}$ (**right**) obtained from the Stokes creeping-flow equation (see Section 8.3). The Stokes equation, however, cannot be solved for complex pore geometries such as those shown here, which is why Section 7.5 develops the Kozeny–Carman relation (7.29) that shows how $<u>$ independent of any geometrical details depends on solid volume fraction φ and specific surface area A_s.

However, the topology of pore spaces in porous media may be quite complex, as seen, for example, in Figure 7.8, rendering solving the Stokes equation out of the question. What we *can* do is to assess how k depends on the solid volume fraction and the specific surface area of a porous medium.

The permeability k has the unit m^2 of an area; that area is proportional to the cross-sectional area of a pore. For example, for a capillary with radius R (Figure 8.7) the permeability is proportional to $k \propto R^2$; for a slit-pore composed of two parallel plates at distance d (Figure 7.9) the permeability scales as $k \propto d^2$. Thus, k is proportional to the square of a characteristic pore diameter. Now the typical dimension of any object, whatever its shape, is proportional to its volume V divided by its area O; for example, for a sphere $V/O \propto$ radius a, and for a cube $V/O \propto$ side l. Hence, for a pore of arbitrary geometry, we can define

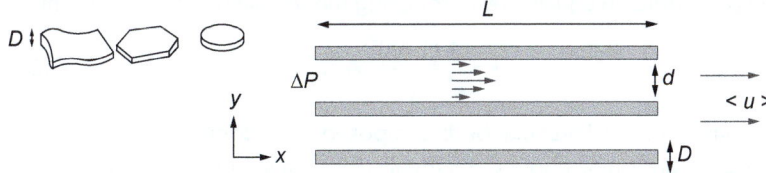

Figure 7.9 Liquid infiltration of a soil of colloidal (clay) platelets with thickness D, modelled by viscous flow with average speed $<u>$ through a parallel stacking of plates at distance d; flow is driven by pressure drop ΔP over distance L. Evaluation of the liquid permeability k of the stacking is the subject of Exercise 7.4.

its characteristic diameter d_p as the liquid volume V_p that fills the pore divided by the solid–liquid area O_s along which the liquid is flowing:

$$d_p = \frac{V_p}{O_s}. \tag{7.25}$$

Let us work out the pore diameter in (7.25) as follows. If the solid phase of the porous medium has a volume V_s, the **specific surface area** A_s of that solid phase is given by:

$$A_s = \frac{O_s}{V_s}. \tag{7.26}$$

Hence, for the characteristic pore diameter we can write

$$d_p = \frac{V_p}{V_s} A_s^{-1} = \frac{1-\varphi}{\varphi} A_s^{-1}, \tag{7.27}$$

where $1-\varphi$ is the pore volume fraction of the porous medium, and φ the fraction of the total porous medium volume that is occupied by the solid phase. The permeability k_p of the pore space is proportional to

$$k_p \propto d_p^2 = \left(\frac{1-\varphi}{\varphi}\right)^2 A_s^{-2}. \tag{7.28}$$

In a liquid permeation experiment, however, we do not measure local fluid flow speeds in pores, but the flow rate averaged over the volume of the whole porous medium (Figure 7.8), including the solid phase. Since inside solids flow rates are zero, the overall liquid permeability $k = (1-\varphi)k_p$:

$$k \propto \frac{(1-\varphi)^3}{\varphi^2} A_s^{-2}. \tag{7.29}$$

This proportionality, known as the **Kozeny–Carman (KC) scaling relation**, tells us how the liquid permeability of a porous medium depends on the volume fraction and the specific area A_s of the solid phase. We see that, according to the KC scaling, an increase in A_s reduces the permeability k—which makes sense because a larger solid–liquid interface entails a higher viscous friction exerted on liquid flowing in the porous medium. An increase in solid volume fraction φ also diminishes k—which is a monotonically decreasing function of φ with the derivative $dk/d\varphi < 0$. We now turn the proportionality sign in (7.29) into an equality by introducing the so-called Kozeny constant C:

$$k = \frac{1}{C} \frac{(1-\varphi)^3}{\varphi^2} A_s^{-2}. \tag{7.30}$$

For porous media and granular beds composed of randomly packed particles, the Kozeny constant is about $C \approx 5$ (Bear, 1988)—a value that is almost independent of particle shape and that also applies to mixed sizes and shapes.

One application of equation (7.30), we note in passing, is the determination of the specific surface area A_s of grains, particles, or powders. A compact of these materials is subjected to a liquid permeation experiment to determine permeability k from the ratio of average flow speed to the applied pressure drop; then the specific surface area of the material in question can be computed via equation (7.30).

The KC scaling (7.29), as we have just seen, is based on the proportionality $\sqrt{k} \propto d_p$ for the characteristic pore size as defined in (7.25). That this proportionality was the right choice follows from the observation that the KC scaling also emerges via the Stokes equation: the scaling is that part of the solution of the Stokes equation which is independent of geometrical details of the porous medium—the details that determine the value for the Kozeny constant C. For example, the KC-scaling is found via numerical solutions of the Stokes equation for flow in random sphere packings (see Heijs and Lowe, 1995). Another example: the solution of Stokes equation for flow in the stacking of parallel plates in Figure 7.9 entails the KC-scaling (Exercise 7.4), and a value of $C = 3$ for the Kozeny constant. For flow in the capillary in Figure 7.10 (Exercise 7.5) we find again the KC-scaling but now with $C = 2$.

The KC-scaling (7.29) implies that permeability k and, hence, colloidal filtration rates are determined by the smallest dimension of the filtrated particles (e.g. the thickness of a platelet (Figure 7.11), which determines the specific surface area A_s). Particles with all dimensions in (or above) the micron range are easy to separate by filtration, but any dimension of particles in the nanometre range yields a high A_s that necessitates high pressures to remove them from a dispersion by filtration (for the impracticality of separating nanoparticles by colloidal filtration see also Exercise 7.11). Sedimentation is not a suitable alternative for separating nanoparticles that are too small for filtration because settling and filtration rates have the same particle-size dependence: sedimentation mobility μ and liquid permeability k in, respectively, equations (7.5) and (7.19) are *both* proportional to particle diameter squared.

It is important to note that we are discussing here filtration and sedimentation of dispersions of *stable*, non-aggregated colloids; coagulation of small particles to large clusters will strongly enhance filtration and sedimentation rates. For example, in the purification of industrial wastewater, heavy metal ions adsorb on iron-oxide nanoparticles that in flocculated state can be easily separated from the waste stream.

7.6 Random particle packings

Particle shape affects the liquid permeability from equation (7.30) of a particle compact in two different ways. First, as already mentioned, particle shape determines the specific area A_s. The second, less obvious, effect of particle shape is that it sets the solid volume fraction φ of the compact. In as far as the compact is a **random close packing (RCP)**—and we will explain in the following paragraph what that is—the statement can be made that packing density φ is uniquely determined by particle shape. Thus, for a chosen particle shape, the liquid permeability can, in principle, be predicted. That, of course, is only possible in as far as the relation between particle shape and

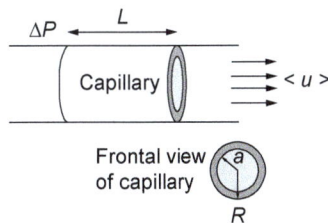

Figure 7.10 Viscous flow through a capillary composed of a cylindrical pore with radius a and solid shell of thickness $R - a$; $<u>$ is the liquid speed averaged over the whole capillary including the solid shell. Flow is driven by a pressure difference ΔP over capillary length L. The liquid permeability of this pore-shell geometry is the subject of Exercise 7.5.

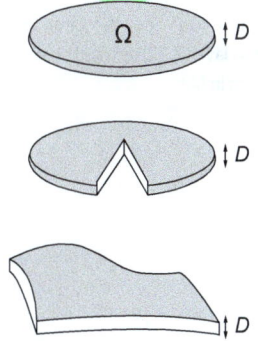

Figure 7.11 A thin pancake (**top**) with area Ω and thickness D has a specific surface area $A_s = \text{area/volume} \approx 2\Omega/D\Omega = 2/D$. This result actually holds for a pancake of arbitrary shape and also remains valid when a slice is cut from the pancake (**middle**) or when it is deformed (**bottom**).

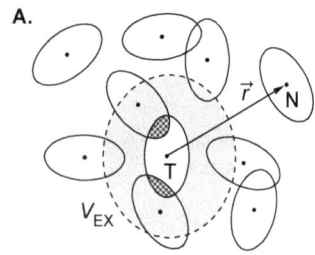

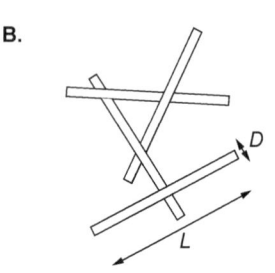

Figure 7.12 (A) A collection of randomly positioned and oriented particles. Vector $\vec{r}$ connects the centres of test particle T and neighbour particle N; it is only when $\vec{r}$ ends in the excluded volume V_{Ex} that the probability that N has an overlap (or contact) with T will be non-zero. (B) Rod positions and contacts become uncorrelated for $L/D \gg 1$; hence thin rods obey the same ideal packing law (7.32) as the ghost particles in (A).

the RCP is known—and that relation has, at the time of writing, only been clearly established for elongated particles in the form of rods and fibres (Philipse, 1996).

An RCP density can be defined as the highest density achievable for a particle packing that has no long-range, crystalline ordering. So the 'randomness' in an RCP refers to long-range randomness; around particles there is local ordering that stems from the excluded volume repulsion; that is, the condition that hard particles cannot overlap—with the exception, as we will see later, of thin-rod packings that are random throughout.

The Bernal sphere packing

The classical instance of an RCP is the Bernal sphere packing, an example of which is shown in Figure 7.13, image labelled $\alpha = 0$. We cannot assign a precise value for the RCP density of the Bernal packing; volume fractions φ from experiments and simulations fall within the density range $\varphi \approx 0.63 - 0.64$. The Bernal packing has been employed to model amorphous solids and **colloidal sphere glasses**—the latter because the Bernal packing is itself a meta-stable glass state as its volume fraction in the range $\varphi \approx 0.63 - 0.64$ exceeds the volume fraction $\varphi = 0.49$ at which colloidal spheres freeze into a crystal. If in a Bernal packing spheres could escape their positions by Brownian motion, the packing would crystallize into an FCC crystal. But spheres are arrested in space by their neighbours which, in turn, are detained by *their* neighbours, etc. The Bernal packing is like a huge three-dimensional traffic jam in which no kinetic pathways are available for spheres to move into an ordered structure of lower free energy. To calculate the RPC density of the jammed spheres is a very complex problem, because positions of spheres at or near contact are highly correlated, due to the excluded-volume repulsion arising from the condition that hard spheres cannot overlap.

Random rod packings

When it comes to packings of anisotropic particles, the shape of most interest is a thin rod—with diameter D and length $L \gg D$—for two reasons. First, the **random rod packing** is the reference model for a broad variety of fibrous materials, including filtration media, cellulose filaments in paper, fibre-reinforced plastics, protein gels, and the xanthane fibres that are widely used as food thickeners. The second reason thin rods deserve special attention is that their RCP density obeys, in marked contrast to the case of spheres, a simple packing law. The underlying reason is that particle contacts in a random rod packing, again in evident contrast to packed spheres, are uncorrelated. To be more precise: correlations between contacts vanish at increasing aspect ratio—a phenomenon that can be qualitatively explained as follows. The surface area of a rod is proportional to DL; a rod–rod contact occupies an area of order D^2, hence contacts occupy a fraction of the rod surface area proportional to $D^2/DL = D/L$. Consequently, in the limit $L/D \to \infty$ this surface fraction vanishes and what remains is a random distribution of ideal point contacts.

Ideal packing law

For particles that have randomly distributed contacts, the relation between particle concentration and particle shape is as follows. Consider a collection of randomly positioned and oriented 'ghost particles' that can freely interpenetrate each other (see Figure 7.12 A). Associated with test particle T in Figure 7.12 is an excluded volume V_{ex}, defined in the following way. When the centre of a neighbour particle is located inside V_{ex}, there is at least one orientation of the neighbour that leads to overlap or contact with T; when the neighbour centre is outside the excluded volume, the contact probability is zero. For an average number density ρ_c of randomly distributed particles, the average number of overlaps, γ_c, with a test particle is the number of particle centres in the excluded volume:

$$\gamma_c = \rho_c V_{ex}. \qquad (7.31)$$

Since the volume fraction φ of particles with volume V_p equals $\varphi = \rho_c V_p$ we obtain an **ideal packing law** of the form

$$\varphi = \frac{V_p}{V_{ex}} \gamma_c. \qquad (7.32)$$

The volume ratio V_{ex}/V_p depends on particle shape and, hence, for a given average γ_c of random contacts, we find from packing law (7.32) how the volume fraction changes with particle shape. For randomly oriented thin rods with diameter D and length $L \gg D$ the volume ratio equals $V_{ex}/V_p = 2L/D$ (Vroege and Lekkerkerker, 1992), so for thin rods:

$$\varphi \frac{L}{D} = \frac{1}{2} \gamma_c; \quad \frac{L}{D} \gg 1. \qquad (7.33)$$

Thin-rod RCP densities

If on average γ_c contacts are needed to arrest a rod in a packing, that number will not change when the rod length increases so γ_c is independent of the rod aspect ratio. Consequently, the RCP density of rods scales with the inverse aspect ratio, as confirmed by the experimental and simulation data in Figure 7.13. From a fit to these data, it follows that $\gamma_c/2 \approx 5$, so the RCP density for thin rods is to a good approximation $\varphi \approx 5D/L$.

This RCP density can be used to evaluate the liquid permeability (7.30) for a medium composed of randomly oriented fibres (see Exercise 7.12). RCP densities are also very useful to estimate the viscosity increase of a dispersion with particle concentration, and to predict at which particle concentrations dispersion viscosities diverge. We return to these rheological consequences of RCP densities in Chapter 8, Section 8.6.

Lastly, as discussed above, the Bernal sphere packing is a meta-stable glass with respect to a crystalline state; the density $\varphi \approx 5D/L$ of random rod packings exceeds the density $\varphi \approx 4D/L$ above which only the liquid-crystalline, nematic phase is thermodynamically stable (Vroege and Lekkerkerker, 1992). Hence, the random packings depicted in Figure 7.1, prepared by sedimentation or colloidal filtration, are actually all colloidal glasses.

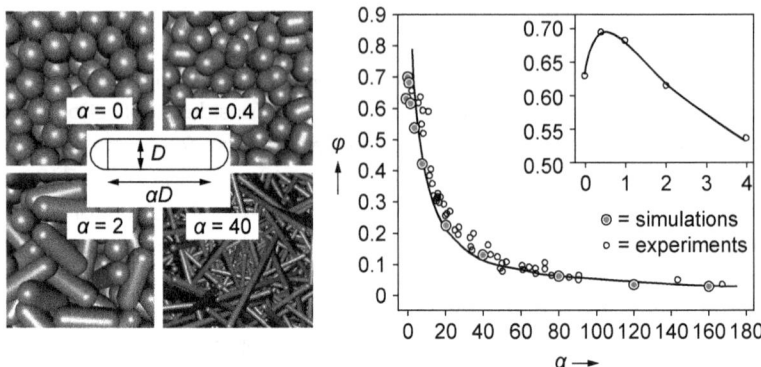

Figure 7.13 Left: Computer-generated random dense packings of spherocylinders for several aspect ratios $\alpha = (L/D) - 1$. **Right**: Volume fractions φ for random packings plotted versus α; solid line is the hyperbola $\varphi\alpha = 5.1$. **Inset**: Magnified plot of simulation data at low aspect ratio; the case $\alpha = 0$ is a random sphere packing with density $\varphi = 0.631$. For $\alpha = 0.4$ the highest random packing density of $\varphi = 0.70$ is achieved. The packing density for $\alpha = 2$ is $\varphi = 0.62$ and already right of the density maximum. For the high aspect ratio $\alpha = 40$, the packing fraction is $\varphi = 5.1/\alpha = 0.13$.

Source: Williams and Philipse, 2003.

Summary

- The mass of sedimenting particles of arbitrary shape can be determined via the Svedberg relation.
- Colloidal sedimentation and filtration rates both increase with particle radius squared.
- Particle charge increases the osmotic pressure exerted by colloids and, consequently, inflates their sedimentation–diffusion equilibrium (SDE) profile.
- Only uncharged particles or macromolecules adopt—in dilute dispersions—a barometric SDE profile from which the buoyant particle mass can be determined.
- Viscous flow in particle compacts and porous media is driven by pressure gradients at a speed that follows from Darcy's law.
- The liquid permeability of a porous structure depends on volume fraction and specific surface area of the solid phase in accordance with the Kozeny–Carman scaling relation.
- Volume fractions φ of random packings produced by sedimentation or filtration are fixed by particle shape; for fibres with aspect ratio L/D, for example, $\varphi L/D =$ constant.

References

J. Bear (1988), *Dynamics of Fluids in Porous Media*. Dover, New York.

S. E. Harding, A. J. Rowe, and J. C. Horton (eds.) (1992), *Analytical Ultracentrifugation in Biochemistry and Polymer Science*. Royal Society of Chemistry, Cambridge.

A. Heijs and C. Lowe (1995), 'Numerical evaluation of the permeability and the Kozeny constant for two types of porous media', *Phys. Rev. E* **5**, 4346.

G. H. Koenderink, S. G. J. M. Kluijtmans, and A. P. Philipse (1999), 'On the Synthesis of Colloidal Imogolite Fibers', *J. Colloid Interface Sci.* **216**(2), 429–431.

A. P. Philipse (1996), 'The Random Contact Equation and Its Implications for (Colloidal) Rods in Packings, Suspensions, and Anisotropic Powders', *Langmuir* **12**, 1127–1138; 5971.

A. P. Philipse, A.-M. Nechifor, and C. Patmamanoharan (1994), 'Isotropic and birefringent dispersions of surface modified silica rods with a bohemite needle core', *Langmuir* **10**(12), 4451–4458.

R. Piazza (2014), 'Settled and Unsettled Issues in Particle Setting', *Rep. Prog. Phys.* **77**, 056602.

K. Planken et al. (2009), 'Ultracentrifugation of Single-Domain Magnetite Particles and the De Gennes–Pincus Approach to Ferromagnetic Colloids in the Dilute Regime', *J. Phys. Chem., B* **113**(12), 3920–3931.

M. Raşa and A. P. Philipse (2004), 'Evidence for a macroscopic electric field in the sedimentation profiles of charged colloids', *Nature* **429**, 857–860.

W. B. Russel (1987), *The Dynamics of Colloidal Systems*. University of Wisconsin press, Wisconsin.

S. Sacanna et al. (2007), 'Observation of a shape-dependent density maximum in random packings and glasses of colloidal silica ellipsoids', *J. Phys.: Condens. Matter* **19**, 376108.

P. G. Saffman, 'On the Settling Speed of Free and Fixed Suspensions', *Stud. Appl. Math.* **52**(2), 115–127.

D. M. E. Thies-Weesie and A. P. Philipse (1994), 'Liquid Permeation of Bidisperse Colloidal Hard-Sphere Packings and the Kozeny-Carman Scaling Relation', *J. Colloid Interface Sci.* **162**(2), 470–480.

G. J. Vroege and H. N. W. Lekkerkerker (1992), 'Phase Transitions in Lyotropic Colloidal and Polymeric Liquid Crystals', *Rep. Prog. Phys.* **55**, 1241.

S. Williams and A. Philipse (2003), 'Random packings of spheres and spherocylinders simulated by mechanical contraction', *Phys. Rev. E* **67**, 051301.

Exercises

7.1 (a) Consider a small cube of porous media with edges Δx through which liquid permeates at constant average speed $<u>$ (Figure 7.6). Derive from a force balance on the cube Darcy's law in its differential form:

$$<u> = -\frac{k}{\eta}\frac{dp}{dx}.$$

(b) Show that for a total pressure drop ΔP across a macroscopic medium extending from $x=0$ to $x=L$ Darcy's law reads in its integral form

$$<u> = \frac{k}{\eta}\frac{\Delta P}{L}.$$

(c) Under part (b) it is assumed that the permeability is a constant. Suppose the permeability varies because the microstructure of the porous medium changes in the x-direction; k is a function of x: $k = k(x)$. Show that the permeability in Darcy's law is then an average permeability of the form:

$$\bar{k} = \frac{L}{\int_0^L k^{-1}(x)\,dx}.$$

7.2 (a) Derive from Newton's second law the speed $v(t)$ for a colloid at time t, settling under gravity; assume that $v(t=0)=0$. Identify the typical time it takes for the colloid to reach the stationary speed v.
(b) Calculate the sedimentation mobility in equation (7.3) for the colloidal C-sphere from Table 3.1 in water $(\eta = 1\text{ mPa s})$.

7.3 Verify that (7.16) is a balance between forces per volume (Nm^{-3}).

7.4 (a) Evaluate the average flow speed $<u>$ and liquid permeability k for soil modelled as a stacking of parallel clay platelets (Figure 7.9) of thickness D at distance d. Start with the formula for the average flow speed in one slit of thickness d. Express k in terms of d.
(b) Rewrite k in terms of the specific surface area A_s from equation (7.26) of the clay plates and compare your finding to the KC relation (7.30). What is the value of the Kozeny constant for the plate stacking in Figure 7.9?

7.5 Demonstrate that the Kozeny constant for the capillary with pore-shell geometry in Figure 7.10 is $C = 2$.

7.6 Check that permeability k in equation (7.18) has the unit m² of surface area.

7.7 Calculate the sedimentation mobility μ in equation (7.5) for particles N, C, and G from Table 3.1, settling under gravity in water $(\eta = 1\text{ mPa s})$.

7.8 Verify the formula for the filtration rate dL/dt in equation (7.22). Hint: employ conservation of volume or numbers of particles (Figure 7.7).

7.9 A colloid is located in a centrifuge cell (Figure 7.3) at a distance $r = 1$ cm from the rotor axis. Calculate the rotor spinning frequency ω (in rpm) for which the colloid weight is ten times as high as in the gravity field.

7.10 Verify expression (7.11) for the potential energy $U(r)$ of a colloid in the centrifuge cell of Figure 7.3. Start with the relation between centrifugal force F and $U(r)$.

7.11 Water $(\eta = 1\text{ mPa s})$ is flowing through a random sphere packing with thickness $L = 1$ cm and a sphere volume fraction $\varphi \approx 0.64$. Estimate the required pressure drop ΔP to maintain an average flow speed of $<u> = 1$ cm/hr for spheres with diameters of, respectively, $R = 5$, 50, and 500 nm. Assume a Kozeny constant of $C = 5$.

7.12 (a) Rewrite the Kozeny–Carman relation for thin, cylindrical rods with diameter D and length $L \gg D$.
 (b) Consider a porous medium composed of randomly oriented rods. What will happen to its permeability if length and diameter of the rods are doubled?
 (c) Estimate the permeability of the medium in (b) for rods with $L = 100$ nm and $D = 5$ nm, assuming a Kozeny constant of $C = 5$.

7.13 (a) Show that the liquid permeability of a dilute array of spheres with radius a with a volume fraction $\varphi \ll 1$ is given by $k_0 = 2a^2/9\varphi$.
 (b) Rationalize why, for given φ, the permeability of the array decreases when spheres get smaller.

7.14 A spherical protein molecule with radius $a = 10$ nm and mass density $\delta_p = 1.15$ g cm^{-3} is immersed in water at 20 °C with a viscosity of 1 mPa s and density $\delta = 1$ g cm^{-3}.
 (a) Calculate the diffusion coefficient of the protein.
 (b) Give an estimate of the protein's sedimentation mobility.
 (c) How far would the molecule diffuse in 1 minute, if we neglected sedimentation?
 (d) How far would the molecule sediment in 1 minute, if we neglected diffusion?
 (e) Calculate the distance the molecule would sediment in 1 minute in a centrifuge, when the molecule is located at distance $r = 10$ cm from the axis of the rotor, spinning around at 2×10^4 rpm.

7.15 An important instance of analytical sedimentation, frequently employed in hospitals and family doctors' practices, is measurement of the settling rate of red blood cells (RBCs) in blood (with added anti-coagulant). Many diseases are accompanied by an increased RBC sedimentation speed. The friction factor f for a disc-like object such as an RBC is larger than for an equivalent sphere with the same volume. We neglect the difference to estimate the settling velocity of an RBC in water under influence of gravity.
 (a) Calculate the equivalent sphere radius for the RBC disc in Exercise 4.3.
 (b) Compute the settling speed of this sphere under gravity in water $\left(\delta = 1.0 \text{ g cm}^{-3}; \eta = 1 \text{ mPa s}\right)$. Take $\delta_p = 1.1$ g cm^{-3} for the mass density of an RBC.
 (c) Calculate the distance this sphere would settle in pure water in one day.

7.16 Table 7.1 lists experimental values for sedimentation mobility μ, diffusion coefficient D, and specific volume $\overline{V}_p$, for various proteins and viruses in water.
 (a) Fill in the missing values in Table 7.1, at locations marked with **X**.
 (b) For every particle type in Table 7.1, μ and D were determined at the same water temperature. Why is that important?

Table 7.1 Sedimentation and diffusion data in water

Substance	$T(°C)$	$\mu / 10^{-13}$ s	$D / 10^{-7}$ cm^{2}s^{-1}	$\overline{V}_p / $ cm^{3}g^{-1}	M
Proteins					
Bovine serum albumin	25	5.01	6.97	0.734	X
Fibrinogen	20	7.9	2.02	X	330,000
Lysozyme	20	1.91	X	0.703	14,400
Viruses					
Bushy stunt virus	20	X	1.15	0.74	10,700,000
Tobacco mosaic virus	20	198	0.46	0.74	X

7.17 Verify the statement in the caption of Figure 7.4 that a balance between particle sedimentation and diffusion flux entails the barometric distribution (7.9).

7.18 Work out equation (7.10) for the barometric profile in (7.9). What result for the osmotic pressure do you obtain?

7.19 A column with height $h = 5$ m contains a random sphere packing with height $h_c = 1$ m and sphere volume fraction $\varphi = 0.64$. The spheres have a total solid volume V and are replaced by the *same* total solid volume V of randomly packed cylinders with aspect ratio $L/D = 20$.
  (a) Estimate the volume fraction φ_c of the random cylinder packing. Hint: Since $L/D \ll 1$ apply the random contact equation (7.33) for thin rods.
  (b) Calculate the height h_c of the cylinder packing.

7.20 Evaluate the average altitude of particles in the barometric height distribution in equation 7.9. Hint: The probability to find a particle at height x is proportional to $\rho_c(x)$.

Find the solutions to these exercises at the end of the book.

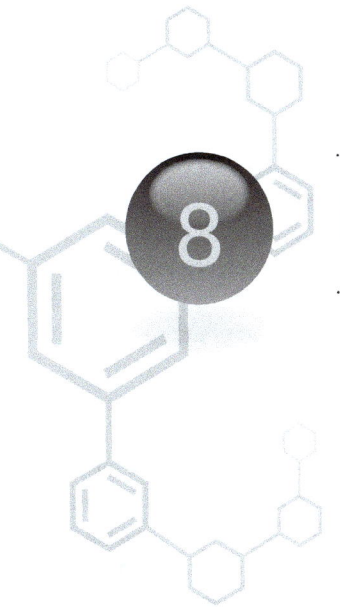

8 Dispersion Flow and Rheology

8.1 Introduction

The science of **rheology** is concerned with the flow and deformation of matter imposed by external mechanical forces. With respect to this deformation two limiting types can be distinguished. In the first type the deformation reverses ('relaxes') spontaneously upon removal of the external force; this reversible deformation is the defining characteristic of an **elastic solid**. The energy needed to create the deformation is stored in such a solid and is fully recovered when that solid relaxes to its initial unperturbed state. At the other extreme, we have liquid matter in which all energy invested to maintain the flow is dissipated as heat. This irreversible energy dissipation is the characterizing property of a **viscous fluid**. Between the two limiting cases of elastic solids and purely viscous liquids are **viscoelastic** systems: fluids and dispersions whose response to applied forces depends on the timescale involved. A concentrated dispersion of starch, for example, exhibits a solid-like resistance towards rapid stirring but yields to a slowly moving spoon. And pitch reacts as an elastic solid if punched, but can be seen to flow over a couple of years down a slope. Also, viscoelastic materials such as mortar, gels, and mayonnaise exhibit a complex combination of liquid-like and solid-like rheological responses. Which type of response prevails depends on the ratio between the time t_R for a material to relax to the time T_{obs} taken to make an observation. This ratio is called the **Deborah number** (De):

$$De = \frac{t_R}{T_{obs}}. \tag{8.1}$$

The number is named after the prophetess Deborah whose hymn in the Old Testament (Judges 5:5) contains the text line that 'the mountains melted before the Lord' which suggests that given infinite observation time even mountains will flow! For $De \gg 1$ a material behaves as an elastic solid; when De is very small it flows like a liquid.

Fluid flow near a wall entails relative motion of adjacent fluid layers that as a result exert frictional forces on each other, proportional to their relative speed and the fluid's **viscosity**. The latter is, loosely speaking, a measure for the stickiness of a liquid: water has a low viscosity and syrup a very high one. Fluids are in stationary motion when the internal viscous forces are balanced by the external forces that drive the flow. This force balance entails the so-called **Stokes equation** for viscous flow. Fluid velocity profiles in pores and capillaries are solutions of the Stokes equation—profiles from which we can compute the total fluid volume that per second, for a given pressure gradient, is transported through a conduit.

Rheology is a vast subject covered by an enormous literature; in this primer we can do no more than provide a first introduction to the topics of viscous flow and dispersion rheology. That introduction starts in Section 8.2 with an outline of fluid viscosity and **Newton's viscosity law**, followed by an account of the significance of the **Reynolds number** (Re). The Stokes equation for viscous flow, developed in Section 8.3, is solved in Section 8.4 for flow in simple pore geometries. Colloids propelled by viscous flow are also subject to the randomizing Brownian motion, and in Section 8.5 we compare Brownian to convective displacements. The presence of colloids enhances a dispersion's viscosity, as we will examine in Section 8.6—which also addresses the marked viscosity increase when colloid concentrations approach random packing densities. Section 8.7 introduces **non-Newtonian fluids**, summarizes the various scenarios for the shear dependence of viscosities, and analyses their rheological response with the **Maxwell model** of viscoelasticity.

8.2 Shear flow, Newton's law, and the Reynolds number

Consider the simple flow condition in Figure 8.1, where a plate with area A glides at constant speed V at distance Δy from a stationary plate located at $y = 0$. The moving plate induces a deformation or a **shear** of the fluid between the two plates, meaning that fluid layers slide along each other at constant, but different speeds—a flow situation also referred to as **laminar flow**. The deformation in Figure 8.1 is expressed in terms of the **shear strain** γ:

$$\gamma = \frac{\Delta x}{\Delta y}. \tag{8.2}$$

The change of the strain with time t, also referred to as the **shear rate** $\dot{\gamma}$, is:

$$\dot{\gamma} = \frac{d\gamma}{dt} = \frac{1}{\Delta y}\frac{d\Delta x}{dt} = \frac{V}{\Delta y} \ (s^{-1}). \tag{8.3}$$

Note that the shear strain is dimensionless, and that shear rate has the unit of reciprocal second. The ratio $V/\Delta y$ is a fluid velocity gradient, which for the geometry in Figure 8.1 happens to be a simple, linear gradient. To also include non-linear gradients, we replace $V/\Delta y$ by the derivative du_x/dy:

$$\dot{\gamma} = \frac{du_x}{dy}, \tag{8.4}$$

where $u_x = u_x(y)$ is the fluid speed in the x-direction at altitude y. A force per unit plane area is a pressure, denoted as p when the force is perpendicular to

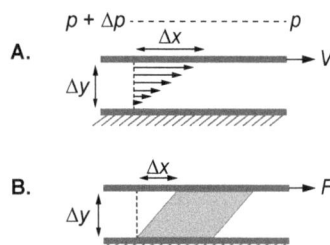

Figure 8.1 (A) A viscous liquid is sheared by an upper plate that moves at constant speed V, parallel to a lower plate mounted on a rigid support. The deformation Δx keeps on increasing in time. The liquid flow is here driven by a pressure gradient $\Delta p/L$: a pressure drop Δp over a distance L in the x-direction. **(B)** An elastic solid sample sandwiched between the two parallel flat plates undergoes a constant deformation Δx due to a force F exerted on the upper plate.

the plane and σ when the force is lying in (tangential to) the plane (Figure 8.1). The latter is also referred to as a tangential pressure, or **stress**:

$$\sigma = \frac{F}{A}. \tag{8.5}$$

Newton's viscosity law

When fluid molecules can freely pass their neighbours, the fluid is easily sheared; when molecules—or any colloids or solute particles—stick together or hinder each other sterically, a larger stress is needed to deform the fluid. This stickiness and hindrance is quantified by the fluid viscosity η: the larger η, the greater the stress required to maintain a certain deformation rate.[1] Accordingly we can define the viscosity as the ratio of shear stress to shear rate:

$$\eta = \frac{\text{shear stress}}{\text{shear rate}} = \frac{\sigma}{\dot{\gamma}}, \tag{8.6}$$

which can be rewritten to

$$\sigma_{xy} = -\eta \frac{du_x}{dy} = -\eta \dot{\gamma}. \tag{8.7}$$

It is this relation between shear stress and shear rate that is known as Newton's viscosity law. The indices xy in σ_{xy} denote an x-directed shear stress that lies in a plane perpendicular to the y-direction. The shear stress, incidentally, opposes the pressure gradient that drives flow in the positive x-direction (Figure 8.1), which is why a minus sign appears in (8.7) since $du_x/dy > 0$. Note that stress is a pressure, with unit Pa; shear rate $\dot{\gamma}$ is a reciprocal second so the viscosity in Newton's law (8.7) has the unit Pa.s = kg m^{-1}s^{-1}. A **Newtonian fluid** is defined as having a viscosity η that is a constant, independent of shear rate. Water, air, alcohol, and milk are familiar examples of Newtonian fluids.

The simple shear flow in Figure 8.1 is an example of viscous flow, also referred to as **Stokes flow**. Purely viscous flow occurs when viscous forces totally dominate the effects of inertia. What that means is the following: to change magnitude or direction of flow velocity $\bar{u}$—on account of Newton's second law—a force has to be exerted on the fluid. That is to say, part of the mechanical work invested to deform the fluid is expended to accelerate fluid mass. However, for small flow rates of sufficiently viscous fluids, this expenditure is negligible compared to the work needed to overcome viscous shear forces.

The Reynolds number

The importance of inertia, or fluid mass, relative to viscous forces, is quantified by the Reynolds number Re, the magnitude of which can be evaluated as follows.

An arbitrary shaped colloid, with mass m and typical dimension L, is at rest, with its centre at position $s = 0$ (Figure 8.2). A force F displaces the particle

---

[1] Overcoming 'stickiness and hindrance' from neighbours by a molecule requires an activation energy E_A, which accounts for the strong, Arrhenius-type temperature dependence $\eta(T) \sim \exp(-E_A/RT)$ of fluid viscosity.

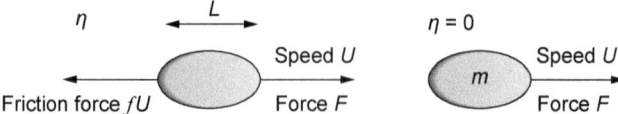

Figure 8.2 Left: A particle with typical dimension L moving at constant speed U through a fluid with viscosity η. **Right:** A particle with mass m advancing at accelerated speed U in a fluid of zero viscosity.

from $s = 0$ to $s = L$. Upon switching on force F, a colloid immersed in a viscous fluid will attain a constant speed U in about $\tau_{MR} = m/f$ seconds. The time τ_{MR} —the momentum relaxation time introduced in Section 3.2—is very much smaller than the 'drift time' $t_L = L/U$ it takes for a stationary moving colloid to reach position $s = L$. In the stationary state F equals the frictional force fU (f is the Stokes friction factor) so the work w_{vis} needed for the displacement is

$$w_{vis} = \int_0^L F ds = \int_0^L fU ds = fUL. \tag{8.8}$$

Now imagine the same colloid submerged in a so-called dry fluid with zero viscosity ($\eta = 0$). Then the force F accelerates the colloid in accordance with Newton's second law $F = m\, dU/dt$ and the work w_{in} required to propel a colloid mass m from $s = 0$ to $s = L$ only stems from the colloid's inertia:

$$w_{in} = \int_0^L F ds = m \int_0^L \frac{dU}{dt} ds = m \int_0^U U dU = \frac{1}{2} m U^2. \tag{8.9}$$

So the invested work equals the colloid's kinetic energy when it has arrived at position $s = L$. The Reynolds number is defined as the dimensionless ratio of inertial to viscous forces:

$$Re = \frac{w_{in}}{w_{vis}} = \frac{mU}{2fL}. \tag{8.10}$$

The Reynolds number can also be expressed as the ratio of the momentum relaxation time $\tau_{MR} = m/f$ to the drift time $\tau_L = L/u$:

$$Re = \frac{\tau_{MR}}{2t_L} = \frac{mU}{2fL}. \tag{8.11}$$

We see that for $Re \ll 1$ the momentum relaxation time τ_{MR} is very much smaller than the displacement time t_L; that is to say, viscosity dominates inertia. It is convenient to rewrite (8.10) as follows. The mass m of an object with mass density δ scales as $m \sim \delta L^3$ whereas the Stokes friction factor scales as $f \sim \eta L$; for example, f is proportional to ηD for a sphere of diameter D and proportional to $\eta \ell$ for a thin rod of length ℓ. Hence, in order of magnitude Re equals

$$Re \sim \frac{\delta UL}{\eta}. \tag{8.12}$$

Clearly, for motion of huge heavy objects $Re \gg 1$, their motion is dominated by their inertia. Colloids and bacteria are sufficiently small to always experience

COLLOID SCIENCE   143

in solutions hydrodynamics at low Reynolds numbers. Unaware of their inertial mass, particles move or are being pushed around in water that to them behaves as viscous syrup—a treacle with which they exchange the random thermal shocks that give rise to the Brownian motion (the diffusive movement that we discussed in Chapter 3).

8.3 The Stokes equation for viscous flow

The velocity flow profile $\vec{u}$ for purely viscous flow at low Reynolds numbers is a solution of the Stokes equation, a differential equation entailed by the balance between pressure and viscous forces on a fluid volume element (Figure 8.3). For the simple flow profile u_x near a flat surface (Figure 8.4A) this force balance is constructed as follows. Flow is driven in the positive x-direction by a gradient in pressure p; the corresponding net normal force (per volume) in the x-direction on a volume element with thickness dx (Figure 8.4B) equals

$$f_N = -\frac{(p+dp-p)A}{Adx} = -\frac{dp}{dx}, \quad (8.13)$$

where A is an area. The net viscous, tangential force (again per volume) on a volume element with height dy (Figure 8.4C) is:

$$f_S = -\frac{(\sigma+d\sigma-\sigma)A}{Ady} = -\frac{d\sigma}{dy}. \quad (8.14)$$

On substitution in (8.14) of Newton's viscosity law (8.7) for the stress $\sigma = \sigma_{xy}$ we obtain for the viscous force

$$f_S = \eta \frac{d^2 u_x}{dy^2}. \quad (8.15)$$

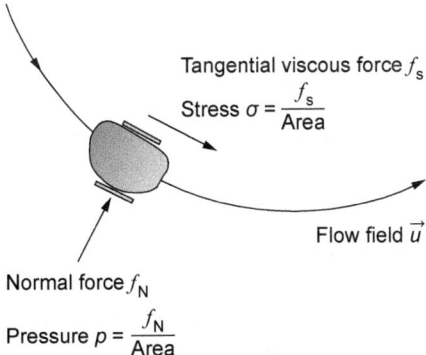

Figure 8.3 A tiny liquid blob in the flow field of a viscous fluid experiences pressures normal to its surface and viscous stresses in directions tangential to the blob surface. For small Reynolds numbers $Re \ll 1$ the blob's inertia has no effect; that is, the net force on the blob is zero and it does not undergo acceleration.

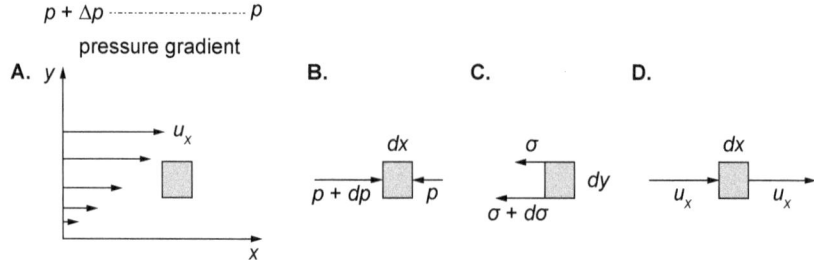

Figure 8.4 (A) Flow profile $u_x = u_x(y)$ near a flat surface located at $y = 0$; the flow is driven by a linear pressure gradient in the x-direction. **(B)** Net pressure dp exerted on a volume element with thickness dx. **(C)** Net viscous stress $d\sigma$ exerted on that same volume element with height dy. **(D)** Fluid enters and leaves the volume element at the same speed such that $du(y)/dx = 0$: the divergence of the flow is zero.

To achieve **stationary flow**, the sum of forces $f_p + f_v$ on any volume element must be zero, which implies

$$-\frac{dp}{dx} + \eta \frac{d^2 u_x}{dy^2} = 0; \quad u_x = u_x(y). \tag{8.16}$$

This is the Stokes or 'creeping flow' equation, for the one-dimensional flow profile in Figure 8.4A. Note that u_x denotes an x-directed flow rate that varies in the y-direction. For an incompressible fluid the Stokes equation has to be supplemented with the so-called continuity condition

$$\frac{du_x}{dx} = 0, \tag{8.17}$$

which expresses that, in a volume element at height y, the fluid mass is constant in time: incompressible fluid enters and leaves the volume element at the same rate (Figure 8.4D). For a general flow field $\vec{u}$ the Stokes equation reads (Acheson, 1992):

$$-\vec{\nabla} p + \eta \nabla^2 \vec{u} = 0; \quad \vec{\nabla} \cdot \vec{u} = 0. \tag{8.18}$$

The continuity condition from (8.17) is generalized here to $\vec{\nabla} \cdot \vec{u} = 0$, which expresses that the divergence of the flow field $\vec{u}$ is zero.

8.4 Flow in simple geometries

In many practical situations fluids and dispersions flow in simple geometries; think of tap water in a conduit pipe, blood in veins lemonade in straws, and colloidal dispersions in the tube of a capillary viscosimeter. In the final case, incidentally, the viscosity of a (Newtonian) fluid is determined from the volume flow rate Q in a tube under a known pressure drop in accordance with the **Hagen-Poiseuille law** (8.33). Stationary viscous fluid flow in a tube—or in any other geometry for that matter—results from a balance between two forces: an external force that drives the fluid forward, and a retarding viscous force exercised by the tube wall on the moving fluid.

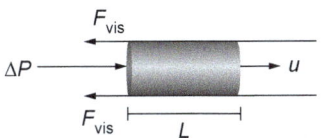

Figure 8.5 A cork moves at constant speed u in a tube because the pressure difference ΔP over cork length L that propels the cork is balanced by the viscous force between tube and cork surfaces that opposes the cork's motion.

Plug flow

Figure 8.5 illustrates this force balance for the simple case of a pressure gradient $\Delta P/L$ that pushes a plug in the form of a cork through a tube. The inner wall of the tube rubs the moving cork and exerts a viscous force F_{vis} on it, which is proportional to the stationary speed u of the cork: $F_{vis} \propto u$. If the pressure gradient in Figure 8.5 increases, cork speed u and F_{vis} increase in proportion. Hence, the force balance on the cork entails the proportionality:

$$u \propto \frac{\Delta P}{L}. \tag{8.19}$$

For the rigid cork in Figure 8.5, all its parts move at the same speed u, a flow situation that is referred to as **plug flow**. A stationary viscous fluid motion will harbour a certain flow profile, to be addressed below, but that does not alter the proportionality in (8.19)—albeit that speed u will have to be replaced by the average flow speed $<u>$:

$$\langle u \rangle \propto \frac{\Delta P}{L}. \tag{8.20}$$

The constant that turns (8.20) into an equality is the liquid permeability k of the geometry through which liquid is flowing, divided by the liquid's viscosity η:

$$\langle u \rangle = \frac{k}{\eta} \frac{\Delta P}{L}. \tag{8.21}$$

We recognize here Darcy's law for unidirectional flow—a law that we met in Chapter 7 in the description of viscous flow in particle packings and porous media.

Flow in a slit

The simplest geometry to consider is a narrow slit formed by two parallel flat plates (Figure 8.6), each of area L^2, at a distance $d \ll L$. For the profile $u_x = u_x(y)$ of the viscous flow in the slit between the plates, the Stokes equation from (8.16) applies:

$$\frac{d^2 u_x}{dy^2} = \frac{1}{\eta} \frac{dp}{dx} = \text{constant}. \tag{8.22}$$

The gradient dp/dx is a constant because for an incompressible fluid $du_x/dx = 0$, so at any position x, fluid is driven by the same pressure gradient. Since $d^2 u_x / dy^2$ is a constant, the fluid flow rate must be a quadratic function of y. Integrating (8.22) twice (Exercise 8.6) indeed yields the parabola

$$u_x = \frac{1}{2\eta} \frac{dp}{dx} \left(y^2 - \frac{1}{4} d^2 \right), \tag{8.23}$$

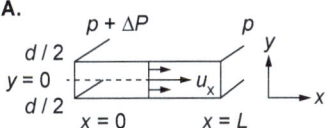

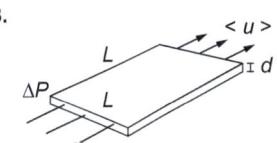

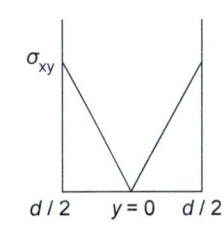

Figure 8.6 **(A)** Parabolic viscous flow profile between two large parallel plates at distance $d \ll L$; the maximal fluid speed occurs at height $y = 0$. **(B)** Driven by pressure gradient $\Delta P / L$, liquid enters and leaves a slit pore at average speed $<u>$; the flowing liquid exerts viscous stress on a total internal pore area of $2L^2[1 + 2d/L] \approx 2L^2$. **(C)** Viscous stress increases from zero at $y = 0$ to its maximum at the pore wall according to equation (8.24).

a solution which satisfies the **stick boundary** or no-slip condition (see also Figure 8.6): $u_x(y = d/2) = 0$ stating that at a plate surface the fluid speed is zero. Insertion of (8.23) in Newton's law (8.7) leads to the viscous stress:

$$\sigma_{xy} = -\eta \frac{du_x}{dy} = -\frac{dp}{dx} y = +\frac{\Delta P}{L} y. \quad (8.24)$$

So the viscous stress increases linearly (Figure 8.6C) from zero at $y = 0$ to its maximal value $\Delta P(d/2L)$: the tangential force (per unit area) the flowing liquid exerts on the walls of the slit pore. The flow speed averaged over the volume between the two plates (see Exercise 8.12) is:

$$\langle u \rangle = \frac{2}{dL} \int_0^{d/2} \int_0^L u_x \, dx \, dy = \frac{d^2}{12\eta} \frac{\Delta P}{L}. \quad (8.25)$$

Here ΔP is the total pressure drop, going in Figure 8.5 from $x = 0$ to $x = L$. Comparison of (8.25) with Darcy's law (8.21) shows that the liquid permeability of the slit pore in Figure 8.6 equals:

$$k = \frac{d^2}{12}. \quad (8.26)$$

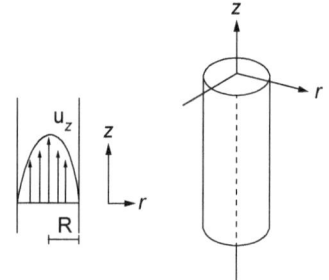

Flow in a cylinder

The axial flow of a Newtonian fluid in a cylinder of radius R (Figure 8.7) has only one velocity component, namely in the z-direction, parallel to the main axis: $u_z = u_z(r)$. The Stokes equation (8.18) reads, for the radial profile,

$$\frac{1}{r} \frac{d}{dr}\left(r \frac{d}{dr} u_z\right) = \frac{1}{\eta} \frac{dp}{dz} = \text{constant}. \quad (8.27)$$

One integration yields the velocity gradient:

$$\frac{du_z}{dr} = \frac{1}{2\eta} \frac{dp}{dz} r + \frac{C}{r}; \, C = 0. \quad (8.28)$$

The integration constant C equals zero, otherwise the velocity gradient and the viscous stress would be infinite on the tube main axis at $r = 0$. From this velocity gradient we obtain the viscous stress via Newton's viscosity law (8.7):

$$\sigma_{zr} = -\eta \frac{du_z}{dr} = \frac{1}{2} \frac{dp}{dz} r. \quad (8.29)$$

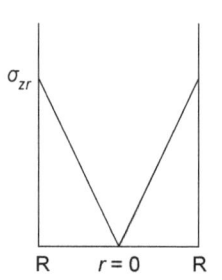

This expression for the viscous stress entails a total viscous force equal to $\pi R^2 \Delta P$ on the inner wall of the tube (Exercise 8.18). The second boundary condition, in addition to absence of an infinite stress at $r = 0$, is the no-slip condition at the tube wall: $u_z(r = R) = 0$. The solution of (8.28) that also satisfies this second condition is:

$$u_z = \frac{1}{4\eta} \frac{dp}{dz} (r^2 - R^2). \quad (8.30)$$

Figure 8.7 Top: Axial flow in a straight tube; the parabolic velocity profile is given by equation (8.30). **Bottom:** Sketch of viscous stress from equation (8.29) which increases from zero in the tube centre to its maximum at the wall.

The average velocity in the tube follows from dividing the integrated flow profile by the tube volume (Exercise 8.12):

$$\langle u \rangle = \frac{1}{\pi R^2 L} \int_0^L \int_0^R u_z \, 2\pi r \, dr \, dz = \frac{R^2}{8\eta} \frac{\Delta P}{L} \qquad (8.31)$$

The similarity to the flat plate result in equation (8.25) is obvious; the different geometry only leads to a different numerical factor in (8.31). Comparison of (8.31) with Darcy's law (8.21) shows that the liquid permeability of the capillary in Figure 8.7 equals

$$k = \frac{R^2}{8}. \qquad (8.32)$$

The permeability of a capillary and that of the slit pore in (8.26) are both proportional to the square of a characteristic pore diameter; it was this proportionality that brought us in Section 7.5 to the Kozeny–Carman relation for the liquid permeability of a porous medium with arbitrary pore geometries.

The average flow speed $<u>$ in (8.31) entails a volume rate of flow through a tube that strongly depends on the tube radius R:

$$Q = \langle u \rangle \pi R^2 = \frac{\pi R^4}{8\eta} \frac{\Delta P}{L}. \qquad (8.33)$$

This is the Hagen-Poiseuille law already mentioned at the start of this Section 8.4; it forms the basis of viscosity measurements on (Newtonian) dispersions: Q follows from the fluid weight that per second leaves a tube, from which η is determined. The R^4 scaling in the Hagen–Poiseuille equation also explains why even minor clogging of arteries may significantly raise blood pressure (Exercise 8.9).

8.5 Diffusion versus convection: the Péclet number

The motion of colloids in a dispersion flowing in geometries like those in Figures 8.6 and 8.7 is a superposition of convection—directed movement imparted by liquid flow—and the undirected, random steps from the colloids' Brownian motion. The latter contributes significantly to displacements of colloids comparable to their own size; traversing large distances, convection is the dominant transport vehicle: we stir to homogenize coffee and milk rather than wait for diffusion to do the mixing.

On the other hand, convection becomes an inadequate means of transport close to a surface, or in sufficiently narrow geometries, where the viscous drag is large. So for small particles or molecules that have to react with a surface—or penetrate a biological cell—the profitable strategy is to cross large distances by convection, followed by Brownian motion for the final sub-micron steps. A typical biological example are viruses that are propelled by the flow of blood or air but that eventually have to arrive at suitable landing places on a new target cell via Brownian motion on a sub-micron scale.

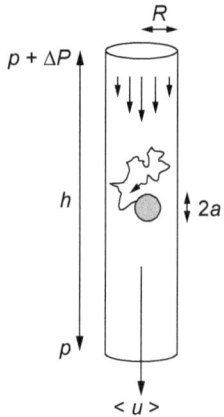

Figure 8.8 Poiseuille flow in a vertical capillary of length h and radius R, driven by pressure difference ΔP caused by the fluid weight. When the Péclet number $Pe \approx 1$, the time needed for a diffusive displacement R of a colloid approximately equals the time needed for the fluid to propel the colloid the same distance.

The Péclet number

A simple case for examining the relative effects of diffusion and convection is provided by a colloid in the Poiseuille flow (Figure 8.8) through a vertical capillary of radius R driven by a pressure difference

$$\Delta P = gh\delta \tag{8.34}$$

for a dilute dispersion with mass density δ in a capillary of length h; g is the gravitational acceleration. The time t_D needed for a colloid of radius a to diffuse a distance equal to the capillary radius R is about:

$$t_D \sim \frac{R^2}{2D} = \frac{3\pi\eta aR^2}{k_B T}. \tag{8.35}$$

We compare this to the time t_C taken by the colloid to travel the same distance by liquid convection. Ignoring the parabolic velocity profile, sketched in Figure 8.8, we just employ the average flow velocity from equation (8.31):

$$\langle u \rangle = \frac{R^2}{8\eta}\frac{\Delta P}{h} = \frac{R^2}{8\eta}g\delta. \tag{8.36}$$

Hence the convection time is approximately

$$t_C \sim \frac{R}{\langle u \rangle} = \frac{8\eta}{Rg\delta}. \tag{8.37}$$

The ratio of diffusion time to convection time is therefore

$$\frac{t_D}{t_C} = aR^3 \frac{3\pi g\delta}{8k_B T}. \tag{8.38}$$

Note that this time ratio also follows from the Péclet number defined as

$$Pe = \frac{\langle u \rangle R}{2D}, \tag{8.39}$$

as can be verified by substitution of (8.36) and the Stokes-Einstein diffusion coefficient for a sphere of radius a in (8.39).

Comparable diffusion and convection

For $Pe \gg 1$ convection dominates whereas for $Pe \ll 1$ diffusion dictates the rate at which a particle migrates. The capillary radius R_B for which $Pe \approx 1$ is about $R_B \approx 1.5$ and $R_B \approx 4$ micron for, respectively, the colloidal C-sphere ($a = 100$ nm) and the nano N-sphere ($a = 5$ nm) from Table 3.1. These estimates for R_B show that for pore radii in the micron range there is a significant transport contribution from both Brownian motion and convection. This puts the radii of blood vessels into perspective. The narrow vessels referred to as 'blood capillaries' are located in body tissues and transport blood from arteries to veins that bring blood back to the heart. Radii of these capillaries are in the range of about 2–5 microns, indicating that species of nanometre size, let alone small molecules, have enough time to reach capillary walls by diffusion during blood flow.

8.6 Concentration-dependent viscosity

The Newtonian viscosity of a liquid is altered, and may even become non-Newtonian (see Section 8.7), when particles are added to form a colloidal dispersion. Suspended colloids increase the viscosity with respect to the solvent viscosity η_0: If, for example, a sphere of radius R is placed in a flow field as sketched in Figure 8.9, forces acting on the upper and lower hemispheres of the particle will make it rotate. The total viscous force F_{vis} opposing sphere rotation is proportional to surface area and solvent viscosity: $F_{vis} \propto \eta_0 R^2$. The torque T required to rotate the sphere equals F_{vis} times the lever arm R (Figure 8.9). Therefore, torque T is proportional to $\eta_0 R^3$. The work w done by torque to produce an angular displacement θ (see Figure 8.9) is:

$$w = T\theta, \tag{8.40}$$

and, hence, is also proportional to $\eta_0 R^3$. Thus, for a number density ρ of spheres, the work w_v spent per unit volume of dispersion on sphere rotations is proportional to

$$w_v = \rho T \theta \propto \rho \eta_0 R^3. \tag{8.41}$$

Due to this expenditure of work we need to apply a higher stress to maintain a certain velocity gradient which, according to Newton's law (8.7), implies a higher viscosity. For the increase of solvent viscosity η_0 to dispersion viscosity η_d we have the proportionalities

$$\eta_d - \eta_0 \propto \rho \eta_0 R^3 \propto \eta_0 \varphi, \tag{8.42}$$

where $\varphi = \rho(4\pi/3)R^3$ is the volume fraction of spheres in the dispersion; note that the volume fraction is small, say $\varphi < 0.01$, because in the argumentation above we considered rotations of single, independent spheres. The constant of proportionality in (8.42) is the **intrinsic viscosity** $[\eta]$—which will be defined in a moment:

$$\eta_d - \eta_0 = \eta_0 [\eta] \varphi. \tag{8.43}$$

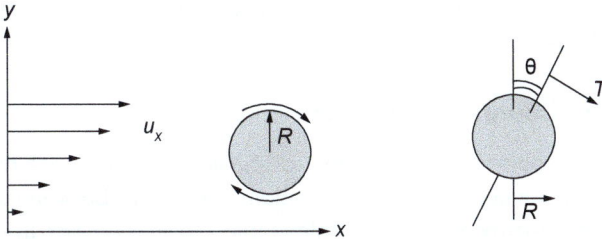

Figure 8.9 Left: A flow field causes a sphere of radius R to rotate. **Right:** The total viscous force F_{tot} on the sphere surface is proportional to $F_{tot} \propto \eta_0 R^2$, and the torque required to rotate the sphere is $T = F_{tot} R \propto \eta_0 R^3$. The work done by this torque entails a viscosity increase.

It is customary to write the concentration dependence of the dispersion viscosity in terms of the **relative viscosity** η_r:

$$\eta_r = \frac{\eta_d}{\eta_0} = 1 + [\eta]\varphi. \tag{8.44}$$

The intrinsic viscosity $[\eta]$ is defined as the limit, at zero concentration, of the **reduced viscosity** η_{red}:

$$[\eta] = \lim_{\varphi \to 0} \eta_{red}; \quad \eta_{red} = \frac{\eta_r - 1}{\varphi}. \tag{8.45}$$

The intrinsic viscosity is a single-particle property that is unaffected by inter-particle interaction—whence the limit $\varphi \to 0$; $[\eta]$ only depends on the shape of the dispersed colloids.

Einstein's viscosity equation

Einstein derived that for spherical particles the intrinsic viscosity equals $[\eta] = 2.5$ (see Fürth (ed.), 1956); thus, Einstein's viscosity equation for sphere dispersions reads

$$\eta_r = \frac{\eta_d}{\eta_0} = 1 + 2.5\varphi. \tag{8.46}$$

One practical application of Einstein's viscosity equation is the determination of the mass density δ_p of spherical particles. The measured viscosity of a dispersion containing a particle weight concentration c leads via (8.46) to a volume fraction φ and the particle mass density $\delta_p = c / \varphi$.

Any deviation from the sphere shape implies, for a given particle volume fraction φ, an increase in dispersion viscosity in comparison to Einstein's equation (8.46) for spheres. When a sphere is deformed to, say, a prolate or oblate particle, the particle's surface area per volume increases, so particles rotating in a shear flow dissipate more energy as heat, which enhances the viscosity. For thin rods of diameter D and length $L \gg D$ the increase in intrinsic viscosity is substantial: for rods with aspect ratio $L/D = 21$ a value of $[\eta] = 6.0$ has been reported (Barnes et al., 1997). Non-spherical particles also randomly pack at volume fractions lower than for spheres, which implies that viscosities rise more steeply with concentration, as will be discussed below. In other words, the non-sphericity of particles enhances viscosities over the whole accessible concentration range, as is illustrated for various particle shapes in Figure 8.10.

The Krieger–Dougherty equation

At increasing particle concentration, dispersion viscosities rise due to hydrodynamic interactions and excluded-volume repulsions between dispersed particles. For a treatment of this complex topic, which is outside the scope of this volume, see Mewis and Wagner (2021). Here we will only address a simple 'effective medium' model that qualitatively accounts for the viscosity increase

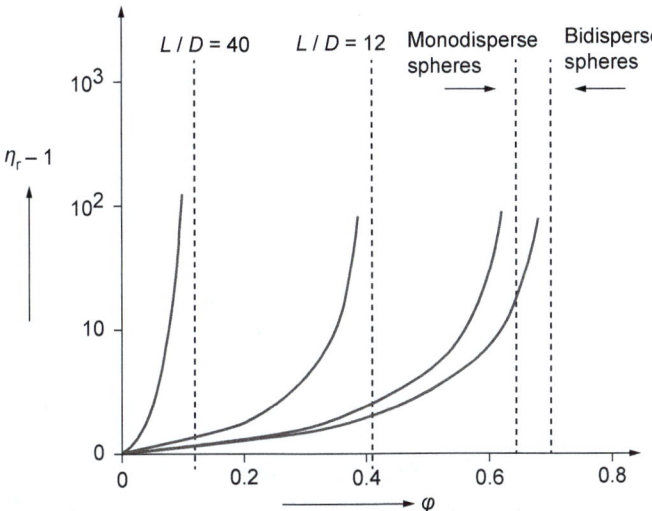

Figure 8.10 Sketch of relative viscosity η_r from the KD equation (8.47), taking maximal packing fractions φ_M equal to random close packing densities φ_{RCP}, versus volume fraction φ. From left to right: Viscosity curves for randomly oriented rods with aspect ratios $L/D = 40$ and $L/D = 12$, monodisperse spheres, and a bidisperse sphere mixture (de Lange Kristiansen et al., 2005). Particularly striking is the enormous viscosity increase when, at given volume fraction φ, spheres are replaced by random rods or fibres; the increase is caused by narrowing of the gap between φ and random close packing densities φ_{RCP} (marked by dashed lines).

of a Newtonian dispersion as function of particle shape and concentration; for examples, see Figure 8.10.

The excluded-volume repulsions referred to above dominate the scene at concentrations that approach the maximal packing fraction φ_M of the dispersed particles—$\varphi_M = 0.74$ for spheres in a crystalline FCC packing and $\varphi_M \approx 0.63\text{--}0.64$ for randomly packed spheres. Moving towards maximal packing, particles increasingly restrict each other's mobility to react to shear forces—a response that completely vanishes when particles are permanently arrested at the density φ_M. In other words, on approaching φ_M the dispersion viscosity diverges. So we can account for at least two limiting cases: volume fractions low enough for the viscosity to be determined solely by the single-particle intrinsic viscosity according to equation (8.44), and the steep viscosity rise when concentrations approach φ_M. These two limits follow from the Krieger–Dougherty (KD) equation for the relative viscosity:

$$\eta_r = \left(1 - \frac{\varphi}{\varphi_M}\right)^{-[\eta]\varphi_M}. \qquad (8.47)$$

The KD equation, as shown in Appendix 8A, presupposes that a colloid experiences the dispersion in which it diffuses as an effective continuous medium, with a viscosity equal to that of the dispersion. This effective medium model ignores the effect of hydrodynamic interactions between colloids; it merely interpolates between (8.44) valid at low concentrations and the rapid viscosity increase near φ_M. So one should not expect more from the KD equation than an estimate of the viscosity increase with particle concentration, albeit with a clear prediction for concentrations at which dispersion viscosities will diverge.

Fibre-dispersion viscosity

The KD equation yields the important insight that anything that lowers the maximum packing fraction of dispersed colloids will, for a given volume fraction, enhance the viscosity. This is particularly evident for randomly distributed colloids, having a φ_M that is about equal to the density φ_{RCP} of the random close packing (RCP)—the packings that we met in Section 7.6 in the context of liquid flow through particle compacts. For example, thin fibres have their φ_{RCP} far below the sphere RCP density. Hence, the viscosity of randomly oriented fibres at given volume fraction is substantially higher (Figure 8.10) than for a sphere dispersion. It is, incidentally, the steep concentration dependence (Figure 8.10) of the dispersion viscosity for disordered cellulose fibres that probably compels the paper-making industry to employ very dilute cellulose dispersions, such that viscosities remain workably low—and aggregate formation of fibres is avoided. An adverse consequence is the high energy costs for the removal of huge water volumes by colloidal filtration—the process that we modelled in Section 7.4.

Sphere mixtures

Mixing spheres of different sizes is a way of increasing the RCP density, so the viscosity of a polydisperse sphere dispersion falls, at given total volume fraction, below the monodisperse case. As an illustrative example (see de Lange Kristiansen et al., 2005): a mixture of large (L) and small (S) spheres, with size ratio $R_L/R_S = 2.6$ and volume fraction ratio $\varphi_S/\varphi_L = 0.28$, has an RCP density of $\varphi_{RCP} \approx 0.69$ which, as can be seen in Figure 8.10, leads to a substantial decrease in the viscosity of a concentrated sphere dispersion.

8.7 Non-Newtonian dispersions

A dispersion is called **non-Newtonian** when its viscosity is shear dependent. Figure 8.11 illustrates typical types of flow curves (i.e. plots of shear stress σ versus shear rate $d\gamma/dt$). The viscosity, indicated by the slope of the dashed lines in Figure 8.11, may increase with $d\gamma/dt$, which manifests **shear thickening** of a dispersion. A concentrated aqueous dispersion of corn starch particles is a well-known example of a shear-thickening fluid: shear stress causes the particles to be jammed into each other, which enhances the viscosity.

When an increasing shear rate lowers the viscosity, the dispersion is said to be **shear thinning**. A familiar example is toothpaste: it rests unmoving on your toothbrush but is easily pressed out of a toothpaste tube. Shear thinning is also observed in dispersions of elongated colloids or fibres (e.g. the xanthane

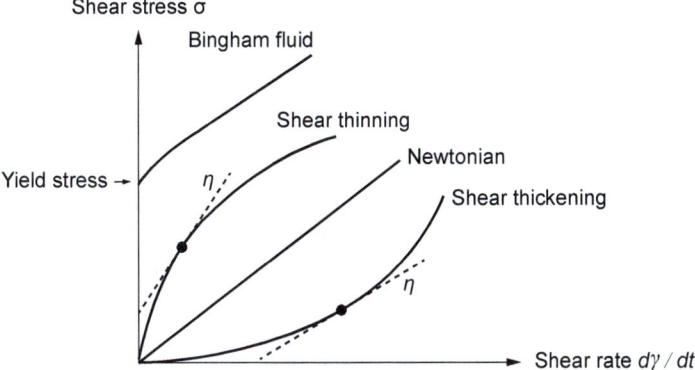

Figure 8.11 Schematic flow curves illustrating various scenarios for the shear rate dependence of the viscosity η—the magnitude of which at any point equals the slope of the tangent at that point.

fibres widely used as food thickeners); the fibres are aligned by a flow field which diminishes the viscosity. We speak of thixotropy when dispersion viscosities recover on cessation of stirring—which manifests rebuilding of microstructures that were sheared apart.

An important non-Newtonian phenomenon is that of **yield stress**: below this stress value the dispersion does not flow. Yield stress is desirable, for example, for 'non-drip' paint: a moving paintbrush will make the paint flow but the flow ceases once the paint has been spread on a wall. Mayonnaise can be stirred with a spoon—that remains standing upright in the unperturbed delicacy. Quite common for colloidal dispersions is **Bingham flow**, where once the yield stress is exceeded the relation between σ and $d\gamma/dt$ becomes linear (Figure 8.11), the slope being the **plastic viscosity** of the dispersion.

Charged colloids

For dispersions of charged colloids, experimental flow curves can change quite dramatically as a function of salt concentration. A concentrated dispersion of stable, charged colloids at low salt concentration is highly viscous, and shear thickening. Addition of salt shrinks electrical double layers surrounding the colloids, which decreases the viscosity. On approach of the salt flocculation concentration, the zero-shear viscosity passes a minimum and rises when particles aggregate. On application of shear, it will be noted that due to flocculation the dispersion has become shear thinning: shear-induced rupture of flocs lowers the viscosity.

Hookean solids

A Newtonian liquid, as we have seen in Section 8.2, is by definition a liquid with a constant, shear-rate-independent viscosity. For such a liquid the displacement *velocity* is linearly proportional to the applied stress. The antithesis of Newtonian liquids are elastic solids that obey **Hooke's law**:

$$\sigma = G\gamma, \tag{8.48}$$

where G is the **elasticity modulus**—which is independent of the shear strain γ. Then the displacement itself is linearly proportional to the stress. Here an obvious model is a spring which linearly increases in length with the pulling force. One striking difference between a viscous Newtonian liquid and an elastic Hookean solid concerns their 'memory'. When a Hookean solid is deformed by a stress, its initial shape will be restored after the stress is removed: the spring jumps back to its initial position. The Hookean solid has a perfect memory in the sense that all work of the external force is stored in reversible elastic deformation. The Newtonian liquid, however, has no memory of its state before the stress was switched on: all work input is irreversibly dissipated as heat.

Maxwell model for viscoelasticity

For a more quantitative picture of viscoelastic response to (changes in) applied stress, we employ the Maxwell model (Figure 8.12), which approximates a viscoelastic material by a spring connected in series to a piston moving in a viscous liquid; the damping piston is also known as a 'dashpot'. Applied mechanical work is first fully employed to pull out the spring; then the piston starts to move and the energy stored in the spring is all gradually dissipated as heat due to viscous friction between piston and fluid. For the spring, an applied stress σ immediately results in a strain γ_E, so for the time dependence of σ and γ_E we have, in view of Hooke's law (8.48):

$$\dot{\gamma}_E = \frac{1}{G}\frac{d\sigma}{dt}, \qquad (8.49)$$

where the subscript E denotes elastic strain. For the piston in the viscous fluid, the time dependence of the strain—the shear rate—follows from Newton's law (8.7):

$$\dot{\gamma}_V = \frac{1}{\eta}\sigma, \qquad (8.50)$$

with the subscript V indicating viscous strain. Since spring and dashpot are connected in series (Figure 8.12) the strains, and equally the strain rates, simply add up to the total rate of shear $\dot{\gamma}$:

$$\dot{\gamma} = \dot{\gamma}_E + \dot{\gamma}_V = \frac{1}{G}\frac{d\sigma}{dt} + \frac{\sigma}{\eta}. \qquad (8.51)$$

Suppose we rapidly pull out the spring (Figure 8.12) and abruptly stop further spring extension at $t = 0$; then strain γ remains constant and $d\gamma/dt = 0$ for $t > 0$. We then find from (8.51) that the decrease $d\sigma/dt$ of the stress in time, by the moving piston (Figure 8.12), is given by:

$$\frac{d\sigma}{dt} = -\frac{G}{\eta}\sigma. \qquad (8.52)$$

From (8.52) we infer that the stress relaxes exponentially to zero (Exercise 8.17):

$$\sigma = \sigma_0 \exp\left[-t/t_R\right]; \quad t_R = \frac{\eta}{G}. \qquad (8.53)$$

COLLOID SCIENCE   155

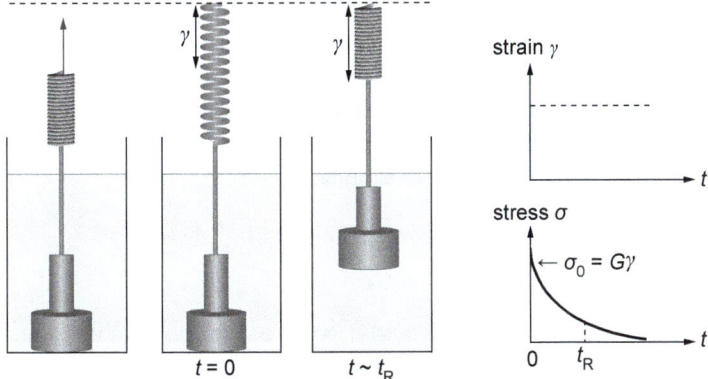

Figure 8.12 A Maxwell element includes a spring modelling elastic energy storage and a piston in a viscous liquid ('dashpot') representing viscous energy dissipation. **Left**: A stress σ_0 rapidly stretches the spring to a strain γ; at $t = 0$ the stretching is abruptly stopped after which the piston starts to move, to gradually relax the stress to zero. **Right**: Sketch of the stress relaxation in time t according to equation (8.53).

Here t_R is the stress relaxation time—which also appears in the Deborah number defined in equation (8.1). At short times $t \ll t_R$ behaviour is mainly elastic, whereas at long times $t \gg t_R$ behaviour is mainly viscous, as it is for a simple Newtonian fluid.

The modelling of viscoelastic fluids by a single Maxwell element is often an oversimplification: in practice colloidal dispersions frequently display complicated rheological behaviour that requires modelling with various Maxwell elements connected in series or in parallel. For this technically rather involved topic, see the treatment in Barnes et al. (1997).

Frequency response

From the analysis of the Maxwell model we inferred that the timescale on which a material is perturbed determines whether Hookean or Newtonian response is dominant. In terms of frequencies one can say that if examined at high frequencies materials behave like solids; in contrast, when studied at sufficiently low frequencies they exhibit viscous properties. A familiar example is glass, an amorphous silicate which acts as a brittle solid under hammer-blow frequencies. However, on very long timescales it slowly deforms as a liquid. The glass blower exploits the fact that flow rates increase at higher temperatures. The microscopic picture here is that glass molecules at room temperature rearrange much too slowly to follow the frequency of hammer blows. At high temperatures, however, the molecules relax fast enough for the glass to respond as a liquid to low-frequency events.

In contrast to silicate molecules in glass, molecules in liquid water relax extremely fast, with $t_R \sim 10^{-12}$ s, so water will only exhibit elasticity at very high frequencies: walking on water requires a footstep frequency of at least 10^{12} Hz. For polymer solutions and colloidal dispersions t_R may be a few seconds, which makes situations with a Deborah number of $De \approx 1$ experimentally accessible. When, for example, a concentrated polymer solution such as salad dressing is stirred, one can clearly observe that the fluid is viscoelastic: when stirring is stopped the dressing bounces back slightly.

Summary

- Newton's viscosity law defines fluid viscosity as the applied shear stress divided by the shear rate that results from this stress.
- In viscous flow frictional forces between sliding fluid layers dominate forces inducing fluid acceleration, such that fluid mass density has no effect on fluid flow.
- The Stokes equation stems from the balance between normal pressure forces and tangential stresses on tiny liquid blobs; its solution is a flow profile in a geometry of interest.
- Energy dissipation by rotating particles in a shear flow causes, for dilute dispersions, a linear increase of dispersion viscosity with particle volume fraction.
- The slope of linear increment of Newtonian viscosity with concentration is the shape-dependent intrinsic viscosity—which markedly increases with particle aspect ratio.
- Viscosities of concentrated dispersions steeply rise on approach of the random packing density of the dispersed particles—a density that is fixed by particle shape.
- Non-Newtonian dispersions harbour divergent scenarios for shear-rate dependence of viscosities; their viscoelastic response is in line with Maxwell's spring-and-dashpot model.

References

D. J. Acheson (1992), *Elementary Fluid Dynamics*. Clarendon Press, Oxford.

H. Barnes, J. Hutton, and K. Walters (1997), *An Introduction to Rheology* (5th edn). Elsevier, Amsterdam.

K. de Lange Kristiansen, A. Wouterse, and A. Philipse (2005), 'Simulation of random packing of binary sphere mixtures by mechanical contraction', *Physica A* **358**, 249.

R. Fürth (ed.) (1956), *A. Einstein. Investigations on the Theory of Brownian Movement*. Dover Publications, New York.

J. Mewis and N. J. Wagner (2021), *Colloidal Suspension Rheology*. Cambridge University Press, Cambridge.

Appendix 8A: The Krieger–Dougherty relation

We seek a concentration dependence $\eta(\varphi)$ of the Newtonian viscosity of a dispersion that entails divergence of the viscosity on approach of the maximal packing fraction φ_M and that, on the other hand, at decreasing volume fractions φ limits to

$$\frac{\eta(\varphi \to 0)}{\eta_0} = 1 + [\eta]\varphi. \tag{8.54}$$

Here $[\eta]$ is the intrinsic viscosity, determined by particle shape, and η_0 the viscosity of the solvent. For the dispersion viscosity at higher concentrations we write:

$$\eta(\varphi) = \eta_0 f(\varphi), \tag{8.55}$$

where $f(\varphi)$ is the function we are looking for. Suppose one colloid in the dispersion is a tracer labelled t; all other particles are host particles with label h. We now assume that the tracer experiences the host dispersion as an effective medium with viscosity $\eta(\varphi_h)$. Such an effective viscosity would manifest itself via the Stokes friction that the tracer experiences when diffusing in the host dispersion, the diffusion coefficient D_t of a tracer sphere with radius R_t being given by

$$D_t = \frac{k_B T}{6\pi \eta(\varphi_h) R_t} \tag{8.56}$$

The implication of (8.56) is that the viscosity of the host dispersion can be determined by measuring the tracer diffusion coefficient, via the dynamic light-scattering method treated in Chapter 11. For the effective medium viscosity we can write:

$$\eta(\varphi_h) = \eta_0 f(\varphi_h). \tag{8.57}$$

When a small volume fraction $\varphi_t \ll \varphi_h$ of tracers is added to the dispersion, the viscosity will approximately increase with a factor $f(\varphi_t)$:

$$\eta(\varphi_h + \varphi_t) \approx \eta(\varphi_h) f(\varphi_t) = \eta_0 f(\varphi_h) f(\varphi_t). \tag{8.58}$$

Substitution of $\varphi = \varphi_h + \varphi_t$ in (8.55) yields:

$$\eta(\varphi_h + \varphi_t) = \eta_0 f(\varphi_h + \varphi_t). \tag{8.59}$$

Comparing (8.58) and (8.59) we must conclude that

$$f(\varphi) = f(\varphi_h + \varphi_t) = f(\varphi_h) f(\varphi_t). \tag{8.60}$$

A function which entails the factorization in equation (8.60) is:

$$f(\varphi) = (1 + A\varphi)^B, \tag{8.61}$$

since

$$f(\varphi_h)f(\varphi_t) = (1 + A\varphi_h)^B(1 + A\varphi_t)^B \sim (1 + A\varphi_h + A\varphi_t)^B, \text{ for } \varphi_t \ll \varphi_h. \quad (8.62)$$

A and B follow from two boundary conditions. First, at infinite dilution we have $f(\phi \rightarrow 0) = 1 + AB\varphi$ which should equal $1 + [\eta]\varphi$ so $AB = [\eta]$. Since $[\eta] > 0$, A and B must have the same sign. Now $B > 0$ in (8.61) would entail viscosities that just keep on increasing on passing φ_M which is physically impossible, so $B < 0$. By taking $1 + A\varphi_{RCP} = 0$ we then impose the divergence $\eta(\varphi \rightarrow \varphi_M) \rightarrow \infty$. Hence $A = -1/\varphi_M$ and $B = -[\eta]\varphi_M$, which on substitution in (8.61) yields the Krieger–Dougherty (KD) relation:

$$\frac{\eta(\varphi)}{\eta_0} = f(\varphi) = \left[1 - \frac{\varphi}{\varphi_M}\right]^{-[\eta]\varphi_M}. \quad (8.63)$$

In constructing the KD relation, no assumptions were made on shape and polydispersity of the colloids involved. Particle shape determines intrinsic viscosity $[\eta]$; shape and polydispersity both govern the value of φ_M. For dispersions of spheres and randomly oriented non-spheres we can equate φ_M to the random-close-packing density φ_{RCP}, as is done in Figure 8.10.

Exercises

8.1 Verify that the Reynolds number Re is dimensionless.

8.2 Estimate for the N, C, and G particles from Table 3.1 the Reynolds number for sedimentation in water due to gravity.

8.3 (a) Suppose you swim at a speed of $U = 2$ m/s; assuming a frontal area of 0.1 m^2 how large is your Re?

 (b) To maintain a constant swim speed you have to invest energy because of (1) viscous friction on your body and (2) displacement of water. Argue which factor is the most important.

8.4 A sphere with diameter d is pulled out (at constant volume) to a thin rod of length L with aspect ratio $L/D = 30$. Argue by which factor the Reynolds number Re goes up or down.

8.5 A force F moves a very large, flat plate with constant speed $u(D)$, at a distance $y = D$ from a parallel wall in water.

 (a) Derive the velocity profile $u_x = u_x(y)$ from the Stokes equation (8.16) and give an expression for the average flow velocity $<u>$.

 (b) Suppose $u(D) = 1$ mm sec^{-1}; $D = 1$ mm, and $\eta = 10^{-3}$ Pa s. How large is F (per unit area)?

8.6 **(a)** Derive the flow profile $u_x = u_x(y)$ by solving the Stokes equation in the form of (8.22) for x-directed flow between two large flat, parallel plates, each with surface area L^2 (see Figure 8.6); assume stick boundary conditions.
  (b) What is the unit/dimension of a viscous stress σ_{xy}?
  (c) Evaluate the profile of the viscous stress σ_{xy} for the flow in Figure 8.6; that is, find σ_{xy} as a function of y.
  (d) Sketch the stress profile; what is the maximum value of the viscous stress?
  (e) Calculate the total viscous force on the inside of the slit pore in Figure 8.6. Argue why your finding is plausible.

8.7 Give the equation for the volume rate of flow Q for the flat plates in Figure 8.6.

8.8 Consider viscous flow in the simple geometries of Figures 8.6 and 8.7 with a 'pure-slip' boundary condition, instead of the stick boundary or no-slip condition that is assumed in Section 8.4 in the derivation of flow profiles. Can you qualitatively explain what happens to the flow profiles when a pure-slip condition applies? Note: A practical example of a situation with a pure-slip boundary is water flowing along a very hydrophobic wall.

8.9 A certain fatty deposit decreases the inner radius of a blood capillary from $R_1 = 5$ microns to $R_2 = 4$ microns. By which factor should pressure p increase to keep the blood volume flow rate at the same level as for a clean blood vessel?

8.10 Why does salt addition decrease the viscosity of a charge-stabilized dispersion?

8.11 Consider the viscosity of a suspension of randomly oriented cellulose fibres (the fibres used, among other things, in paper-making). What will happen to the viscosity if, at given fibre concentration, the fibres reposition themselves parallel to the fluid flow? Why?

8.12 Verify the integrations to obtain the average flow speed $<u>$ in **(a)** equation (8.25) and **(b)** equation (8.31).

8.13 A dispersion of stable silica spheres with radius $a_1 = 1\mu m$ is mixed with a dispersion of silica spheres with radius $a_2 = 0.1\mu m$ at constant volume fraction $\varphi = 0.5$. How will the mixture viscosity compare to the viscosity of the separate dispersions? Why?

8.14 An aqueous dispersion contains spherical proteins with a specific volume of $\bar{v} = 1$ cm^3/g and a weight concentration $c = 10$ g/L. Calculate the relative viscosity of the protein solution.

8.15 Show that for low sphere volume fractions the KD equation (8.47) reduces to Einstein's viscosity equation (8.46).

8.16 Does Einstein's viscosity equation (8.46) also apply to polydisperse spheres?

8.17 Verify that the exponential stress relaxation (8.53) is a solution of differential equation (8.52).

8.18 (a) Derive the viscous stress profile sketched in Figure 8.7.
(b) Evaluate the total viscous stress on the inside of the capillary in Figure 8.7 and compare this total stress to the net external force $F_{ext} = \pi R^2 \Delta P$ that drives the flow. What is your conclusion?

8.19 Calculate the capillary radius R_B (in Figure 8.8) for which $Pe \approx 1$ for the colloidal C-sphere (radius $a = 100$ nm) and the nano N-sphere ($a = 5$ nm) from Table 3.1 ($T = 298$ K).

Find the solutions to these exercises at the end of the book.

9 Magnetic Dispersions

9.1 Introduction

There is a class of inorganic colloids that have the intriguing feature of responding to an applied magnetic field. Instances of this response are extensive field-induced structure formation inside a dispersion, or the shape change of the dispersion as a whole to a pattern of spikes, as shown in Figure 9.1A. The attribute of these colloids that causes their responsiveness to a field is their magnetic **dipole moment**—a moment that may be a permanent one, or a transitory one that is induced by the external magnetic field. Distributions of such magnetic, or magnetizable, colloids in liquid carriers are referred to as **magnetic dispersions**, and the aim of this chapter is to provide an introduction to the magnetic properties of these captivating fluids.

Magnetic dispersions do not occur in nature, so they have to be prepared in the laboratory. Here one could think of a straightforward preparation route to realize a liquid phase with magnetic properties, and that is simply the melting of a **ferromagnetic** metal. A potential drawback here is that the melting point of the ferromagnetic metals iron, nickel, or cobalt is fairly high—iron, for example, melts only at 1,538 °C—and that would leave us with the problem of handling very hot magnetic liquids. That problem, however, will never arise because the melting route towards liquid magnets is impassable: all ferromagnetic materials known to humankind lose their ferromagnetism at the so-called **Curie temperature**—and that temperature is invariably below the material's melting point. Iron, for example, has a Curie point of 770 °C, which is 768 °C below its melting point. At and above the Curie point the ordered magnetic dipole moments in iron spontaneously transform to a disordered collection of randomly oriented dipoles. Iron, in other words, becomes **paramagnetic** which represents a much weaker form of magnetism than the ferromagnetism exhibited by iron below the Curie temperature.

Molten magnets do not exist, but fluids that mimic them, nevertheless, do: they are the magnetic dispersions that come in the form of **ferrofluids**;

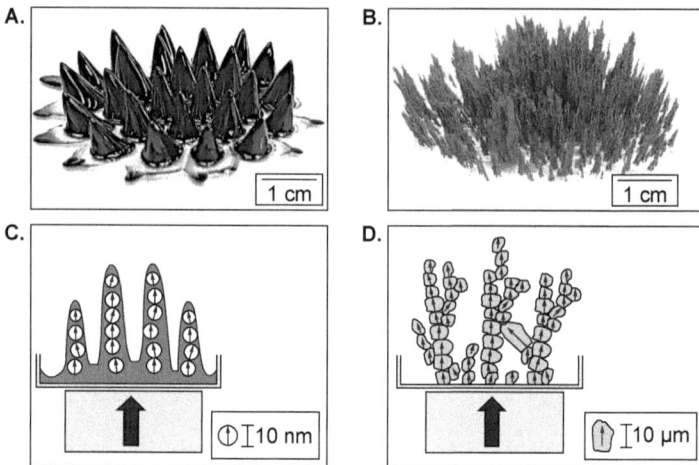

Figure 9.1 (A) A perpendicular field from a permanent magnet (not visible here) applied to a ferrofluid pool produces a static pattern of spikes that mutually repel as they are magnetized in the same direction (Rosensweig, 1985). **(C)** The ferrofluid spikes in **(A)** retain their fluid character, in contrast to the micron-size iron grains in **(B)** that snap into bristled, solid-like structures, schematically depicted in **(D)**, formed by strongly attractive, magnetized grains.
Source: Photography in (A) and (B) by Ben Erné.

Section 9.3 reviews their intriguing properties and applications. An important characteristic of a ferrofluid is its **magnetization curve** which displays the fluid's magnetization as a function of an applied magnetic field; these curves are examined in Section 9.4, followed by a consideration of the disappearance ('relaxation') of magnetization after the external field has been switched off. For applications of ferrofluids, for reasons that will be explained, it is essential that magnetic particles do not aggregate; this is the context for the discussion of interactions between dipolar colloids in Section 9.5. In contrast to the magnetite particles in ferrofluids that each comprise a magnetic **single domain (SD)**, the so-called **magnetic microbeads** are colloids that are filled with a large number of SDs. Dispersions of these beads are the subject of Section 9.6. First, however, we will recap in Section 9.2 some characteristics of magnetic materials—which will be needed to better comprehend the behaviour of the magnetic dispersions that will be discussed in later sections.

9.2 Magnetism in solids and solutions

We will summarize here magnetic properties that relate to the way in which atomic magnetic moments are arranged in a material; Figure 9.2 summarizes various options for this dipole ordering. In **ferro-** and **ferrimagnetic** materials, atomic magnetic dipoles spontaneously align to form a magnetized domain. In ferromagnetic materials, dipoles in one such domain all point in the same direction;

Spontaneous domain formation

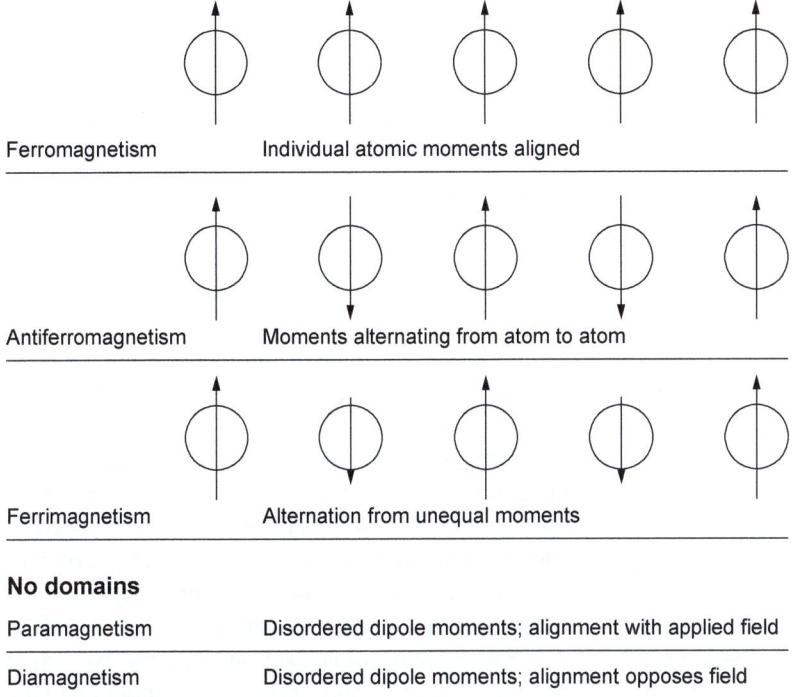

No domains

Paramagnetism	Disordered dipole moments; alignment with applied field
Diamagnetism	Disordered dipole moments; alignment opposes field

Figure 9.2 Overview of various types of magnetic behaviour. Ferrofluids exhibit superparamagnetism.

ferromagnetism is exhibited by iron, cobalt, and nickel and many of their alloys. In a ferrimagnetic substance such as magnetite, Fe_3O_4 or $FeO.Fe_2O_3$, a minority of dipoles is oriented opposite to the other parallel dipoles; this mix of oppositely directed moments occurs in every crystalline unit cell. Since the number of dipoles in the two directions are different, they do not cancel each other and the ferrimagnetic material remains strongly magnetic, although less than a ferromagnetic metal.

The situation when all magnetic moments point in the same direction, as with for example the moments in a ferromagnetic domain, is referred to as **saturation magnetization**. The saturation value of the magnetization, denoted m_{sat}, is defined as the maximal net dipole moment per unit volume of magnetic material, and has the unit of ampere per metre (A m^{-1}). It follows that the magnetic dipole moment m_d (A m^2) of a ferromagnetic domain with volume V_p equals

$$m_d = m_{sat} V_p. \tag{9.1}$$

When looking at a ferromagnetic specimen as a whole, the magnetic moment may be very much less than the saturation moment, the reason being that the material is composed of multiple magnetic SDs that are magnetized in various directions such that magnetic dipole moments cancel each other,

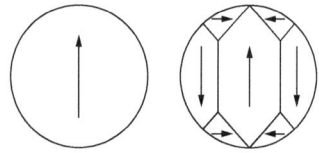

Figure 9.3 Left: Uniformly magnetized sphere of ferromagnetic material. The arrow represents the direction of saturation magnetization of this single-domain particle. **Right**: Sphere of ferromagnetic substance segmented into domains, with arrows indicating the direction of saturation magnetization in each domain.

partly or completely. An example of a multi-domain structure in a ferromagnetic material is shown in Figure 9.3. In essence, a ferromagnetic substance breaks up into domains to minimize the magnetic field energy, which would be considerable if the whole material were magnetized in one direction. The material, however, does not divide itself into domains indefinitely because it requires energy to create the **domain walls** which separate the domains; such a wall is a region in which the dipole moment gradually rotates from the direction in one domain to the dipole direction in its neighbour domain.

For magnetite the maximal size of an SD is about $d \approx 40$ nm; for magnetite crystals larger than that, subdivision of magnetic domains occurs, their number increasing with increasing volume of the magnetite particle. In a magnetite ferrofluid average particle diameters are typically in the range $d \approx 10-15$ nm, with a polydispersity of around 35 and with the implication that all particles in the size distribution contain an SD, with a magnetic dipole moment given by equation (9.1).

The behaviour, called **paramagnetism**, results from the tendency of molecular magnetic moments to align with an external magnetic field; thus, inside a paramagnetic substance the external field is enhanced. The property of paramagnetism is displayed, for example, by liquid oxygen and solutions of rare-earth salts. Other examples of paramagnetic systems are the $FeCl_2$ and $FeCl_3$ solutions from which the ferrimagnetic magnetite particles are prepared, as described in Appendix 9A. Paramagnetic molecules, incidentally, have their magnetic moment in a random orientation, and moments align independently from each other as the interaction between paramagnetic molecules is negligible. In the ferrofluids to be discussed in Section 9.3, we have an analogous situation: magnetic particles in these fluids align their moment parallel to an applied field but the magnetic interaction between the particles is negligible. In short, the colloids interact magnetically with the field but not with each other. Since, however, the magnetic moments of magnetic colloids are, for given strength of the applied field, so much larger than for paramagnetic molecules, ferrofluids are called **superparamagnetic**.

The weakest type of magnetic behaviour is **diamagnetism**. The atoms in diamagnetic materials do not possess a permanent magnetic moment. The feeble diamagnetic response to an external field is the induction of a small magnetization in a direction at 180° to the applied field, such that inside the material the field is diminished. Weak as it may be, of all types of magnetism, diamagnetism is probably the most widespread; it occurs for example in the inert gases, in most metals and nonmetals, and in many organic compounds.

9.3 Ferrofluids

Ferrofluids are concentrated, stable dispersions of small (3–15 nm) particles that each carry a permanent magnetic moment. Most ferrofluids are based on particles composed of magnetite, an iron oxide with an atomic composition represented by Fe_3O_4 or $FeO.Fe_2O_3$. Magnetite is a ferrimagnetic material that oxidizes to maghemite (γ-Fe_2O_3)—an oxidation that is harmless as it does not alter the magnetic properties of the particles involved. The colloids are usually sterically stabilized by a grafted oleic acid layer and dispersed in nonpolar solvents, such as cyclohexane or any other oil (see Figure 9.4); the oleic acid layer prevents particles from aggregating by the Van der Waals attraction, or the magnetic attraction that is further investigated in Section 9.5.

Ferrofluid preparation

Traditionally, non-aqueous ferrofluids are prepared via extensive ball-milling of magnetite rocks in an organic solution of adsorbing surfactants—an instance of the comminution method that we encountered earlier in Chapter 1, Section 1.1. These magnetite minerals are composed of multiple magnetic domains that in the milling process are fractured into small (3–15 nm) particles that each contain one SD, or a portion thereof. So, the particles in a ferrofluid possess fully saturated magnetism, with a magnitude of the dipole moment given by equation (9.1). Instead of this comminution technique, which may take weeks, a fast condensation route may be employed in which magnetite particles precipitate upon alkalization of an $FeCl_2$ / $FeCl_3$ solution; Appendix 9A provides a recipe for the synthesis of oleic acid-coated magnetite particles. This synthesis exemplifies particle formation by the homogeneous nucleation in a supersaturated solution that we studied in Chapter 2; supersaturation is induced here by the formation of almost insoluble iron oxides when the pH of iron chloride solution is suddenly raised by addition of ammonia.

Molten magnet mimicry

On a microscopic scale, a ferrofluid comprises dense swarms of nanomagnets that tumble and zigzag around by rapid Brownian motion. When this agitated magnetic crowd starts to migrate in a field gradient, the dispersion begins to move as a whole; it displays the behaviour that molten ferromagnets would also exhibit. A remarkable example of the molten-magnet character of ferrofluids is shown in Figure 9.1A: when a ferrofluid with a free surface is exposed to a sufficiently strong magnetic field, its initially flat surface changes to a periodic pattern of spike-like shapes; this phenomenon of spike formation is known as the **Rosensweig instability**. This instability is extensively discussed in Rosensweig (1985); here a brief account will suffice. Magnetic particles line up along magnetic field lines to lower their energy; protrusions are formed that repel each other as they are composed of ferrofluid elements that are magnetized in the same field direction. Growth of protrusions is opposed by gravity that pulls ferrofluid downwards, and by the surface tension that resists the increase in the fluid's surface area. Surface tension is also responsible for the smooth shape of the spikes; the spikes formed by iron filings are much more irregular—and quite rigid as they are composed of strongly magnetized micron-sized iron grains (Figure 9.1B).

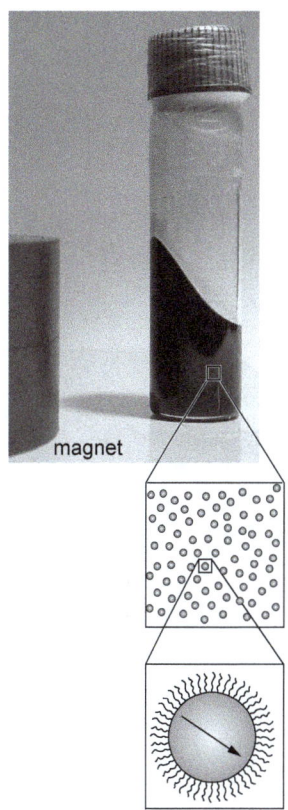

Figure 9.4 Ferrofluid adopts a skewed meniscus near a magnet. The fluid in the picture is a concentrated, stable dispersion of magnetite particles in cyclohexane; an illustration is shown of a magnetite particle (typical diameter about 10 nm) with an embedded permanent magnetic moment and a surface grafting of oleic-acid molecules.

Source: Figure prepared by Ben Erné.

Ferrofluid applications

The molten-magnet behaviour of ferrofluids gives rise to a variety of applications (Socoliuc et al., 2022). The fluids are employed, for example, for permanent sealing of rotary shafts, where a magnet keeps an oily ferrofluid (with an ultra-low vapour pressure) in place, enclosing a rotating shaft. Other applications include ferrofluids for sealing rotating hard discs in computers, and ferrofluid drops in loudspeakers and hearing aids to provide lubrication and better discharge of heat. An emerging application of ferrofluids is the so-called magnetic density separation: the partition of materials from plastics or electronic waste, on the basis of their effective density in magnetic liquids exposed to a vertical field gradient (van Silfhout, 2020).

We note here in passing that, apart from their appearance in ferrofluids, magnetic nanoparticles are also the subject of current research that focuses on biomedical applications (Pankhurst et al., 2003) such as magnetically controlled delivery of drugs attached to magnetic particles, and the use of magnetic particles as contrast agents in magnetic resonance imaging (MRI). In MRI scans of the human body, the contrast of an organ of interest can be enhanced by the selective attachment of chemically modified magnetic nanoparticles to that organ.

Colloidal stability

We return to magnetic particles dispersed in ferrofluids, to consider a crucial prerequisite for the application of these fluids as 'molten magnets': the long-term colloidal stability of a fluid. This stability requires that the sum of magnetic attraction and the Van der Waals attraction between two contacting particles is smaller than thermal energy $k_B T$ such that Brownian motion keeps particles separate, as exemplified by the magnetite particles depicted in Figure 9.5A. For magnetic contact attractions of a few $k_B T$, dipolar structures may spontaneously form, as shown in Figure 9.5C. These reversible structures grow or dissolve upon, respectively, increasing or decreasing particle concentration. For the large magnetite crystals extracted from **magnetotactic bacteria** depicted in Figure 9.5D, magnetic attractions are so strong that the crystals irreversibly snap into rigid chains and ring-like structures.

When structures or aggregates of particles form in a ferrofluid, they will rapidly move towards a magnet, as illustrated in Figure 9.6 (left), or sediment under gravity: the initially homogeneous dispersion no longer mimics a molten ferromagnet as it de-mixes into solvent and a sediment of clusters. Since the particle magnetic moment increases with particle volume according to (9.1), colloidal stability requires sufficiently small particles; for magnetic iron oxides the average particle diameter is typically around $d \approx 10$ nm; for particles of that size an oleic acid layer on their surface sufficiently weakens the contact attraction between particles to ensure colloidal stability. In Section 9.5 we will examine the interaction between magnetic particles in more detail.

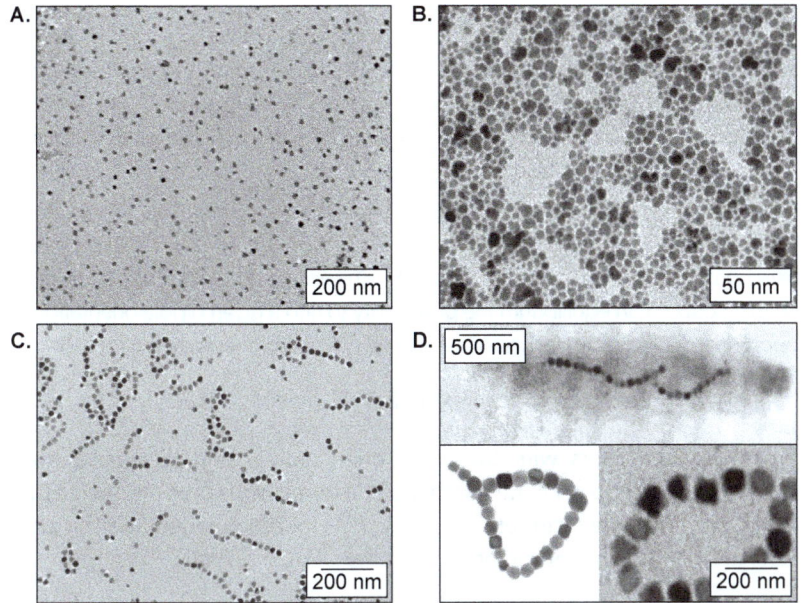

Figure 9.6 Right: In a colloidally stable ferrofluid high concentrations of free nanoparticles remain homogeneously distributed in the skewed fluid near a (neodymium) magnet. **Left**: Magnetized particle clusters in a flocculated ferrofluid 'sediment' onto the magnet; an unstable fluid no longer mimics a molten ferromagnet and de-mixes into solvent and a particle compact.

Source: Picture courtesy of A. van Silfhout.

Figure 9.5 (A) *In situ* cryo-TEM image of oleic acid-coated magnetite particles (average diameter $d = 14$ nm and $\lambda \approx 0.3$) exerting dipolar attractions too weak to induce aggregation. **(B)** TEM image of particles ($d \approx 7$ nm $\pm$ 35%) from an aqueous, charge-stabilized ferrofluid. **(C)** Cryo-TEM image of magnetite particles ($d = 24$ nm and $\lambda \approx 2$) associating to reversible, flexible dipolar chains (Klokkenburg et al., 2006). **(D) Top**: TEM image of a magnetotactic bacterium with a backbone of single-domain magnetite crystals; $d \approx 50$ nm and $\lambda \approx 40$. **Below**: Magnetite crystals extracted from magnetotactic bacteria snap into dipolar necklaces by strong magnetic contact attractions (Philipse and Maas, 2002). The clustering observed in **(B)** is enforced by drying a stable, concentrated dispersion on a TEM grid; cryo-TEM in **(A)** and **(C)** directly images particles as they are in dispersion.

Source: Images courtesy of H. Meeldijk, **(A)**, **(C)**, **(D)**; A. van Silfhout **(B)**.

9.4 Equilibrium magnetization of superparamagnetic dispersions

The permanent magnetic moments of SD magnetite particles can be aligned by an external magnetic field, much like compass needles that orient in the Earth's magnetic field. The latter has no significant effect on magnetite particles: to orient the 'nanocompass needles' of a ferrofluid, the applied field must be considerably stronger than that of the Earth. The alignment by the nanoneedles parallel to the lines of an external magnetic field is opposed by thermal rotations of the needles; they make random angular steps, as a result of which the direction of the moments adopts a certain angular distribution. For weakly magnetic particles in a weak field, like the magnetic nanoneedles in the Earth's magnetic field, this distribution will be of almost spherical symmetry; for strong dipoles in a strong field, the angular distribution adopts the shape of a narrow peak in the field direction.

Equilibrium dipole orientations

We will now evaluate the equilibrium distribution of polar angles θ between dipoles and field lines in the range from $\theta = 0$ to $\theta = \pi$ (Figure 9.7). The energy E of a dipole $\vec{m}_d$ in an applied field $\vec{H}$, scaled on the thermal energy $k_B T$, equals

$$\frac{E}{k_B T} = -\frac{\mu_0 \vec{m}_d \cdot \vec{H}}{k_B T} = -\alpha \cos\theta; \quad \alpha = \frac{\mu_0 m_d H}{k_B T}. \tag{9.2}$$

Here α comprises the maximal energy $\mu_0 m_d H$ achieved for dipoles parallel to the field such that $\cos\theta = 1$. In equilibrium the probability to find a given dipole orientation is proportional to the Boltzmann factor associated with the dipole energy E in (9.2):

$$\text{Boltzmann factor} = \exp[-E/k_B T] = \exp[\alpha \cos\theta]. \tag{9.3}$$

Since dipole vectors with the same orientation θ end in annular area $A = 2\pi \sin\theta \, d\theta$ (Figure 9.7), the number of these vectors is proportional to this area A. Therefore, the distribution function $P(\theta)$ for dipole orientations is proportional to the Boltzmann factor in (9.3) times $A = 2\pi \sin\theta d\theta$:

$$P(\theta) = C \exp[\alpha \cos\theta] \times 2\pi \sin\theta d\theta. \tag{9.4}$$

The constant of proportionality C follows from the normalization requirement (Exercise 9.5)

$$\int_0^\pi P(\theta) d\theta = 1, \tag{9.5}$$

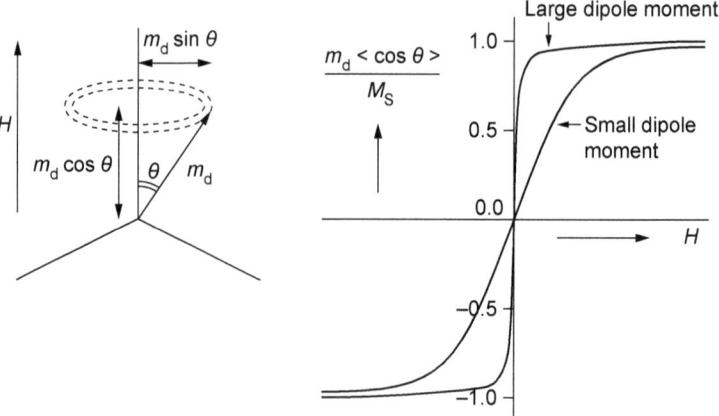

Figure 9.7 Left: Dipole moment m_d has its lowest energy when parallel to the external magnetic field H; due to rotational Brownian motions, dipole moments adopt an equilibrium distribution in the angular range $\theta = 0$ to $\theta = \pi$. **Right**: Sketch of ferrofluid magnetization curves; $m_d <\cos\theta>$ is the average magnetization per volume in the field direction, and saturation magnetization M_s is reached when $<\cos\theta> = 1$. The slope near the origin increases with particle volume because moment m_d scales with particle volume (see equation (9.1)), and the slope rises with m_d in accordance with equation (9.11).

and the normalized angular distribution function reads:

$$P(\theta) = \frac{\alpha}{4\pi \sinh(\alpha)} \exp[\alpha \cos\theta] 2\pi \sin\theta d\theta. \tag{9.6}$$

It should be noted that this distribution function holds for magnetic particles that do not interact: the dipole moments of individual particles orient in the external field completely independent of each other.

Magnetization curves

A magnetization curve is a plot of the average magnetization of a sample as a function of the applied magnetic field. We shall derive here an expression for the magnetization curve for a dispersion of non-interacting dipolar particles, with the primary aim of assessing how the magnetic particle moment m_d can be determined from a measured magnetization curve. The contribution of a dipole to the net magnetization induced by the external field H is the component $m_d \cos\theta$ of the dipole moment parallel to the field (see also Figure 9.7). The average value of this component is $m_d <\cos\theta>$, where the average cosine of the angle between dipoles and field equals:

$$\langle \cos\theta \rangle = \frac{1}{C} \int_0^\pi \cos\theta\, P(\theta) d\theta = \Lambda(\alpha), \tag{9.7}$$

as follows from substitution of (9.6) in (9.7) and performing the integration (Exercise 9.6); the Greek uppercase letter Λ denotes the so-called **Langevin function** defined by:

$$\Lambda(\alpha) = \coth(\alpha) - \frac{1}{\alpha}; \quad \coth(x) = \cosh(x)/\sinh(x) = (e^x + e^{-x})/(e^x - e^{-x}). \tag{9.8}$$

For a liquid dispersion with a number density ρ of colloids with an embedded permanent magnetic moment, the net magnetization M per volume of dispersion is:

$$M = \rho m_d <\cos\theta> = \rho m_d \Lambda(\alpha). \tag{9.9}$$

At strong external fields such that $\alpha \gg 1$, the Langevin function approaches $\Lambda(\alpha) = 1$ signifying that all dipoles are aligned parallel to the field. Then the colloidal dispersion has reached its maximal or **saturation magnetization** M_s:

$$M_s = \rho m_d, \quad \text{for } \alpha \gg 1. \tag{9.10}$$

In a weak external field ($\alpha \ll 1$) thermal angular displacements of dipoles are significant. From the weak-field limit of the Langevin function (Exercise 9.4)

$$\Lambda(\alpha) = \frac{\alpha}{3}, \quad \text{for } \alpha \ll 1, \tag{9.11}$$

it follows that the initial linear slope of the magnetization curve (Figure 9.7), or the initial **magnetic susceptibility**, is given by:

$$\chi_i = \frac{M}{H} = \rho \frac{\mu_0 m_d^2}{3k_B T}. \tag{9.12}$$

Magnetic moment determination

Knowledge of the colloid number density ρ would allow determination of the particle magnetic moment m_d in equation (9.12) from the experimentally determined susceptibility. Colloid number densities, however, are seldom accurately known: the experimental concentration measure is a colloid *weight* concentration and its conversion to a number density requires the particle mass density or specific volume which are not easy to determine. Fortunately, we can eliminate the colloid number density by scaling the measured initial susceptibility on the saturation magnetization from (9.10):

$$\frac{\chi_i}{M_s} = \frac{\mu_0 m_d}{3k_B T}. \tag{9.13}$$

This determination of magnetic moments, it should be noted, only works when distributions of particle magnetic moments are sufficiently narrow. In a polydisperse mixture of magnetic moments, the large dipole moments will align at low fields and, hence, dominate χ_i whereas weak moments will only contribute on approach of the saturation magnetization.

Relaxation of dipole orientations

When the applied external field—which was employed to measure a magnetization curve—has been switched off, the aligned particle magnetic moments revert to their initial, random distribution. This reversion is called **magnetic relaxation** and the typical time it takes for moment orientations to randomize is called the magnetic relaxation time. The magnetization of a ferrofluid can, in fact, relax via two different mechanisms, both having in common that they rely on thermal rotation of dipole orientations. One mechanism is the rotational Brownian motion of the particles as a whole in the liquid; the second is thermal rotation of the magnetic vector *inside* a particle. In a frozen ferrofluid, particles are immobilized and only the second mechanism is in action. The orientational decline of $<\cos\theta>$ by Brownian rotations is a single-exponential decay (as derived, for example, in Philipse, 2018) in time t:

$$\langle \cos\theta \rangle = \exp[-t/\tau_{RR}]; \quad \tau_{RR} = \frac{1}{2D_r}. \tag{9.14}$$

Here τ_{RR} is the characteristic time for **Brownian relaxation** and D_r the rotational diffusion of the colloid in question. For a sphere with volume $V_p = (\pi/6)d^3$ the rotational diffusion coefficient and the associated relaxation time are given by:

$$D_r = \frac{k_B T}{\eta \pi d^3} = \frac{k_B T}{6\eta V_p}; \quad \tau_{RR} = \frac{3\eta V_p}{k_B T}, \tag{9.15}$$

where η is the solvent viscosity. In a single-domain crystal the magnetic vector is directed along the 'easy axis' of magnetization and the vector must cross an energy barrier to adopt other orientations. The height of the barrier equals KV_p, where K is the **anisotropy constant** of the material. The characteristic time τ_{NR} for **Néel relaxation**—that is, for magnetization fluctuations in the crystal—is:

$$\tau_{NR} = \frac{1}{f_0} \exp[KV_p/k_B T], \qquad (9.16)$$

where f_0 is a frequency of order 10^9 Hz. The rotational relaxation time τ_{RR} increases linearly with particle volume V_p whereas the Néel relaxation time grows exponentially with V_p. The consequence is that for magnetite particles larger than $d \approx 15$ nm Brownian rotations totally dominate magnetic relaxation.

9.5 Interactions between dipolar magnetic colloids

Magnetic particles in a ferrofluid, as we already observed in Section 9.3, must be small enough to avoid their aggregation by magnetic attraction. Here we will examine where the size dependency of magnetic interparticle interactions actually comes from, and how these interactions can be computed for a given particle size.

For uncharged spheres in an organic solvent or oil there are, in addition to the magnetic interaction that will be dealt with shortly, two other interactions between particles that are operative, namely the Van der Waals attraction and the so-called hard-sphere repulsion (in essence the Born repulsion that prohibits interpenetration of hard-sphere surfaces). What these two interactions have in common is that, at least for spheres, they are *isotropic*: viewed from a sphere centre the interaction energy with the neighbour sphere is the same in every direction, depending only on the centre-to-centre distance. The peculiar consequence of an embedded magnetic dipole moment in the spheres is a dipolar interaction energy between them that is *anisotropic*, namely either attractive or repulsive, depending on the orientation of the dipoles relative to each other.

Potential interaction energy

The interaction energy U for two dipoles with magnetic moment m_d (see Figure 9.8) at centre-to-centre distance r is (Rosensweig, 1985):

$$\frac{U(\vec{r}, \text{or})}{k_B T} = \lambda \left(\frac{d}{r}\right)^3 f_{or}; \; r \geq d. \qquad (9.17)$$

Here 'or' stands for the orientational dependence of the dipolar interaction, a dependence that is quantified by the function f_{or}:

$$f_{or} = \hat{s}_1 . \hat{s}_2 - 3(\hat{s}_1 . \hat{r})(\hat{s}_2 . \hat{r}). \qquad (9.18)$$

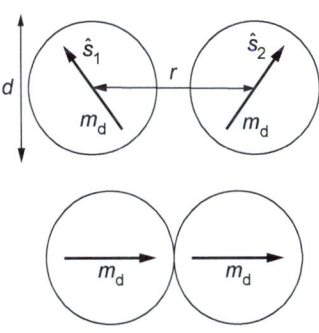

Figure 9.8 Top: Two spheres of diameter d with an embedded, permanent dipole moment m_d with orientations specified by unit vectors $\hat{s}_1$ and $\hat{s}_2$. **Bottom**: Dipolar attraction is maximal for two dipoles in head-to-tail contact: $w_{max} = -2\lambda$, where λ is the coupling parameter defined in equation (9.19). Thermal fluctuations of dipoles weaken contact attractions to the contact value of the potential of mean force (see equation (9.23)).

In this orientational part of the dipolar interaction, the carets indicate unit vectors for dipole orientations and for vector $\hat{r}$ (see Figure 9.8). The parameter λ in (9.17) measures the strength of the dipolar interaction contact energy at centre-to-centre distance $r = d$:

$$\lambda = \frac{\mu_0 m_d^2}{4\pi k_B T d^3} = \left(\frac{\mu_0 m_{sat}^2}{24 k_B T}\right) V_p ; \quad m_d = m_{sat} V_p, \tag{9.19}$$

where $V_p = (\pi/6)d^3$ is the volume of a sphere with diameter d, and $\mu_0 = 4\pi \times 10^{-7}$ J/A^{2}m is the vacuum permeability.

Head-to-tail dipoles

For dipoles in a head-to-tail configuration, unit vectors in (9.18) all point in the same direction such that their in-products all equal unity, hence $f_{or} = 1 - 3 = -2$, so the interaction between two head-to-tail dipoles that follows from (9.17) is an attraction with magnitude:

$$\frac{U(r)}{k_B T} = -2\lambda \left(\frac{d}{r}\right)^3 ; \quad r \geq d. \tag{9.20}$$

Since $\lambda \propto d^3$ we see that the amplitude of this attraction is proportional to $U \propto d^6$; this strong particle-size dependence of dipolar attractions underpins the statement made in Section 9.3 that magnetite particles in a ferrofluid should be small, typically $d \approx 10$ nm, to avoid agglomeration. The maximal attraction occurs for two dipoles in head-to-tail configuration at contact (Figure 9.8):

$$\frac{U_{max}}{k_B T} = -2\lambda, \text{ for } r = d. \tag{9.21}$$

This result entails that the maximal dipolar attraction increases with particle diameter as $U_{max} \propto d^3$. The chaining of relatively large ($d \approx 50$ nm) SD magnetite crystals in Figure 9.5D forms an illustrative example of strong dipolar contact attractions at work.

The contact energy of particles will be weakened if particles are covered with a dense oleic acid layer (see also Figure 9.4); for a layer of thickness Δ, the centre-to-centre contact distance increases from d at centre-to-centre distance r to $2\Delta + d$. On substitution of $r = 2\Delta + d$ in (9.20) we then obtain for the diminished contact attraction:

$$\frac{U_{max}}{k_B T} = \frac{-2\lambda}{(1 + 2\Delta/d)^3}, \text{ for } r = d + 2\Delta. \tag{9.22}$$

Interaction free energy

The pair interaction energy U in (9.17) is a **potential energy**, namely the work needed to bring two dipoles at fixed orientations from infinity to centre-to-centre distance r. The **free energy** of interaction takes the unavoidable thermal fluctuations of dipole orientations into account, which can be achieved by taking the Boltzmann-weighted average of (9.17) over all dipole orientations (as worked out in detail in Chan et al., 1985). The result is a so-called interaction potential of mean force $w(r)$ which for weak dipoles equals:

$$\frac{w(r)}{k_B T} \sim -\frac{\lambda^2}{3}\left(\frac{d}{r}\right)^6, \quad \text{for } \lambda < 1 \text{ and } r \geq d + 2\Delta. \tag{9.23}$$

The dipolar pair potential (9.17) is, depending on the dipole orientations, repulsive as well as attractive, but the potential of mean force (9.23) is always attractive; the reason is that attractive dipole configurations are weighted with a much larger Boltzmann factor than repulsive arrangements. Note also the distance decay $w(r) \propto r^{-6}$ which is reminiscent of molecular Van der Waals attractions (Appendix 5B) which decline in the same way—that is no coincidence, since molecular Van der Waals attractions also comprise Boltzmann-weighted averages over fluctuating (electrical) dipole orientations. The contact interaction free energy from (9.23) is $w(r = d) \sim -\lambda^2 k_B T/3 \, (\lambda < 1)$ which is significantly weaker than the head-to-tail contact attraction from (9.21). This weakening is caused by thermal fluctuations by which dipole orientations rotationally 'diffuse out' of the linear head-to-tail configuration.

If we model an SD magnetite particle by an uncharged, hard sphere of diameter $d + 2\Delta$ with an embedded dipole moment, the interaction potential for this so-called **dipolar hard sphere (DHS)** reads:

$$\frac{w(r)}{k_B T} \begin{cases} = \infty & 0 \leq r < d + 2\Delta \\ = -\dfrac{\lambda^2}{3}\left(\dfrac{d}{r}\right)^6 & r \geq d + 2\Delta \text{ and } \lambda < 1. \end{cases} \tag{9.24}$$

The hard-sphere part of the DHS potential is a steep repulsion that prevents spheres from interpenetrating below the contact distance $r = d + 2\Delta$. The dipolar part of the DHS potential represents only a weak magnetic attraction; (9.23) is obtained under the assumptions that λ is small—which makes (9.24) applicable to weakly magnetic particles, such as the ones depicted in Figure 9.5A.

Magnetic versus Van der Waals attraction

In addition to a magnetic attraction there is also, of course, the ever-present Van der Waals attraction between magnetic colloids. The relative importance of both attractions depends on the particle volume. The Van der Waals attraction between two spheres of diameter d at centre-to-centre distance r, follows from the attractive part of the DLVO potential between two spheres in equation (5.59) in Chapter 5:

$$w(r) = -A_H \frac{d}{24(r-d)}. \tag{9.25}$$

Most striking is the strong increase of the amplitude of the magnetic attraction in the DHS potential (9.24), proportional to d^6, in comparison to the amplitude of the Van der Waals attraction, which is roughly proportional to d.

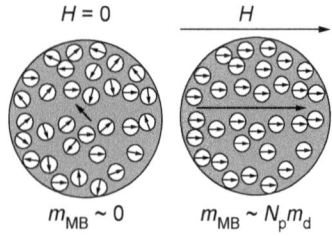

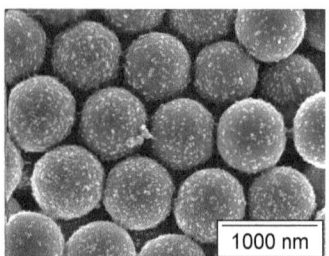

Figure 9.9 Top: A micro-bead containing N_p superparamagnetic particles with dipole moment m_d has a net dipole moment m_{MB} that increases from virtually zero in zero field to its maximum $N_p m_d$ in a saturating external field. **Bottom**: SEM image of magnetizable spheres in the form of silica spheres covered with single-domain cobalt-ferrite nanoparticles.
Source: Claesson et al. (2007).

The implication is that upon increase of particle size, the balance fairly quickly tips to the magnetic attraction: for magnetite particle diameters larger than about $d \approx 10$ nm the contribution of the Van der Waals attraction to the interparticle interaction is negligible.

9.6 Magnetizable dispersions

Ferrofluid particles incorporate only one SD with magnetic moment m_d; so-called **magnetic micro-beads** are micron-sized colloids that contain a large number N_p of SD nanoparticles (Figure 9.9). In zero magnetic field, magnetic moments of embedded nanoparticles stay randomized by thermal Néel rotations such that a bead's dipole moment m_{MB} is effectively zero. An external field, however, aligns the SD dipoles (Figure 9.9) such that a bead's dipole moment increases to $m_{MB} \approx N_p m_d$. This substantial induced dipole moment is widely exploited in the separation of biomolecules from complex solutions and body fluids (Ruffert, 2016). Beads are functionalized with, for example, antibodies that specifically bind to biomolecules floating in solution. In zero field, beads remain dispersed in solution where they collect the desired biomolecules; on application of an external field gradient (from a simple hand-held magnet) the magnetized beads aggregate, rapidly migrate to the magnet, and are subsequently drawn out of solution.

Magnetophoresis

The migration of micro-beads and aggregates thereof, induced by a magnetic field, is an instance of **magnetophoresis**. For it to occur the magnetic field must have a gradient; a homogeneous magnetic field only rotates dipole moments (the rotations that we investigated in Section 9.4) but does not bring moments into motion because the magnetic energy of dipoles is independent of their position. In a magnetic field gradient, however, magnetic dipoles move towards an increasing field strength. The magnetic force F on a particle dipole moment m_d exerted by the component dH/dx of the field gradient in the x-direction is given by

$$F = m_d \mu_0 \frac{dH}{dx}. \tag{9.26}$$

This force can, at sufficiently low concentrations, be equated to the viscous force $f_d v_d$ on a single, free SD particle, to obtain for the magnetophoresis speed v_d:

$$v_d = \frac{m_d \mu_0}{f_d} \frac{dH}{dx}, \tag{9.27}$$

where f_d is the SD particle's Stokes friction factor. So from an experimental determination of the magnetophoresis rate one can obtain the particle's magnetic moment m_d, provided the friction factor f_d is known, for example, from the diffusion coefficient of an SD colloid. For a fully magnetized bead, with friction factor f_{MB} and magnetic moment $m_{MB} = N_p m_d$, the magnetophoresis speed v_{MB} is:

$$v_{MB} = \frac{N_p m_d \mu_0}{f_{MB}} \frac{dH}{dx} = N_p \frac{f_d}{f_{MB}} v_d. \qquad (9.28)$$

Owing to the larger number $N_p \gg 1$ of SD particles it contains, a micro-bead migrates much faster in a given field gradient than one SD particle. Fully magnetized beads will aggregate into chains of, say, z beads. The chain friction factor will be somewhat larger than that of a single bead but this modest increase in hydrodynamic resistance will be overshadowed by the large dipole moment $zN_p m_d$ per chain, which will lead to a significant increase of the net rate of magnetophoresis.

Magnetophoresis equilibrium profiles

The accumulation of large, magnetized aggregates or chains on a magnet (see also Figure 9.6, left) is hardly counteracted by Brownian motion: their diffusion coefficient is simply too small for that. For small SD particles, however, 'back diffusion' is able to compete with magnetophoresis, a competition that entails a **magnetophoresis–diffusion equilibrium (MDE)** concentration profile in which convection and diffusion are in balance. The convective particle flux j_m of SD particles due to magnetophoresis follows from (9.27) as:

$$j_m = \rho v_d = \rho \frac{m_d \mu_0}{f_d} \frac{dH}{dx}, \qquad (9.29)$$

where $\rho = \rho(x)$ is the particle number density at position x. The counteracting diffusive flux j_D follows from Fick's first law:

$$j_D = -D \frac{d\rho}{dx}; \quad D = \frac{k_B T}{f_d}. \qquad (9.30)$$

Here D is the Stokes–Einstein diffusion coefficient of the SD particles. In an MDE the fluxes from (9.29) and (9.30) balance each other, which leads to the differential equation:

$$\frac{d\rho}{\rho} = -\frac{m_d}{k_B T} \mu_0 dH, \qquad (9.31)$$

the solution of which is the MDE profile

$$\frac{\rho(x)}{\rho_0} = \exp\left[-\frac{m_d \mu_0}{k_B T} H(x)\right]; \quad \rho_0 = \rho(x = 0). \qquad (9.32)$$

Here we have explicitly indicated that both particle concentration ρ and field strength H depend on position x in the MDE profile. Note that in general (9.32) is not a single exponential; only for a linear increase of the field strength:

$$H(x) = \beta x; \quad \beta > 0, \qquad (9.33)$$

the MDE profile is a single exponential

$$\frac{\rho(x)}{\rho_0} = \exp\left[-\frac{m_d \mu_0 \beta}{k_B T} x\right]. \tag{9.34}$$

This is the 'magnetic equivalent' of the sedimentation–diffusion equilibrium profile that was treated in Section 7.3.

Summary

- Atomic dipole ordering determines the spectrum of magnetic receptiveness of materials, varying from small, negative diamagnetic to the large positive ferromagnetic susceptibility.
- Ferrofluids are concentrated, stable dispersions of single-domain magnetic particles that respond to a magnetic field gradient much like a molten ferromagnet would do.
- Magnetite synthesis is essentially alkalization of iron chloride solutions leading to single-domain particles, with size polydispersity due to overlap of nucleation and growth phases.
- The magnetization of superparamagnetic ferrofluids by external fields, strongly amplified compared to paramagnetic gases, yields experimental values of magnetic colloid moments.
- Magnetic relaxation entails randomization of aligned moments by Brownian particle rotation, and internal Néel contribution vanishes exponentially with increasing colloid size.
- Interaction between uncharged dipolar colloids comprises excluded-volume repulsion and magnetic attraction, which for growing particle size rapidly dwarfs Van der Waals attraction.
- Rapid micro-bead magnetophoresis, resulting from all embedded dipoles having the same field-induced orientation, enables highly selective biomolecular separation from solution.

References

D. Chan et al. (1985), 'The stability of colloidal suspensions of coated magnetic particles in aqueous solution', *IBM J. Res. Develop.* **29**(1) January, 11–17.

E. Claesson et al. (2007), 'Measurement of the zero-field magnetic dipole moment of magnetizable colloidal silica spheres', *J. Phys.: Condens. Matter* **19**, 36105.

M. Klokkenburg et al. (2006), 'Quantitative real-space analysis of self-assembled structures of magnetic dipolar colloids', *Phys. Rev. Letters* **96**, 037203.

Q. A. Pankhurst et al. (2003), 'Topical Review: Applications of magnetic nanoparticles in biomedicine', *J. Phys. D. Appl. Phys.* **36**, R167–R181.

A. P. Philipse (2018), *Brownian Motion: Elements of Colloid Dynamics*. Springer Nature, Cham.

A. Philipse and D. Maas (2002), 'Magnetic Colloids from Magnetotactic Bacteria', *Langmuir* **18**, 9977–9984.

R. Rosensweig (1985), *Ferrohydrodynamics*. Cambridge University Press, New York.

C. Ruffert (2016), 'Magnetic bead—Magic Bullet', *Micromachines* **7**(2), 21.

V. Socoliuc et al. (2022), 'Ferrofluids and bio-ferrofluids: looking back and stepping forwards', *Nanoscale* **14**(13). Royal Society of Chemistry.

G. van Ewijk, G. J. Vroege, and A. P. Philipse (1999), 'Convenient preparation methods for magnetic colloids', *J. Magn. Magn. Mater.* **201**, 31–33.

A. van Silfhout (2020), *Magnetic Sedimentation in Aqueous Ferro Fluids for Magnetic Density Separation*. Thesis, Utrecht University.

Appendix 9A: Small-scale ferrofluid synthesis

The following ferrofluid synthesis is based on van Ewijk, Vroege, and Philipse (1999).

Magnetite precipitation

Dissolve simultaneously $FeCl_2 \cdot 4H_2O$ (3.29 g, 16.5 mmol) and $FeCl_3 \cdot 6H_2O$ (8.68 g, 32.1 mmol) in 380 ml demineralized water. Add, under vigorous stirring at room temperature, 25 ml ammonia (25 per cent); immediately a dark, magnetic precipitate forms, which is collected with a permanent magnet and, after decantation of the supernatant, mixed with 40 ml 2M HNO_3. The nitric acid brings the pH below the isoelectric point of iron oxide, and **repeptizes** the precipitate.

Magnetite oxidation

After stirring for 5 minutes, the oxidation of magnetite to maghemite is completed by adding 60 ml of an aqueous 0.35 M $Fe(NO_3)_3$ solution and subsequent refluxing of the stirred solution at its boiling point for one hour. On a permanent magnet, the maghemite settles as a reddish sediment. After decanting the supernatant and washing the precipitate twice with 100 ml 2 M HNO_3 (decant the acid as much as possible), the precipitate is redispersed in 50 ml demineralized water to a stable, black maghemite sol with a typical solid weight concentration of 5–6 g/l.

Oleic acid coating

The maghemite particles can now be grafted with oleic acid on a small scale at room temperature. To that end, 2 ml of the aqueous sol is diluted with 50 ml demineralized water, coagulated by adding a few drops of ammonia (25 per cent) and sedimented on a magnet. After decanting the supernatant and washing with 50 ml water, 100 ml water is added to the gently stirred precipitate, followed by the addition of 6–8 ml oleic acid. Within a few minutes, all maghemite colloids migrate into the oil phase where, after separation from the colourless aqueous phase, they are washed three times with 10 ml ethanol to remove water and any excess surfactant. After drying in a nitrogen flow, the oleic acid-coated maghemite particles are easily redispersed in a few millilitres of cyclohexane to form a stable dispersion, which can be manipulated with a magnet.

Size polydispersity

An electron microscope image of the thus synthesized particles will look typically like the TEM image from Figure 9.5D, and display colloids with an average diameter of around 7–10 nm and a size polydispersity of about 30–40 per cent. This significant spread in particle size is due to the circumstance that particle formation already starts before pH gradients (resulting from the addition of ammonia) have disappeared by stirring. So, there is considerable overlap of nucleation and growth in the magnetite precipitation process, which broadens the size distribution.

Exercises

9.1 (a) Calculate λ in (9.19) for magnetite ($m_s \approx 430\,\text{kA m}^{-1}$) at room temperature, for single-domain particles with diameters 1, 10, and 30 nm.
  (b) For which diameter is the maximal contact attraction in (9.21) equal to $k_B T$?

9.2 By which factor does the parameter λ in (9.19) decrease if magnetite particles of diameter d are covered by an oleic acid layer of thickness Δ?

9.3 (a) Calculate the rotational relaxation time τ_{RR} for magnetite colloids of diameter $d = 10$ nm in a ferrofluid at 25 °C in kerosene with viscosity $\eta = 0.002\,\text{kg.m}^{-1}.\text{s}^{-1}$.
  (b) Estimate the energy barrier $KV_p/k_B T$ in Néel rotation that yields a Néel relaxation time τ_{NR} equal to the magnitude of τ_{RR} from question (a).
  (c) What is the value of the anisotropy constant for the magnetite particles in (a) that yields the energy barrier from (b)?

9.4 Verify (9.12).

9.5 Find normalization constant C from (9.4) and (9.5).

9.6 Verify the calculation of $<\cos\theta>$ in (9.7).

9.7 Verify the weak-field limit of the Langevin function in (9.10).

9.8 Estimate the energies of the following assemblies of dipolar spheres. Which of them are the most stable?

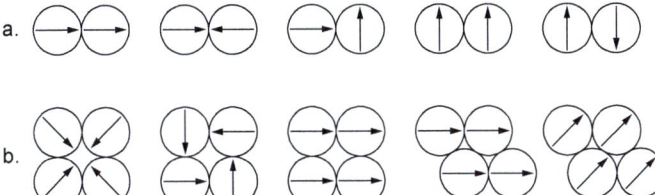

9.9 Consider spheres with diameter $d = 10$ nm composed of iron oxide with a bulk magnetization $m_{sat} = 430$ kA m^{-1} at 300 K.
  (a) Calculate the dipole moment m_d of one particle.
  (b) Evaluate in units of $k_B T$ the interaction energy of one sphere with an external field H for $H = 1, 10, 100, 1{,}000$ kA m^{-1}.
  (c) Sketch a magnetization curve, using your answers from (b), with the vertical axis scaled on the saturation magnetization.
  (d) A dispersion of the iron-oxide spheres has a measured saturation magnetization of 100 kA m^{-1}. Calculate the sphere volume fraction in the dispersion.
  (e) How large is the initial magnetic susceptibility of the dispersion in (d)?

9.10 Consider the MD-equilibrium profile $\dfrac{\rho(x)}{\rho_0} = \exp[-\dfrac{m_d \mu_0}{k_B T} H(x)]$ from equation (9.32).
  (a) Calculate the value of $k_B T / \mu_0 m_d$ for the particles from Exercise 9.9.
  (b) What is the physical meaning of this value?

9.11 (a) Derive the second virial coefficient B_2 for the DHS potential in equation (9.24). First evaluate the hard-sphere part, B_2^{HS}, of the second virial coefficient; then evaluate B_2^{MAG}, the magnetic part, assuming particle magnetic moments are weak.
  (b) What effect does the magnetic particle moment have on the osmotic pressure of a dispersion?

Find the solutions to these exercises at the end of the book.

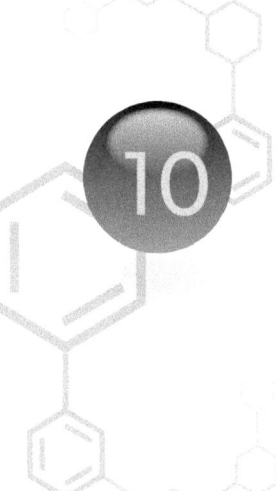

10 Association Colloids and Emulsions

10.1 Introduction

When you vigorously stir a mixture of olive oil and vinegar, the turbid salad dressing that results is an example of an **emulsion**. Depending on the relative amounts of oil and vinegar at the start of the emulsification process, you have either a water-in-oil emulsion with vinegar droplets in a continuous oil phase, or an oil-in-water emulsion with oil droplets floating in a continuous water (vinegar) phase. The salad dressing is, like *any* emulsion, thermodynamically unstable: the dressing exhibits **coalescence**, the spontaneous fusion of droplets, ultimately leading to phase separation of the emulsion into the initial vinegar and oil phases.

The emulsion could be kinetically trapped by adding gelling agents that strongly increase the viscosity of the continuous phase, which delays coalescence. Alternatively, droplets could be stabilized by addition of surface-active agents or **surfactants** that adsorb on droplet surfaces. This adsorption reduces the oil–water **interfacial tension** and, hence, diminishes the driving force for coalescence. Incidentally, employing surface-active proteins as stabilizer will turn the dressing into mayonnaise—an emulsion, if well prepared, in which coalescence is imperceptibly slow.

On washing your hands after all this cookery in a bowl of water with a bar of soap, you will notice that the water turns bluish: you have made a solution of **micelles**—aggregates of surfactant molecules. The bluish appearance, incidentally, of a micellar solution has the same origin as the blue colour of the sky during the day, namely the Rayleigh light scattering (a topic to which we will return in Chapter 11). Whereas an unstabilized salad dressing demixes within a few hours, micellar solutions do not change in an appreciable time because they are thermodynamically stable. Micelles belong to the category of **association colloids**, defined as structures of colloidal dimensions that spontaneously form in solution by self-assembly of surfactant molecules. Other examples of association colloids are vesicles and biological cell membranes—structures that are much more complex than the micelles to which we restrict our attention here.

The phenomenon of surface tension is shown in Section 10.2 to be the cause of the solubility of small droplets, and the ensuing Ostwald ripening. Surface tension being the primary cause of emulsion instability motivates us to explore in Section 10.3 ways to undo this surface tension effect—via adsorption of surfactants and small particles on droplet surfaces. In Section 10.4 we switch the focus from multi-molecular entities in the form of unstable emulsions to the spontaneously formed multi-molecular micelles. We address, among other things, the thermodynamics of micelle formation and the equilibrium micellar size distributions. This chapter is a first introduction to emulsions and micellar dispersions that serves as a stepping stone to the in-depth treatment of these systems in Leermakers et al. (2005) and Walstra (2005).

10.2 Interfacial tension and the Laplace pressure

The boundary between two phases such as two immiscible liquids, or a liquid and a solid phase, is referred to as an **interface**; the term **surface** is usually reserved for the cases when one phase is a gas. Molecules in a surface or interface represent a higher state of energy than molecules in the bulk. To understand why, let us look at the surface of liquid water. A water molecule in the bulk experiences the Van der Waals attraction exerted by its neighbour molecules from all sides. A molecule at the surface, however, lacks about half of these neighbours and, consequently, experiences a net pull into the water bulk. Thus, energy is needed to transfer a water molecule from that bulk to the surface, so increasing surface area requires energy.

The tensile force of surface tension

The extra energy required to form a surface or interface is quantified by the surface tension or interfacial tension γ which equals the reversible work needed to increase the interface with a unit area, for example 1 cm². The thermodynamic definition of surface tension is

$$\gamma = \left(\frac{\partial G}{\partial A}\right)_{p,T,n}, \tag{10.1}$$

which expresses that γ equals the variation in Gibbs free energy G with surface area A, at constant pressure p and temperature T, and constant number of moles n. Let us apply this definition to an oil–water interface created by immersing an oil film, carried by a rectangular frame, in water, as sketched in Figure 10.1. The film will, when left alone, spontaneously contract to decrease the oil–water interface. Suppose the film with constant width L shrinks from a length x to $x-dx$; then, everything else being constant, the Gibbs energy decrease by the contraction follows from (10.1) as

$$dG_{\text{CON}} = \gamma dA = -\gamma 2L dx, \tag{10.2}$$

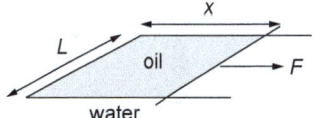

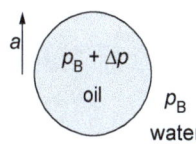

Figure 10.1 **Top**: An oil film of width L immersed in water in a rectangular frame, one side of which can be moved by an external force F—which counteracts spontaneous film contraction. **Bottom**: An oil droplet of radius a immersed in water; here it is the internal Laplace pressure Δp, in excess of bulk pressure p_B, that opposes spontaneous reduction of the oil–water interface.

where the factor 2 accounts for the upper and lower side of the oil film. To counteract the spontaneous contraction a force F is applied (see Figure 10.1) that compels the film to stretch. The Gibbs energy increase due to elongating the film a distance dx equals the reversible work done by force F:

$$dG_{STR} = Fdx. \qquad (10.3)$$

In equilibrium, the film area remains unchanged because the total change in Gibbs energy is zero: $dG_{CON} + dG_{STR} = 0$, from which it follows that

$$\gamma = \frac{F}{2L}. \qquad (10.4)$$

Thus, surface tension can also be seen as a force per unit length (with dimension Nm^{-1})—a tensile force that is oppositely directed to the external force F sketched in Figure 10.1.

Excess Laplace pressure under concave surfaces

We now consider, instead of a flat oil–water interface, the spherical interface between an oil droplet and the continuous water phase. The spontaneous tendency of the interface to shrink is, instead of the external force for the flat interface, now counteracted by an *internal* force, namely the **Laplace pressure** Δp inside the droplet. This Δp is actually the pressure in the droplet in excess of the pressure p_B in the surrounding bulk phase (Figure 10.1). Imagine an infinitesimal, spontaneous decline in droplet radius from a to $a-da$; then the decrease in droplet surface area is $8\pi ada$ (Exercise 10.6), which corresponds to the Gibbs energy decrease:

$$dG_{surf} = -\gamma 8\pi ada. \qquad (10.5)$$

The compression of the droplet amounts to an increase in Gibbs energy equal to the volume work done on the droplet. Since, due to the infinitesimal radius decrease, the Laplace pressure remains virtually constant, this volume work equals Δp times the volume change:

$$dG_{vol} = \Delta p 4\pi a^2 da. \qquad (10.6)$$

When the spontaneous surface reduction and the forced droplet volume compression are in balance, equilibrium is present, as signified by a constant droplet radius and a zero change in the total Gibbs energy:

$$dG = dG_{vol} + dG_{surf} = 0. \qquad (10.7)$$

On substitution of (10.5) and (10.6) in the equilibrium condition (10.7) we obtain the Laplace equation for the excess pressure in a droplet:

$$\Delta p = \frac{2\gamma}{a}. \qquad (10.8)$$

Note in the derivation of the Laplace equation we have tacitly assumed that surface tension γ is independent of droplet radius a, and equal to its bulk value. This assumption is valid as long as droplet radii are much larger than molecular size.

Coarsening by Ostwald ripening

The dependence in (10.8) of the excess pressure inside a droplet on the droplet size has an important consequence—which we will investigate for an emulsion of oil droplets in water. Consider the fate of a small emulsion droplet next to a larger one; from (10.8) it follows that in the small droplet oil molecules are at a higher pressure than in the larger drop. Hence molecules from the small droplet have the spontaneous tendency to join the larger droplet. This joining can occur via diffusion in the continuous phase, or by the merger of droplets in close contact, a process referred to as coalescence. In other words, the emulsion exhibits **coarsening**: its average droplet size increases because large droplets grow at the expense of disappearing small ones. Emulsion coarsening is an instance of Ostwald ripening, the process that we investigated earlier in Section 2.5, in the context of particle nucleation and growth in a supersaturated solution.

10.3 Stabilized emulsions

Emulsification of oil and water requires the creation of a huge interfacial area between two mutually repelling liquids. As a practical example, consider the emulsion known as mayonnaise: it contains 80 volume per cent olive oil, in the form of droplets with an average diameter of about 5 microns. So a 500 ml pot of mayo contains about 479 m^2 of interface between olive oil and water, an area almost 13,000 times larger than the area of the pot itself (Exercise 10.1).

A high energy input (from mixing, high-speed stirring, or ultra-sonification) is needed to realize such large interfacial areas. This energy input increases the interfacial Gibbs free energy of the system which, in turn, tends to lower this free energy by spontaneous area reduction (Figure 10.2) via Ostwald ripening or droplet coalescence. Ultimately, the system macroscopically de-mixes into two liquids, the one with the lower mass density floating on top of the other.

Stabilizing surfactants

One strategy to kinetically retard destabilization of emulsions, as mentioned in the introduction, is to employ emulsifiers in the form of surface-active agents or surfactants. The latter are amphiphilic molecules with a hydrophilic head that prefers the water phase, and a water-repelling hydrophobic tail that favours the oil phase. Due to their amphiphilicity surfactants position themselves in the oil–water interfacial area, as sketched for an oil droplet in water in Figure 10.3A. Surfactants lower the tendency of droplets to coalesce because they decrease the interfacial tension γ_{ow} of the oil–water interface. Hence, the Gibbs free energy decrease $\Delta G = \gamma_{ow} \Delta A$ associated with an interfacial area decrease ΔA, the ΔG that drives coalescence, becomes smaller.

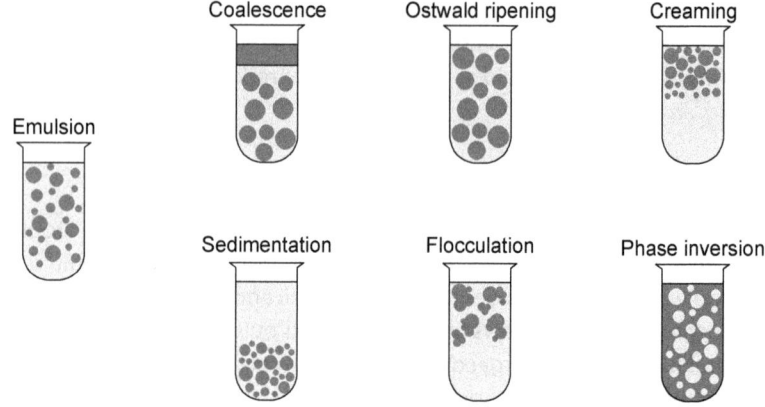

Figure 10.2 **Left**: Illustration of a kinetically stable emulsion. **Right**: Emulsion instability materializes itself in various ways: droplet coalescence, coarsening by Ostwald ripening, creaming, sedimentation, flocculation, and phase inversion.

Source: Figure courtesy of Riande Dekker, Van der Waals–Zeeman Institute.

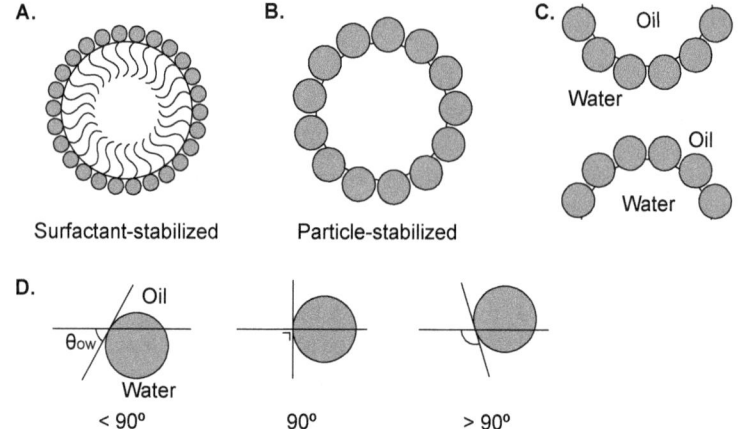

Figure 10.3 **(A)** Emulsion droplet kinetically stabilized by surfactants, with their hydrophobic tail and hydrophilic head residing, respectively, in oil and water. **(B)** Droplet stabilized by adsorbed particles forming a steric barrier between oil (o) and water (w) phase. **(C)** Particle wetting properties determine whether a w/o or an o/w emulsion is formed. **(D)** An o/w emulsion forms for oil–water contact angles $\theta < 90°$ whereas angles $\theta > 90°$ create a w/o emulsion. The adsorption energy is maximal for angles $\theta \approx 90°$ occurring for amphiphilic particles having no preference for either oil or water.

Source: Figure adapted from Dekker (2022).

Pickering stabilization

Instead of surfactants, solid particles can also be employed as stabilizers, in so-called **Pickering emulsions**. Here particles form a physical barricade at the oil–water interface (Figure 10.3B) that puts a stop to coalescence. The efficiency of Pickering stabilization strongly depends on the wetting properties of the particles involved. The optimal situation is partial wettability of particles by both phases, leading to contact angle at the oil–water interface of (close to) 90°. For hydrophobic particles the contact angle exceeds 90° which favours water-in-oil emulsions; slightly hydrophilic particles induce contact angles below 90° which is apt for stabilizing oil-in-water emulsions; see also Figure 10.3C and D.

Frequently applied stabilizers for Pickering emulsions are silica particles (Figure 10.4), which have the advantage that their wetting properties can be tuned via grafting of hydrophobic or hydrophilic molecules onto the silica surface. Other inorganic stabilizers include gibbsite minerals and clay platelets. In the food industry bio-organic stabilizers are also employed, such as proteins, starch, and the linear polysaccharide chitosan; the stabilizing agents in mayonnaise are egg yolk particles; and in margarine (Figure 10.4) the Pickering stabilizers are solid fat crystals that keep water droplets separated in a liquid oil phase.

Gibbs adsorption energy

Two crucial features of particle adsorption at the oil–water interface are the following. First, the adsorption is spontaneous as the change in adsorption free energy G_A is negative: $\Delta G_A < 0$. Second, the Gibbs adsorption energy for a particle is usually much larger than the particle's thermal energy $k_B T$. Hence, particle adsorption cannot be undone by thermal fluctuations: Pickering emulsions are kinetically stabilized by irreversibly adsorbed particles.

We will evaluate the adsorption energy for the case of a hydrophilic solid sphere with radius a that migrates from water to the oil–water interface (Figure 10.5), at a position where a sphere cap of height h protrudes into the oil phase, with wetting angle θ. The transfer comprises a free energy price for the particle because water in contact with area A_{cap} of the hydrophilic sphere cap is replaced by oil:

$$\Delta G_{price} = A_{cap}(\gamma_{SO} - \gamma_{SW}). \tag{10.9}$$

Here γ_{SO} and γ_{SW} are the interfacial tensions between the solid phase and, respectively, oil and water. On the other hand, the cape's base area A_{base}

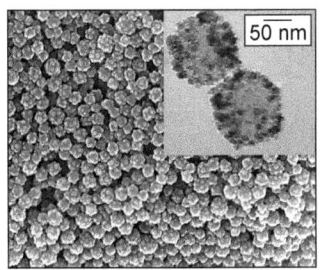

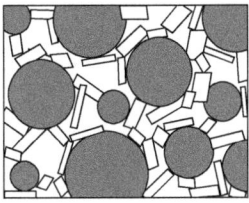

Figure 10.4 Top: SEM images of Pickering emulsion composed of (polymerized acrylic) oil droplets stabilized by silica nanoparticles. Bottom: Illustration of margarine: An emulsion of water droplets in edible oil with fat crystals keeping water droplets separated. Average droplet size is about 10 microns. In your mouth fat crystals melt, such that Pickering stabilization is lost and tasteable water droplets merge—and mix with your saliva.

Source: Figure courtesy of Sacanna et al. (2007).

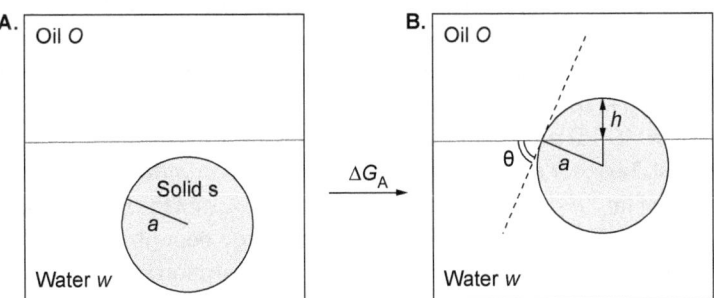

Figure 10.5 A hydrophilic sphere in water (**A**) adsorbs at the oil–water interface, with a sphere cap of height h (**B**) protruding into the oil phase. The adsorption is spontaneous: $\Delta G_A < 0$; the Gibbs energy decrease is caused by removal of an oil–water interfacial area equal to the base area $A_{base} = \pi a^2(1-\cos^2\theta)$ of the sphere cap.

shuts off an area A_{base} of oil–water interface with surface tension γ_{WO} which decreases the Gibbs energy with an amount:

$$\Delta G_{gain} = -A_{base}\gamma_{WO}. \tag{10.10}$$

The three surface tensions in (10.9) and (10.10) are related to the three-phase contact angle θ via **Young's equation**:

$$\gamma_{SO} = \gamma_{SW} + \gamma_{WO}\cos\theta, \tag{10.11}$$

which in essence represents a force balance on the three-phase contact line between oil, water, and the solid sphere. From equations (10.9), (10.10), and (10.11) we find for the total Gibbs energy change ΔG_A for the particle transfer to the oil-water interface:

$$\Delta G_A = \Delta G_{price} + \Delta G_{gain} = (A_{cap}\cos\theta - A_{base})\gamma_{WO}. \tag{10.12}$$

The areas of the oil-protruding sphere cap and its base are (see Exercise 10.3):

$$A_{cap} = 2\pi a^2(1-\cos\theta);\ A_{base} = \pi a^2(1-\cos^2\theta), \tag{10.13}$$

which on substitution in (10.12) yields:

$$\Delta G_A = -\pi a^2 \gamma_{WO}(1-\cos\theta)^2. \tag{10.14}$$

This is the adsorption free energy for a hydrophilic sphere, with a contact angle in the range $0° \leq \theta \leq 90°$; repeating the derivation for a hydrophobic sphere moving from the oil to the oil–water interface we also find (10.14), except with a term $(1+\cos\theta)^2$ instead of $(1-\cos\theta)^2$. We can gather all this in one formula by writing for the particle's adsorption Gibbs free energy:

$$\Delta G_A = -\pi a^2 \gamma_{WO}(1-|\cos\theta|)^2. \tag{10.15}$$

The adsorption energy as function of the wetting angle, sketched in Figure 10.6, comprises a deep cusp, with a minimum at $\theta = 90°$: the wetting angle at which the removed oil-water interfacial area reaches its maximum value, namely the base area πa^2 of a half-sphere.

Equation (10.15) predicts that for small nanoparticles ΔG_A may already be substantial. Take, for example, a toluene–water system with interfacial tension $\gamma_{WO} = 36\,mNm^{-1}$. For a nanoparticle with radius $a = 1nm$ the maximal adsorption energy is $\pi a^2 \gamma_{WO} = -27.5\,k_BT$: more than large enough to safeguard an amphiphilic particle (for which $\theta \approx 90°$) against desorption by thermal fluctuations. The adsorption energy drops to $\Delta G_A \approx -k_BT$ for $\theta \approx 37°$, meaning that adsorption is only reversible around or below this wetting angle.

The fact that ΔG_A is negative even at contact angles close to $0°$ and $180°$ (Figure 10.6) implies that particles remain 'surface active' and self-assemble at an oil–water interface, over a wide range of hydrophobicity.

COLLOID SCIENCE

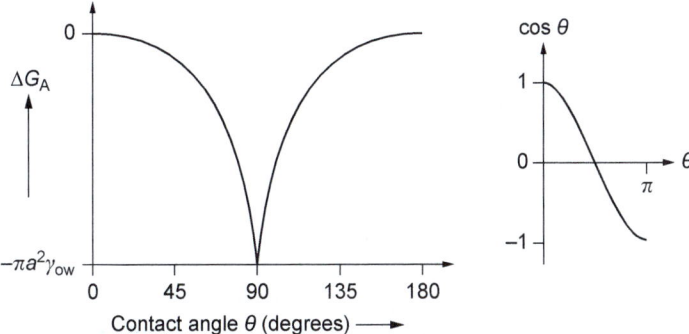

Figure 10.6 Sketch of the variation in the Gibbs energy G_A, according to equation (10.15), for adsorption of a sphere of radius a on an oil–water interface with tension γ_{ow}. At contact angle $\theta = 90°$ the Gibbs energy is maximal, as the eliminated interfacial oil–water is maximal, namely the base area πa^2 of a half-sphere.

A.
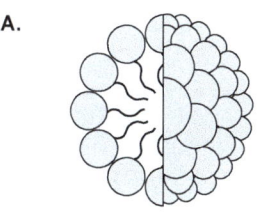

B.
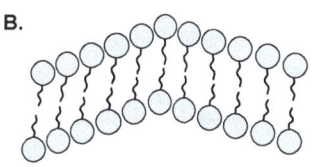

C.
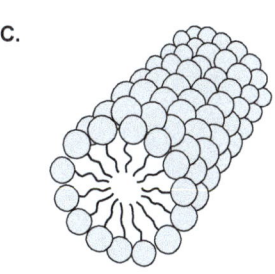

Figure 10.7 Example of micellar structures: **(A)** spherical, **(B)** lamellar, and **(C)** cylindrical. Head groups are shown as spheres, connected to a surfactant tail. The shape of micellar aggregates is determined by the size of the surfactant's head group and the length of its hydrophobic chains. For an account of surfactant geometry and micellar shape see Israelachvili (1992).

10.4 Micellization

The formation of colloids by growth in a molecular solution, which we studied in Chapter 2, is an activated process wherein only clusters above a certain critical size grow irreversibly by uptake of surrounding molecules or monomers. Micelles, in contrast, spontaneously form in the absence of any nucleation barrier, and ultimately evolve to colloids with reproducible size and shape that are controlled by thermodynamics. The solution concentration of surfactants at which they start to self-assemble to micelles is known as the **critical micelle concentration (CMC)**. It appears to be the case that initially formed micelles are more or less spherical associates containing in order of magnitude 100 molecules. At higher concentrations these roughly spherical 'droplets' merge to other morphologies such as discs, cylinders, and laminar forms, as sketched in Figure 10.7.

Soaps and surfactants

Molecules forming micelles are **amphipathic**: one part of the molecule is highly soluble in the solvent concerned, the other part is insoluble. The solvent may be an organic one, or water. For micelles in aqueous systems the amphipathic molecule is a surfactant comprising a polar head group and a hydrocarbon chain. Ionic surfactants possess a charged head group. Familiar examples of ion surfactants are the classical **soap** molecules which comprise the sodium or potassium salts of fatty acids (carboxylates) and sodium dodecyl sulphate (which has an anion as head group), a component of many cleaning and hygiene products. Non-ionic surfactants are block copolymers AB, composed of a hydrophilic A-chain connected to a water-repelling B-chain. These surfactants are widely used in applications such as pharmaceuticals, cosmetics, and detergency.

Detergency

Let us consider briefly the latter application and see how micelles can work as a **detergent** in laundering and washing. Fatty and oily substances—main ingredients of dirt—strongly adsorb as thin films on textile fibres. Surface-active molecules diminish the adhesion between fat and textile and 'roll up' fat to droplets that—with their surface covered by polar head groups that protrude into the water phase—remain floating in solution, without redeposition on textile and surfaces to be cleaned. The twofold role of micelles in washing appears to involve, first, being the supply source of the single surfactants that are consumed in adsorption processes, and, second, contributing to cleansing via **solubilization**—that is, the uptake of non-polar, fatty substances in the hydrocarbon-like, hydrophobic cores of micelles.

Micelle–monomer equilibria

We continue now with a closer look at the process of micellization to address its most remarkable feature, and that is the abrupt change in monomer concentration around the CMC from a linear concentration increase to a flat plateau value that represents a virtually constant monomer concentration (see Figure 10.8). This quite sudden transition can be explained by the Debye model as being a consequence of the thermodynamic equilibrium between micelles and free monomers. Suppose a micelle is composed of n monomeric soap molecules S; then, neglecting any effect of non-ideality, the micelle–monomer equilibrium can be represented by

$$nS \rightleftarrows S_n, \qquad (10.16)$$

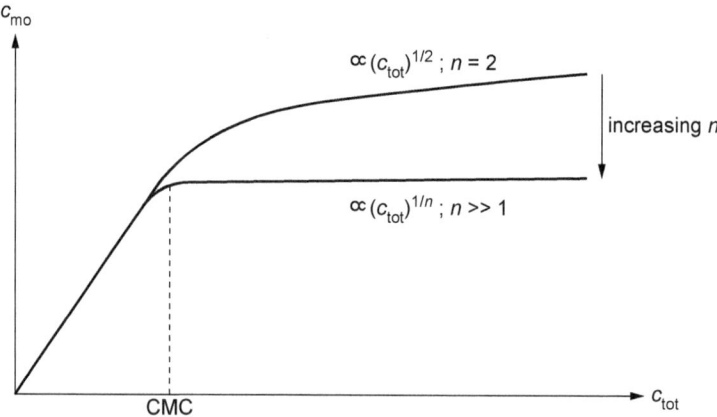

Figure 10.8 Sketch of surfactant–monomer concentration c_{mo} versus the total concentration c_{tot} of added surfactants. Around the CMC micelles start to form composed of n monomers. For the case of dimers ($n = 2$) the monomer concentration beyond the CMC increases with the square root of c_{tot} (Exercise 10.8). For large n, added monomers virtually all end up in newly formed micelles such that above the CMC, c_{mo} is almost constant.

with concentrations determined by the equilibrium constant K:

$$K = \frac{c_{mi}}{(c_{mo})^n}, \qquad (10.17)$$

where c_{mi} is the micelle concentration and c_{mo} the concentration of free soap molecules. For a total concentration c_{tot} of added soap molecules, mass conservation requires that:

$$c_{tot} = c_{mo} + nc_{mi}. \qquad (10.18)$$

From (10.17) and (10.18) we can eliminate the micelle concentration to obtain:

$$c_{tot} = c_{mo} + Kn(c_{mo})^n. \qquad (10.19)$$

The number n of monomers in a micelle is substantial, typically $n \sim 50–100$, so to a very good approximation we can ignore c_{mo} with respect to $Kn(c_{mo})^n$, to find for the monomer concentration that is in equilibrium with the micelles:

$$c_{mo} = \left(\frac{c_{tot}}{nK}\right)^{1/n}, \text{ for } n \gg 1. \qquad (10.20)$$

Clearly the increase of monomer concentration as function of the total added soap concentration is very slight, since c_{tot} is raised to a power $1/n$ that is close to zero. Therefore the following scenario emerges (as sketched in Figure 10.8): below the CMC the monomer concentration linearly increases until the CMC is reached (which, incidentally, is not a sharp concentration but a rather narrow window of soap concentrations in which micelles start to form). A further increase in total surfactant concentration leads almost exclusively to an increase in the number of micelles and an only marginal rise in the concentration of single molecules.

Micellar size distributions

In a micellar solution the micelles do not all have the same size n; the association of surfactants to micelles is a reversible process; micelles form and fall apart and, hence, the size of the micelles is distributed. The shape of this distribution follows from the condition of thermodynamic equilibrium between surfactants and micelles. This condition reads in terms of the chemical potentials of the species involved:

$$\mu_n = n\mu_1. \qquad (10.21)$$

The chemical potential μ_n of a micelle of size n equals n times the chemical potential μ_1 of a surfactant monomer. Assuming both micelles and monomers are present at low concentrations, the chemical potentials may be written as

$$\mu_i = \mu_i^0 + k_B T \ln x_i, \qquad (10.22)$$

where μ_i^0 is the standard chemical potential ($\mu_i = \mu_i^0$ for $x_i = 1$) and $k_B T$ the thermal energy; x_i is a mole fraction, equal to x_1 and x_n for, respectively, free surfactants and micelles. Substitution of (10.22) for $i = 1$ and $i = n$ in the equilibrium condition (10.21) leads to the following equilibrium distribution of micellar size:

$$x_n = (x_1)^n \exp\left(\frac{-\Delta\mu^0}{k_B T}\right); \quad \Delta\mu^0 = \mu_n^0 - n\mu_1^0. \tag{10.23}$$

Here $\Delta\mu^0$ is the chemical potential difference for the surfactant association $nS \rightarrow S_n$ under standard conditions. We note in passing that the thermodynamic equilibrium constant K_{thermo} for reversible particle association follows from (10.23) as

$$K_{thermo} = \frac{x_n}{(x_1)^n} = \exp\left(\frac{-\Delta\mu^0}{k_B T}\right), \tag{10.24}$$

a dimensionless constant that should be clearly distinguished for the 'practical' equilibrium constant K in (10.17) which has the dimension of (concentration)$^{1-n}$.

Linear micelles

Let us further work out the equilibrium distribution (10.23) for the simple case of attractive monomers that reversibly combine to a flexible, linear aggregate, as sketched in Figure 10.9. For a linear chain of particles we can put

$$\Delta\mu^0 = (n-1)w_{rev}, \tag{10.25}$$

where w_{rev} is the reversible work needed to create any of the $n-1$ bonds in the chain; the assumption here is that the interaction between monomers is short-ranged such that only interactions between contacting monomers contribute to the chain's free energy. On substitution of (10.25) into (10.23) we arrive at

$$x_n = (x_1)^n \exp\left(\frac{(1-n)w_{rev}}{k_B T}\right), \tag{10.26}$$

which is a size distribution in which the number of n-mers decreases exponentially with aggregate size n. It can be expected that the chain length distribution in (10.26) applies not only to monomers but to any reversible aggregation of weakly attractive sub-particles. In Chapter 9 we came across an example of such aggregation by magnetite particles that associate into flexible dipolar chains, as depicted in Figure 9.5C. The chain length distribution of the chains in Figure 9.5C is indeed an exponential one.

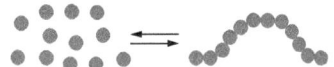

Figure 10.9 Example of weakly attractive monomers in equilibrium with a worm-like, linear aggregate of size $n = 12$.

Summary

- An emulsion is a dispersion of liquid droplets in a continuous fluid medium that can be stabilized against coalescence by surfactants or Pickering particles that reside in the liquid–medium interface, thereby reducing interfacial tension.

- The tendency of interfaces to contract due to surface tension is, for convex interfaces, counteracted by the Laplace pressure. This pressure's radius dependence accounts for the Ostwald ripening of polydisperse droplets.
- Surface-active molecules (monomers), having passed their solubility limit, self-assemble to multi-molecular micelles. Thermodynamic equilibrium between monomers and micelles entails that virtually all monomers added beyond the critical micelle concentration end up in a micelle.
- The reversible micelle formation produces distributions of micellar sizes; for linear (worm-like) chains the number fraction of chains decreases exponentially with the chain length.

References

R. Dekker (2022), *Emulsion Stability and Rheology*. PhD Thesis, Utrecht University.

J. Israelachvili (1992), *Intermolecular and Surface Forces* (2nd edn). Academic Press, San Diego.

F. Leermakers, J. C. Eriksson, and J. Lyklema (2005), 'Associaton Colloids and their Equilibrium Modelling', in *Fundamentals of Colloids and Interface Science*, vol. 5, J. Lyklema (ed.). Elsevier, Amsterdam, pp. 4.2–4.121.

S. Sacanna et al. (2007), 'Thermodynamically stable pickering emulsions', *Phys. Rev. Letters* **98**, 158301.

P. Walstra (2005), 'Emulsions', in *Fundamentals of Colloids and Interface Science*, vol. 5, J. Lyklema (ed.). Elsevier, Amsterdam, pp. 8.2–8.92.

Exercises

10.1 (a) Verify the intefacial area of 479 m² between olive oil and water in the 500 ml pot of mayonnaise mentioned in Section 10.3.
(b) Calculate the ratio of this oil area to that of the cylindrical pot, assuming the pot's height equals its radius R.

10.2 Calculate (in units of k_BT) the Gibbs energy for desorption of a sphere with radius $a = 3$ and 30 nm from an oil–water interface, assuming $\gamma_{OW} = 0.03\,\text{N}\,\text{m}^{-1}, \theta = 60°$, and $T = 298\,\text{K}$.

10.3 Verify the formulas in (10.13) for the area of a sphere cap and the area of its base.

10.4 Plot ΔG_A (in units of k_BT) as function of wetting angle θ for adsorption of particles with radius $a = 5$ nm at a toluene–water interface for which $\gamma_{WO} = 36\,\text{mN}\,\text{m}^{-1}$.

10.5 Creaming is the process by which oil droplets rise to the surface, where they form a concentrated emulsion—the creaming process is illustrated in Figure 10.2. A dilute oil-in-water emulsion contains

oil droplets with mass density $\delta_p = 0.8\,\text{g/ml}$ in water with viscosity $\eta = 10^{-3}\,\text{Pa s}$ and mass density $\delta = 1\,\text{g ml}$.
  (a) Calculate the creaming speed for droplets of radius $R = 1$ and $10\,\mu\text{m}$.
  (b) Estimate how many hours it takes for creaming to complete in a bottle with a height of 30 cm.

10.6 A droplet of radius a shrinks infinitesimally to one with radius $a - da$. What are the sign and magnitude of the corresponding Gibbs free energy change dG_{surf}?

10.7 Calculate the pressure in atmospheres needed to press water ($\gamma_{air-water} = 72 \times 10^{-3}\,\text{Nm}^{-1}$ at $25\,°\text{C}$) out of a porous glass membrane containing parallel uniform pores with a diameter of $0.10\,\mu\text{m}$.

10.8 Refer to equation (10.19); consider the case $n = 2$ where soap molecules only form micelles in the form of dimers.
  (a) In this case how does monomer concentration c_{mo} depend on the total concentration c_{tot} of added soap molecules?
  (b) Sketch this dependence in a plot of c_{mo} versus c_{tot}, and explain how this plot modifies upon increasing the number of soap molecules in a micelle to a value $n \gg 1$.

10.9 (a) Find an expression for the osmotic pressure π exerted by the mixture of monomers and dimers from Exercise 10.8, assuming both species are ideal.
  (b) How large is π when $n \gg 1$—that is, when micelles are composed of a very large number of monomers?
  (c) How large is pressure π in the two limiting cases for the dimerization equilibrium constant $K \to 0$ and $K \to \infty$?

Find the solutions to these exercises at the end of the book.

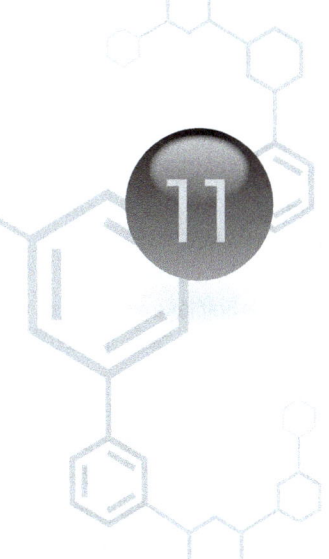

11 Colloid Characterization by Light Scattering and Electrophoresis

11.1 Introduction

Key characteristics of particles in a colloidal dispersion are their shape, size, and electrical surface charge. Knowledge of particle size is important because it allows us to, for example, assess how fast particles will diffuse around in solution; recall from Chapter 3 on Brownian motion that diffusion coefficients depend on particle size via the Stokes–Einstein equation (3.39). Particle size also determines, as we saw in Chapter 7, the rate at which colloids sediment, and the speed at which particles can be removed from solution by colloidal filtration. For determining colloidal size, two widely applied optical scattering methods will be introduced here. Section 11.2 addresses **static light scattering** (SLS), which yields a particle size from the time-averaged intensity of light scattered by a dispersion. Section 11.3 introduces **dynamic light scattering** (DLS), which determines the particle diffusion coefficient and, hence, particle size via the time-dependent intensity fluctuations around the time-averaged value.

The evaluation of the electrical charges of colloids, dealt with in Section 11.4, is important because of the many significant consequences of surface charge and its ensuing electrical double layer (EDL) surrounding colloids. In Chapter 5 we saw how colloidal stability results from the overlap between two EDLs. From the surface potential Ψ_0 of the colloids, the height of the stabilizing repulsive barrier from the DLVO potential in equation (5.58) can be estimated—an estimate that is obstructed by the circumstance (as we will see in Section 11.4) that electrophoresis provides only an approximate value for Ψ_0. Another consequence of colloidal charge is the significant increase in osmotic pressure exerted by a dispersion—a pressure rise that we studied in Chapter 4 via the Donnan model. Furthermore, mention was made in Chapter 8 of the considerable viscosity increase going from neutral to charged colloids, owing to the viscous energy dissipation by the EDL's accompanying particles in a flow field.

Particle charge also enhances osmosis, the spontaneous invasion of solvent into a dispersion (the process that we encountered in Chapter 4).

Telling osmosis examples are the voluminous urine uptake by super-absorbing baby nappies which is caused by the highly charged polyelectrolytes in the nappies' fabric, and the shrinkage of red blood cells (RBCs) in solutions of high salinity. Knowledge of the charge on the proteins in an RBC, for example, would permit a prediction for the salinity above which this shrinkage will occur.

Before entering the introduction to light scattering, let us first take a brief look at size determination by microscopy, which comprises the rating of particle dimensions on photographic images. The method, at first sight, looks straightforward in comparison to the experimentally and theoretically involved scattering techniques. Microscopy, however, has several drawbacks when it comes to particle sizing. The image resolution of optical microscopy is limited in order of magnitude to optical wavelengths, and the number of visualized particles is relatively small in comparison to SLS and DLS which sample large particle numbers in macroscopic dispersion volumes. Furthermore, the electron beam in an electron microscope can cause particles to deform or melt (as is the case of polymeric colloids); and electron microscopy may reveal clustering of particles that is actually caused by sample preparation, namely drying of dispersion droplets on a grid—an instance of this clustering artifact can be seen in Figure 9.5B for the case of magnetite colloids. Moreover, SLS and DLS are also the obvious methods to characterize the soft colloids we studied in Chapter 10, namely micelles and emulsion droplets, because they only exist in liquid media.

11.2 Static light scattering

When a colloidal dispersion is illuminated, the fluctuating electric fields of the incoming light waves interact with electrons in the materials of which colloids and solvent are composed. This interaction brings electrons into a vibratory motion in the form of oscillating electrical dipoles that emit ('scatter') light waves in all directions. Here, and in what follows, it is assumed that colloids are irradiated by monochromatic light that is not absorbed by them: photons of one colour impinging on particles are either transmitted or scattered. Consider now the total intensity of this scattered light that radiates through the surface $4\pi r^2$ of a spherical shell, around an oscillating dipole centred at $r = 0$. If no light is lost due to adsorption, the intensity radiating through a shell is the same for *any* radial distance r. Hence, the scattered intensity I_s decreases with distance in proportion to $I_s \propto 1/r^2$. This brings us to the definition of the **Rayleigh ratio** R:

$$R_\theta = \frac{r^2 I_s \text{(per unit volume of dispersion)}}{I_0}. \tag{11.1}$$

Here I_0 is the intensity of the incident light. The Rayleigh ratio measures the static light scattering (SLS) intensity from a unit volume of dispersion, relative to the incident light intensity. The subscript θ in (11.1) signifies that the SLS intensity generally depends on the scattering angle θ—the angle between incident and scattered light, as indicated in Figure 11.1.

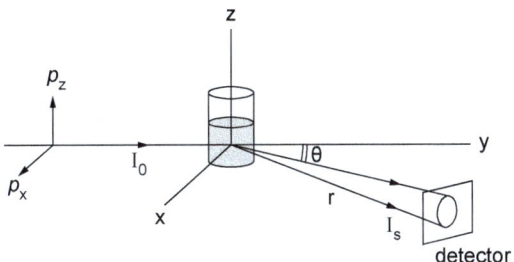

Figure 11.1 Diagram of measuring geometry for light-scattering experiments. The axis of the cell containing dispersion parallels the z-axis lying in the plane of the paper. For a scattering angle θ the scattered light intensity I_s is detected at distance r from the scattering object—a tiny volume of dispersion. The detector moves in the xy-plane perpendicular to the paper. Incident unpolarized light of intensity I_0 comprises vertically and horizontally polarized beams (each of intensity $I_0/2$) represented by the mutually perpendicular vectors p_z and p_x.

Rayleigh scattering by small particles

To comprehend basic features of light scattering it is best to start with scattering from small colloids with dimensions that are much smaller (by a factor of twenty) than the wavelength of light—corresponding to particle dimensions of less than about 20–25 nm. Light waves emitted by dipoles in such small scatterers are all in phase, so each colloid actually radiates as one huge oscillating dipole, emitting radiation that is also known as **Rayleigh scattering**. The detailed mathematical derivation of the Rayleigh ratio (presented in Kerker, 1969) would be out of place here; instead, we will just report the outcome of the derivation, which is Rayleigh's equation (11.2), and discuss its physical meaning, together with some important characteristics of Rayleigh scattering that this equation entails.

For incoming light that is polarized in the vertical z-direction (see Figure 11.1), the scattered light intensity is independent of scattering angle θ; the associated Rayleigh ratio R_V is given by

$$R_V = Kcm; \quad K = \frac{2\pi^2 n^2}{\lambda_0^4}\left(\frac{dn}{dc}\right)^2. \tag{11.2}$$

Here m is the mass of the small Rayleigh scatterers; c is their weight concentration, which is low enough for the effect of interparticle interactions to be negligible. The optical parameter K comprises the wavelength λ_0 of incident light, and the refractive index n of the *solution*. The characteristic wavelength dependence λ_0^{-4} of Rayleigh scattering stems from the amplitude E_s of the electric field radiated by oscillating dipoles. Electromagnetic theory shows that this amplitude is proportional to $E_s \propto \lambda_0^{-2}$ (Kerker, 1969). The scattered intensity is in turn proportional to $I_s \propto E_s^2$, which implies that the wavelength dependence of the intensity is $I_s \propto \lambda_0^{-4}$.

This λ_0^{-4} dependence entails that blue light is scattered more strongly than red light—hence the blue colour of light scattered by the sky during the day, and the bluish appearance of dilute milk and other dilute emulsions and suspensions of small particles. The wavelength dependence in (11.2) also explains the redness of transmitted light by the setting sun, and the use of more weakly scattered yellow lamp light when cars are driven through the fog.

Refractive indices and optical contrast

The solution's refractive index increment dn/dc in the Rayleigh equation (11.2) shows that particles only scatter light when their refractive index n_p differs from the index n_0 of the *solvent* in which they are dispersed—that is to say, when the **optical contrast** $n_p - n_0$ is non-zero. The magnitude of the contrast, and hence the particle refractive index n_p, can be determined from a measured value of dn/dc, a determination which you will be asked to investigate in Exercise 11.3. When the optical contrast is small enough, the refractive index n of the *solution* changes linearly with particle concentration c, such that dn/dc of the solution is given by

$$\frac{dn}{dc} = \frac{n - n_0}{c}. \tag{11.3}$$

The appearance of the square $(dn/dc)^2$ in equation (11.2) ensures that only the absolute value of the optical contrast matters: a particle with an optical contrast of, say, $n_p - n_0 = -0.1$ scatters with the same intensity as one for which $n_p - n_0 = +0.1$.

Rayleigh scattering and particle volume

When a colloidal dispersion is sufficiently dilute, the scattering is simply the sum of contributions from individual colloids—which is why the SLS in (11.2) depends linearly on particle concentration c. Since both weight concentration c and particle mass m in (11.2) depend on particle volume V_p via

$$c = \rho \delta_p V_p; \quad m = \delta_p V_p, \tag{11.4}$$

where ρ is the number density of particles with mass density δ_p, the SLS intensity is proportional to $R_\theta \propto V_p^2$, which translates to $R_\theta \propto a^6$ for spheres of radius a. The strong particle size dependence of SLS entails that dust motes and particle aggregates are strong scatterers that easily dominate the contribution from the small Rayleigh particles. Hence, these contaminants should be removed by filtration or centrifugation (processes investigated in Chapter 7) from a dispersion prior to an SLS measurement. The size dependence also implies that the larger particles in a size distribution contribute the most to the total SLS intensity, such that SLS yields an apparent particle size that may be considerably larger than the number-averaged size obtained from counting particles on electron micrographs. Apparent light-scattering radii are further discussed in Appendix 11A.

The intensity of Rayleigh scattering, as argued above, strongly depends on particle volume. Particle *shape*, however, is immaterial: equation (11.2) applies to small particles of arbitrary shape. Indeed, when the light waves emitted by a collection of oscillating dipoles are all in phase, any shape change of this collection does not alter the total intensity of the scattered light.

Particle mass from Rayleigh scattering

According to equation (11.2) Rayleigh scattering can, in principle, be employed to determine the absolute mass m of dispersed particles. The determination requires measurement of the refractive index n of a dilute dispersion as function of particle concentration c which yields a value for dn/dc. Then, for a chosen wavelength λ_0, the optical constant K is known, and an absolute value for particle mass (and also, of course, molar mass $M = N_{AV}m$) follows from a measured value of the Rayleigh ratio R_V for a chosen concentration c (Exercise 11.7).

Scattering of unpolarized light

The Rayleigh ratio R_H for the case of an incident beam that is polarized in the horizontal x-direction (Figure 11.1) is given by

$$R_H = R_V \cos^2\theta = Kcm\cos^2\theta, \quad (11.5)$$

with R_V and K as given by (11.2). The term $\cos^2\theta$ appears in (11.5) because for horizontally polarized light, a detector measures an angle-dependent scattering intensity. For scattering angles $\theta = 0°$ and $\theta = 180°$ (respectively, forward and back scattering) where $\cos^2\theta = 1$, the detector perceives, so to speak, the full amplitude p_x (see Figure 11.1) of the scattered field. When $\theta = 90°$, $\cos^2\theta = 0$, only the vertical component contributes: the scattered light at this angle is vertically polarized. The total scattered intensity for unpolarized light, as measured by the total Rayleigh ratio R_0, is simply the sum of the two polarization contributions:

$$R_0 = R_V + R_H = Kcm(1+\cos^2\theta); \quad K = \frac{2\pi^2 n^2}{\lambda_0^4}\left(\frac{dn}{dc}\right)^2. \quad (11.6)$$

This equation describes, for example, light scattering by air molecules bathing in unpolarized sunlight; the angular pattern of the light scattered by these molecules is sketched in Figure 11.2.

Light scattering by larger particles: the form factor

Light scattered by particles larger than Rayleigh scatterers exhibits a θ-dependence in addition to the $1 + \cos^2\theta$ term in (11.6). This additional term, which forms the basis of particle size determination by SLS, is based on the following. For a colloid larger than a 'Rayleigh particle', its electrons are not at any moment subjected to the same incident electric field. Hence, the light scattered from different regions in the colloid's interior is out of phase and interference occurs that reduces the scattering intensity, in an amount that depends on scattering angle θ. The theory that incorporates interference effects is usually referred to as the Rayleigh–Gans–Debye (RGD) theory, which applies when

$$(n_p - n_0)L \ll \lambda, \quad (11.7)$$

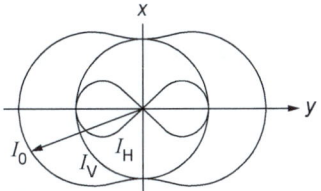

Figure 11.2 Intensity of light scattered by a small particle at the origin from an unpolarized incident beam, as a function of the scattering angle. Intensities I_V of vertically polarized scattered light and I_H of horizontally polarized light with corresponding Rayleigh ratios in, respectively, equations (11.2) and (11.5) add up to the dumbbell-shaped curve I_0—the Rayleigh ratio (11.6). Shown here are intensity distributions in the horizontal xy-plane of Figure 11.1; three-dimensional distributions are found by rotating the diagram around the y-axis.

where L is the maximal distance in a particle. When requirement (11.7) is fulfilled, light in a colloid is scattered only once; when, for example, the optical contrast $n_p - n_0$ is too large, multiple scattering in a particle occurs, which invalidates RGD theory. The outcome of the RGD theory is a modification of the Rayleigh ratio R_0 in (11.6) with the so-called particle **form factor** $P(Q)$:

$$R(Q) = R_0 P(Q); \quad 0 \leq P(Q) \leq 1. \tag{11.8}$$

Here Q is the so-called wave vector that is defined by

$$Q = \frac{4\pi n}{\lambda_0} \sin(\theta/2). \tag{11.9}$$

The form factor $P(Q)$ has the property that $P(Q = 0) = 1$; at finite wave vectors $P(Q)$ causes the appearance of maxima and minima in the angular distribution of scattered light. For example, the form factor of an optically homogeneous sphere of radius a,

$$P(Q) = \left[3 \frac{\sin Qa - Qa \cos Qa}{(Qa)^3} \right]^2, \tag{11.10}$$

yields the pronounced maxima and minima that can be seen in Figure 11.3. The minima occur at wave vectors Q_m that make the numerator in (11.10) zero (Exercise 11.5). The first minimum occurs at $Q_m a = 4.5$, so the magnitude of the wave vector Q_m at which this minimum occurs in an SLS experiment yields the radius a of the light-scattering spheres in dispersion.

Small-angle scattering: the Guinier equation

In addition to the location of a scattering minimum, particle size can also be determined from scattering intensities at small wave vectors. A Taylor expansion of (11.10) for small Qa values (Exercise 11.9) results in

$$P(Q) = 1 - \frac{1}{3}(Qa)^2, \text{ for } Qa \ll 1. \tag{11.11}$$

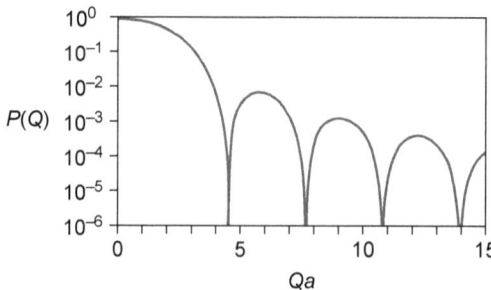

Figure 11.3 Plot of the form factor of a sphere in equation (11.10).

In terms of the radius of gyration of a sphere (Exercise 11.6), $R_g = a(3/5)^{1/2}$, equation (11.11) modifies to

$$P(Q) = 1 - \frac{1}{3}(QR_g)^2, \text{ for } QR_g \ll 1, \qquad (11.12)$$

a result that is known as the **Guinier equation**. The important point to note here is that, although we found (11.12) via the form factor of a sphere in (11.10), the Guinier equation (11.12) is actually valid for *any* particle shape: the gyration radius R_g in (11.12) may belong to particles as divergent as polymer coils, proteins, fibres, and clay platelets. As a consequence, scattering curves for particles having the same gyration radius but different shapes will coincide at very small angles. Employing the expansion $\ln(1-x) = -x + \ldots$ for $x \ll 1$, we can rewrite the Guinier equation as

$$\ln P(Q) = -\frac{Q^2 R_g^2}{3}, \text{ for } Q_{Rg} \ll 1, \qquad (11.13)$$

which implies that the logarithm of scattered intensity decreases linearly with the wave vector squared. The slope of the resulting linear curve, also known as a Guinier plot, yields the particle's radius of gyration. An example of such a Guinier plot for light scattering by latex colloids can be seen in Figure 11.4. Importantly, for determining particle size via the Guinier equation we do not need absolute values of scattering intensities (the values that are required for establishing the mass of Rayleigh scatterer via equation (11.2)). For particle sizing the Q-dependence of the SLS intensity suffices. In other words, we can ignore constant factors and plot the logarithm of the scattered intensity versus (the square of) wave vector Q, as is done, for example, in Figure 11.4.

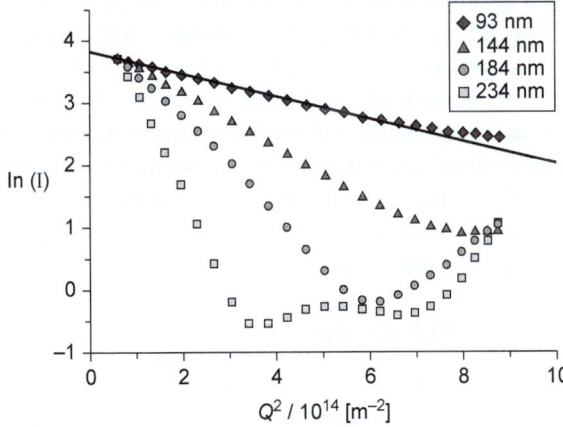

Figure 11.4 The logarithm of the scattered intensity I (in arbitrary units) as function of squared wave vector Q^2, for aqueous dispersions of fluorinated latex spheres in water, having the indicated radii. The slope of the drawn line yields a radius of gyration, in accordance with Guinier equation (11.13).

Source: SLS data taken from G. H. Koenderink et al. (2001).

We end our brief introduction to SLS with two provisos. First, the focus here is on SLS as a characterization tool for single non-interacting particles in dilute dispersions. Concentrated systems are outside the scope of this primer, but it deserves mention that scattering techniques have been widely applied as research tools for studies on particle interactions, structure factors, and phase transitions. For an excellent review of these studies, see Vrij & Tuinier (2005). A second proviso for the application of SLS is that particles should not absorb light, and that the optical contrast between colloids and solvent is low enough for a dispersion to be sufficiently translucent. This translucency may be checked by shining a hand-laser through a small glass vial of dispersion; if the laser beam remains a thin sharp line, scattering by the colloids is sufficiently weak. If, however, the beam fans out to some extent, the dispersion is too turbid for SLS to apply. Aqueous dispersions of inorganic oxides and hydroxide colloids are generally either opaque or coloured, or both; then characterization could employ scattering by neutrons or X-rays, for which those dispersions are much more transparent than for light. The size determination via the form factor (11.10) and the Guinier equation (11.13) works exactly the same as for light scattering, as explained in Ottewill (1982).

11.3 Dynamic light scattering

In the previous section we established how particle size can be determined via SLS through the angular dependence of the scattered light intensity I_S. Now this I_S is actually the average value of an intensity signal that vacillates in time. Interestingly, particle size follows not only from the static value of I_S but, as we will see here, also from the intensity fluctuations around this static average that are detected at a given scattering angle. The monitoring and analysis of these fluctuations is referred to as dynamic light scattering (DLS). The application of DLS for particle size determination requires, just as for SLS, dilute dispersions of weakly scattering, non-absorbing colloids. In addition, DLS determines the size of particles (as we will see below) via their diffusion coefficient $D = k_B T/f$. It is important that colloids are dispersed in a solvent with Newtonian viscosity η, because when that is the case a hydrodynamic sphere radius a_H simply follows from the Stokes factor $f = 6\pi\eta a_H$ that accounts for the viscous force on a sphere moving in a Newtonian fluid.

Fluctuating diffraction patterns

In contrast to SLS, DLS requires laser light because its coherence implies that the wave motion is in phase across a plane of propagating light—waves all march in step, so to speak. This coherence ensures that phase differences between scattered waves stem only from differences in positions of scattering centres. For a fixed array of stationary particles the mentioned phase differences produce static diffraction patterns that contain bright spots due to constructive interference, much like the diffraction peaks in the scattering of X-rays by a crystal. In a colloidal dispersion, however, particles are in continuous Brownian motion, leading to arrangements of spots that change continuously—dances of light speckles that are as enduring as Brownian motion itself. Shine a laser pointer through a cloud of chalk dust produced by clapping two blackboard erasers together, and you will see on a white wall the twinkling of light scattered in the forward direction. The time duration of these twinkles is what interests us: if particles diffuse faster, then twinkling is swifter, so somehow these intensity fluctuations implicate a value for the particles' diffusion coefficient. Let us see how this comes about.

Speckle fluctuation times

The time τ_c is the typical fluctuation time of the speckle pattern at the detector and is, consequently, also the typical time needed for colloids to diffuse a sufficient distance to bring about these fluctuations. That sufficient diffusion distance is the inverse of the wave vector Q defined in (11.9): at small Q (small scattering angle, large wavelength) colloids have to diffuse a large distance Q^{-1} for intensities to significantly fluctuate; at large Q that distance is correspondingly smaller. Recall from Chapter 3 on Brownian motion that the mean-squared displacement (MSD) in the x-direction for a particle with diffusion coefficient D equals

$$<x^2> = 2Dt. \qquad (11.14)$$

When in (11.14) the diffusion time t equals the fluctuation time τ_c, then the particle MSD equals $<x^2> = Q^{-2}$, from which it follows that

$$\tau_c = \frac{1}{Q^2 D} \qquad (11.15)$$

For typical values of the wave vector Q and particles of sub-micron size, τ_c is in the range of microseconds to milliseconds. A value for τ_c can be determined with DLS as follows.

Correlations between intensity fluctuations

In a DLS set-up a fluctuating speckle illuminates a photo-multiplier that feeds a 'photon correlator' which, by a process detailed elsewhere (see Berne and Pecora, 1976), determines $g(\tau)$, the so-called temporal correlation function of the scattered light field. What this function $g(\tau)$ represents is the relation—the 'correlation'—between intensity at time t and intensity at a later time $t+\tau$. To unpack this correlation effect, we write intensity $I(t)$ as the sum of the time-averaged intensity $<I>$ and a time-dependent fluctuation $\Delta(t)$; see Figure 11.5:

$$I(t) = <I> + \Delta(t): <\Delta(t)> = 0. \qquad (11.16)$$

The time average of the fluctuations is by definition zero, as indicated in (11.16). Now multiply fluctuation $\Delta(t)$ with the fluctuation at time $t + \tau$ and take the time average of this product—that is to say, evaluate $<\Delta(t)\Delta(t+\tau)>$.

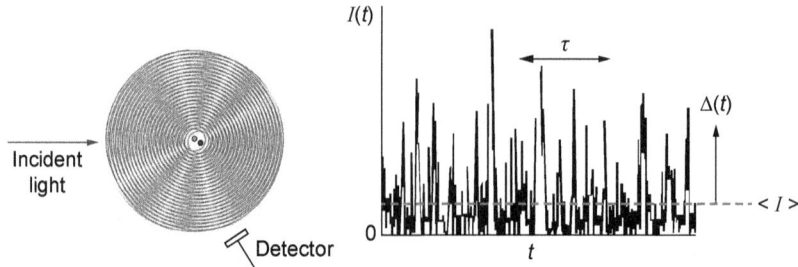

Figure 11.5 Left: Model for the static interference between scattered light waves coming from two motionless particles. **Right**: When the colloids diffuse, the interference pattern continuously varies and a detector, at fixed scattering angle, records a vacillating scattered intensity $I(t)$. $\Delta(t)$ is a fluctuation—a deviation at time t from the time-averaged intensity $<I>$. Separated by a time interval $t \gg \tau$, these fluctuations are random, unconnected events; at times $t \ll \tau$ they are highly correlated. From the time correlation between intensity fluctuations, a particle diffusion coefficient can be deduced.

When the 'delay time' τ is much larger than the characteristic time τ_C from (11.15), fluctuations are independent such that:

$$<\Delta(t)\Delta(t+\tau)> = <\Delta(t)><\Delta(t+\tau)> = 0, \text{ for } \tau \gg \tau_C. \quad (11.17)$$

Absence of correlation is signified by the correlation function being zero: $g(\tau \gg \tau_{CO}) = 0$. When $\tau \ll \tau_C$ particle diffusion has not changed the interference pattern significantly yet, entailing connected fluctuations that in the limit $\tau \to 0$ even coincide:

$$<\Delta(t)\Delta(t+\tau)> = <\Delta^2(t)>, \text{ for } \tau \to 0. \quad (11.18)$$

Note that the time average in (11.18), being an average of quadratics, is always positive. Maximal correlation implies that the correlation function equals unity: $g(\tau = 0) = 1$.

The simplest application of DLS is to a dispersion of identical spheres at concentrations low enough for the spheres to diffuse independently from each other. For such freely diffusing spheres the correlation function $g(\tau)$ is the single exponential:

$$g(\tau) = \exp(-\tau / \tau_C). \quad (11.19)$$

A plot of $\ln g(\tau)$ against the delay time τ yields a value for the characteristic timescale τ_C in (11.15). From the Stokes–Einstein diffusion coefficient $D = k_B T / 6\pi\eta a_H$ the hydrodynamic particle radius then follows as:

$$a_H = \frac{k_B T}{6\pi\eta} Q^2 \tau_C. \quad (11.20)$$

In practice the time τ_C is measured at various scattering angles (various values of Q) and diffusion coefficient D is obtained from a plot of $1/\tau$ against Q^2.

On particle sizes from DLS and SLS

DLS measures the effective radius a_H of the 'hydrodynamic unit' that diffuses around in solution. Due to solvent adsorption at the particle surface, a_H could be a nanometre or so greater than the radius of the 'bare' particle found from SLS—a difference that in most cases will be insignificant in view of the typical uncertainty of 4–5 per cent in DLS and SLS radii. What *does* lead to significant differences between particle sizes obtained from SLS and DLS is the occurrence of particle polydispersity, the reason being that SLS and DLS sample different moments of the particle size distribution. Apparent particle radii found through SLS and DLS on dispersions of polydisperse spheres are evaluated in Appendix 11A.

11.4 Electrophoresis

Having discussed particle size determination from scattering experiments, we now focus on another characterization of colloids, namely the evaluation of electrical charge on the particles via **electrophoresis**. The crux of an electrophoresis experiment is as follows. A uniform electrical field with strength E brings charged particles into motion, a movement that is counteracted by the viscous friction between particles and adjacent solvent, as schematically illustrated in Figure 11.6. When electrical and frictional forces are in balance, particles attain a stationary speed v_{el} from which we deduce the particles' **electrophoretic mobility**, defined as $\mu_{el} = v_{el}/E$. Our next task is then to convert this measured mobility[1] to a particle charge, or an electrical potential at or

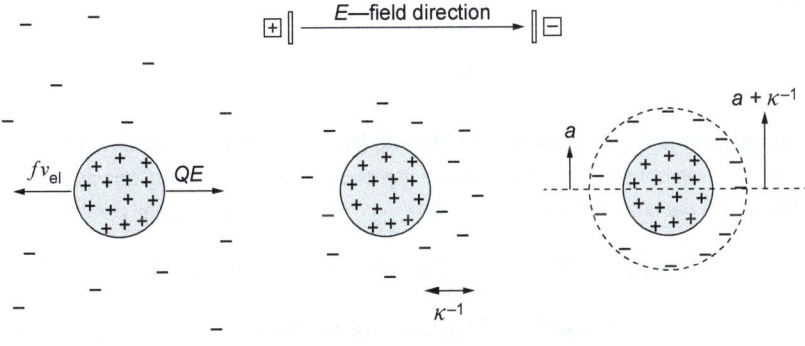

Figure 11.6 Spheres of radius a carrying charge Q migrate in a homogeneous electric field E. **Left**: In a salt-free solution with Debye length $\kappa^{-1} \gg a$, the colloid drifts as a bare charge Q at speed $v_{el} = QE/f$ (f is the sphere's friction factor) in the field direction. **Middle**: A sphere enveloped by a diffuse ion cloud with a thickness κ^{-1} comparable to a; for simplicity only counter-ions are shown. **Right**: The diffuse counter-ion cloud is replaced by a concentric shell bearing a negative charge equal to $-Q$ at distance κ^{-1} from the sphere surface. The viscous force on counter-ions in the EDL is approximated by the Stokes friction factor of a sphere of radius $a+\kappa^{-1}$, leading to the mobility in equation (11.39).

[1] One often reads about 'zeta potentials measured by electrophoresis', but that is rather misleading: the primary experimental data are mobilities from which model-dependent zeta potentials are deduced.

near the particle surface. To work out this conversion we first need to formulate the relation between the particle surface charge Q and surface potential Ψ_0. This relation follows from the electrical potential profile around a charged colloidal sphere.

Electrical potential profile around a charged sphere

The electrical potential $\Psi = \Psi(r)$ at distance r from the centre of a charged sphere is the solution of the Poisson–Boltzmann (PB) equation. For small potentials we can employ the PB equation in the so-called Debye–Hückel (DH) approximation:

$$\frac{1}{r^2}\frac{d}{dr}\left(r^2 \frac{d\Psi}{dr}\right) = \kappa^2 \Phi; \quad \Phi = \frac{ze\Psi}{k_B T} < 1. \tag{11.21}$$

Here Φ is the dimensionless electric potential. For a 1:1 electrolyte at $T = 298\,K$, $\Phi = 1$ for $\Psi = 25.7\,mV$. Thus the DH approximation implies that at 25 °C electrical potentials are less than about 25 mV. In the PB equation (11.21) κ is the inverse of the Debye screening length κ^{-1}, the length scale that we first met in Chapter 5 in our discussion of the PB equation in the form of (5.10). We repeat here the definition of the Debye length:

$$\kappa^{-1} = \left(\frac{\varepsilon\varepsilon_0 k_B T}{2\rho_s (ze)^2}\right)^{1/2}, \tag{11.22}$$

where ρ_s is a number density of a z–z electrolyte in a solution with dielectrical constant $\varepsilon\varepsilon_0$. The solution of (11.21) is in terms of the electrical potential Ψ (Exercise 11.16):

$$\Psi(r) = c_0 \frac{e^{-\kappa r}}{r}. \tag{11.23}$$

So the potential decays exponentially with increasing radial distance r. To find the integration constant c_0 we can make use of the electro-neutrality condition as follows. The net charge density $\rho^* = ze(\rho_+ - \rho_-)$ reads, again for small potentials:

$$\rho^* = ze\rho_s \left(e^{-\Phi} - e^{+\Phi}\right) = -2ze\rho_s \Phi = -\varepsilon\varepsilon_0 \kappa^2 \Psi, \text{ for } \Phi \ll 1. \tag{11.24}$$

Here we have employed the Boltzmann distributions $\rho_\pm = \rho_s \exp[\mp\Phi]$ and linearized them for small potentials to $\rho_\pm = \rho_s(1 \mp \Phi)$. The charged sphere and its EDL are in total electrically neutral; hence the sum of the charge Q on the colloid plus the integrated net charge density in the surrounding solution must be zero:

$$Q + \int_a^\infty \rho^* 4\pi r^2 dr = Q - 4\pi\varepsilon\varepsilon_0 \kappa^2 \int_a^\infty \Psi r^2 dr = 0. \tag{11.25}$$

On substitution of the potential Ψ from (11.23) in (11.25) and carrying out the integration (Exercise 11.17) we find eventually for integration constant c_0:

$$c_0 = \frac{Q e^{\kappa a}}{4\pi\varepsilon\varepsilon_0 (1 + \kappa a)}. \tag{11.26}$$

Hence the potential profile around the sphere becomes:

$$\Psi(r) = \frac{Q}{4\pi\varepsilon\varepsilon_0 r} \frac{e^{-\kappa(r-a)}}{(1+\kappa a)}. \quad (11.27)$$

In this radial Debye–Hückel potential profile, we recognize Coulomb's law $\Psi(r) = Q/4\pi\varepsilon\varepsilon_0 r$ for the electrical potential around a sphere carrying charge Q, multiplied by an exponential decay due to screening of this charge Q by the surrounding electrolyte solution.

Surface potential versus surface charge

The relation we are after, that between surface potential and surface charge, follows from (11.27) as:

$$\Psi_0 = \Psi(r=a) = \frac{Q}{4\pi\varepsilon\varepsilon_0 a} \frac{1}{(1+\kappa a)}. \quad (11.28)$$

We recall that κ^{-1} is the typical distance over which the electrical potential decays, from the particle's surface potential Ψ_0 to the potential $\Psi = 0$ in the bulk. When κ^{-1} is very large in comparison to particle size a, the surface potential (11.28) reduces to the value given by Coulomb's law:

$$\Psi_0 = \frac{Q}{4\pi\varepsilon\varepsilon_0 a}, \text{ for } \kappa a \ll 1. \quad (11.29)$$

In the limit $\kappa a \ll 1$—approached at low salt concentrations—the EDL is so much expanded that counter-ions, instead of staying in the vicinity of a colloid, are more or less evenly distributed ('delocalized') in solution (Figure 11.6). In the limit $\kappa a \ll 1$ we have a 'bare', isolated charged colloidal sphere that migrates in a homogeneous electrolyte solution, unhindered by any EDL. We will, for this simplest instance of electrophoresis, work out the electrophoretic mobility of the bare charged colloid.

Electrophoresis of an isolated charged sphere: the Hückel equation

Suppose the colloid carries Q coulombs of surface charge and moves at stationary speed v_{el} due to an electric field E. The balance between electrical force QE and frictional force fv_{el} on the colloid yields an electrophoretic mobility of the form:

$$\mu_{el} = \frac{v_{el}}{E} = \frac{Q}{f}, \text{ for } \kappa a \ll 1. \quad (11.30)$$

Note that in a similar fashion we defined in Chapter 7 the sedimentation mobility $\mu = v/g$ as the stationary speed v per unit of gravitational field. The friction factor f, theoretically unknown for most particle shapes, can be eliminated via a DLS measurement of the diffusion coefficient $D = k_B T/f$. Then the particle charge follows from what can be called the electrophoretic analogue of the Svedberg relation (7.4):

$$Q = k_B T \frac{\mu_{el}}{D}, \qquad (11.31)$$

showing how, in principle, particle charge Q can be determined from DLS and electrophoresis without knowledge of particle size and shape. For a sphere with hydrodynamic radius a_H in a medium with viscosity η we obtain from (11.30):

$$\mu_{el} = \frac{Q}{6\pi\eta a_H}. \qquad (11.32)$$

It is generally assumed that electrophoresis does not directly probe the surface potential in (11.29) but, instead, the electrical potential at a certain **slipping plane** or **shear plane**: an imaginary envelope around the colloid separating stagnant liquid (moving along with the colloid) at the colloid surface from freely flowing bulk liquid. The electrical potential in this plane is referred to as the **zeta potential** ζ. Under the reasonable assumption that the slipping plane coincides with the hydrodynamic particle radius a_H—the radius that, after all, incorporates immobilized solvent—we have, instead of (11.29):

$$\zeta = \frac{Q}{4\pi\varepsilon\varepsilon_0 a_H}. \qquad (11.33)$$

A combination of (11.32) and (11.33) yields the so-called **Hückel equation** for the relation between mobility and zeta potential:

$$\mu_{el} = \frac{2}{3}\frac{\varepsilon\varepsilon_0\zeta}{\eta}, \quad \text{for } \kappa a \ll 1. \qquad (11.34)$$

The inequality $\kappa a \ll 1$ in (11.34) signals that the Hückel equation only holds for sphere radii much smaller than the Debye screening length. To give some numerical examples: in an aqueous solution containing 10^{-3} mol dm^{-3} NaCl, the Debye length is $\kappa^{-1} = 9.7$ nm, so for nanoparticles the Hückel limit $\kappa a \ll 1$ is accessible. RBCs, in contrast, have a diameter of several microns such that $\kappa a \gg 1$, making equation (11.34) inapplicable to electrophoresis of RBCs.

Colloids dressed with a double layer: the concentric-shell model

As the salt concentration rises, colloids gradually leave the Hückel limit and become 'dressed' with an EDL, with characteristic thickness κ^{-1}. In the EDL the concentration ρ_+ of counter-ions exceeds co-ion concentration ρ_-; the electric field pulls the excess counter-ions (with number density $\rho_+ - \rho_-$) in the direction opposite to the electrical force on the colloid. Now if a colloid carries Q coulombs of surface charge then, because of overall electro-neutrality, the excess counter-ions jointly carry a charge of $-Q$ and, consequently, experience a total electrical force $-QE$. Suppose counter-ions migrate at speed v_{el} and experience jointly a frictional force $f_{ions}v_{el}$, where f_{ions} is the ions' joint friction factor, then the mobility of the ions is given by:

$$\mu_{el,ions} = \frac{v_{el}}{E} = -\frac{Q}{f_{ions}}. \tag{11.35}$$

The ions' collective friction factor f_{ions} can be estimated via what we will call the 'concentric-shell model'. In Exercise 5.9 you were asked to compute the average distance between excess counter-ions in an EDL and the surface of the charged colloid; the outcome was that, within the DH approximation, this average equals the Debye length κ^{-1}. Imagine now that actually *all* excess counter-ions are located at a distance κ^{-1} from the sphere surface (Figure 11.6, right). Thus, the counter-ions are constrained to diffuse on a two-dimensional surface, namely a concentric shell with radius $a_H + \kappa^{-1}$. Note that this relocation of counter-ions to this shell does not change the total electrical force $-QE$ on them; the ions are merely replaced to estimate the hydrodynamic friction they experience. That estimate follows from the assessment that the shell's viscous resistance conforms to the usual Stokes friction factor for a sphere, namely $6\pi\eta(a_H + \kappa^{-1})$. Hence, we can put $f_{ions} \approx 6\pi\eta(a_H + \kappa^{-1})$, leading to:

$$\mu_{el,ions} \approx -\frac{Q}{6\pi\eta(a_H + \kappa^{-1})}. \tag{11.36}$$

The mobility of the bare colloidal sphere is given by equation (11.32): $\mu_{el} = Q/6\pi\eta a_H$, so the net mobility of sphere plus shell is therefore

$$\mu_{el} \approx \frac{Q}{6\pi\eta a_H} - \frac{Q}{6\pi\eta(a_H + \kappa^{-1})} = \frac{Q}{6\pi\eta a_H(1 + \kappa a_H)}. \tag{11.37}$$

The zeta potential follows, within the DH approximation, from (11.28) as

$$\zeta = \frac{Q}{4\pi\varepsilon\varepsilon_0 a_H}(1 + \kappa a_H)^{-1}. \tag{11.38}$$

Combining (11.37) and (11.38) we find for the relation between mobility and zeta potential:

$$\mu_{el} = \frac{2}{3}\frac{\varepsilon\varepsilon_0 \zeta}{\eta}, \text{ for } \kappa a < 1, \tag{11.39}$$

which precisely equals the Hückel equation (11.34). So within the validity of the concentric-shell model for viscous friction on ions, the EDL does not affect the colloid's mobility. To rationalize this, compare the charge Q for the bare sphere in (11.33) to the charge in (11.38) on a colloid dressed with an EDL. For given zeta potential the latter can accommodate more surface charge than a bare colloid, making the electrical force on the dressed colloid $(1 + \kappa a_H)$ times larger than the force on the bare colloid. However, the viscous force on the colloid also increases with a factor of $(1 + \kappa a_H)$ so the stationary colloid speed remains the one given by the Hückel equation.

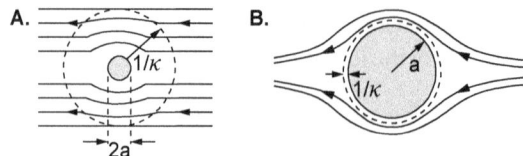

Figure 11.7 Schematic of the lines that represent the electric field and liquid flow in the vicinity of a charged colloid during electrophoretic motion, in the Hückel limit of small κa (**A**) and in the Smoluchowski limit of large κa (**B**).

The Smoluchowski limit at high salt concentrations

Equation (11.39) is valid only for low surface potentials—the DH approximation that underlies (11.38)—and also presupposes that $\kappa a < 1$, or the salt concentration is low enough for the Debye length to be larger than the colloid radius. For small Debye lengths such that $\kappa a \gg 1$, we encounter in our calculation of the mobility a complication that arises from the applied electric field as follows. In the Hückel limit $\kappa a \ll 1$, as we have seen, colloids are surrounded by a homogeneous salt solution; on this neutral solution the applied external field has no effect, which is indicated schematically by the parallel field lines in Figure 11.7A; the electric field (almost) looks as if no colloid is present. In the so-called Smoluchowski limit of large $\kappa a \gg 1$, however, an insulating, nonconducting particle distorts the field in most of the double layer region such that electric field lines curve around the colloid, as illustrated in Figure 11.7B. This curvature, in the end, does not complicate the derivation of the mobility, as we can simply ignore it: field and flow lines run parallel to the colloid surface through a thin EDL, such that their effect is the same as for a flat EDL.

Electro-osmosis and the Smoluchowski equation

We start examining electrophoretic particle mobility in the $\kappa a \gg 1$ limit with the following consideration. So far in our discussion of electrophoresis we have looked at motion of charged particles relative to the surrounding liquid. When, instead of residing on a mobile colloid, electrical charge is located on a stationary wall, it is, on application of an electric field, the liquid that moves relative to the solid. This liquid flow, referred to as **electro-osmosis**, arises from the pull of the electric field on ions in the liquid: as they move they drag liquid along with them. The essential point is now that for thin EDLs ($\kappa a \gg 1$) there is no physical difference between this electro-osmosis and electrophoresis. So if we manage to evaluate the electro-osmotic liquid speed V near a charged wall, we directly obtain the electrophoretic particle speed—this being equal in magnitude to V but opposite in direction.

Far away from the stationary surface liquid motion occurs at stationary speed V (Figure 11.8). In the slipping plane near the particle surface the solvent velocity is zero, so in the y-direction perpendicular to the surface a velocity gradient is present which entails a net viscous force on a volume element in the vicinity of the surface. This volume element is located in the EDL and, hence, contains net electric charge. The applied electric field E exerts a force on this net charge, which balances with the viscous force to yield, as we will see in a moment, the stationary bulk fluid speed V. The electrical force F_E on a volume element at distance y from a charged surface equals

$$\frac{F_E}{A} = E\rho^*(y)dy, \qquad (11.40)$$

where $\rho^*(y)$ is the net charge density (in coulombs per volume) in the volume element; A is the area of the element's sides. The tangential viscous force per area A is the viscous stress σ which according to Newton's viscosity law (discussed in Chapter 8, Section 8.2) equals:

$$\sigma_v = \frac{F_v}{A} = -\eta\frac{dv_x}{dy}. \qquad (11.41)$$

Here $v_x = v_x(y)$ is the fluid speed in the x-direction at height y; η is the solvent viscosity. Consequently, the net viscous stress exerted on the volume element (see also Figure 11.8) is the stress in its lower side at height y minus that in the volume's upper side at altitude $y + dy$:

$$\sigma_{v,net} = -\eta\frac{dv_x}{dy}\bigg|_y + \eta\frac{dv_x}{dy}\bigg|_{y+dy} = \eta\frac{d^2v_x}{dy^2}dy. \qquad (11.42)$$

In the stationary state the fluid volume element moves at constant speed because the total force (per unit area) on the element is zero: $F_E A^{-1} + \sigma_{v,net} = 0$, which entails that

$$E\rho^*(y) = -\eta\frac{d^2v_x}{dy^2}. \qquad (11.43)$$

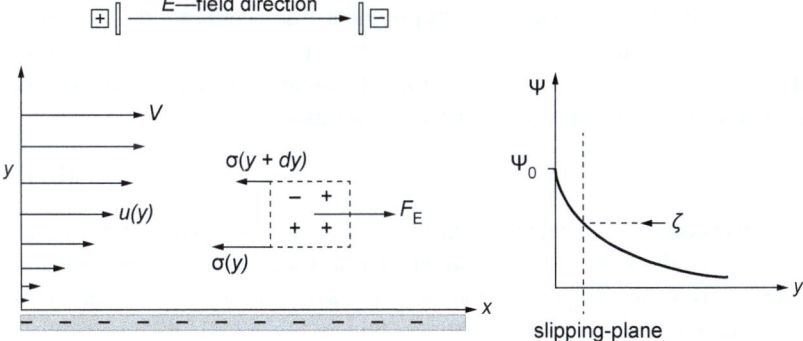

Figure 11.8 Left: Sketch of an electro-osmosis experiment. A homogeneous, x-directed electric field exerts a force F_E on a solution volume element at height y in the EDL near a static charged surface at $y = 0$. The electric field shears the EDL, causing tangential liquid displacements that entail a net viscous force on the volume element, the balance of which with F_E yields the electro-osmotic fluid speed V in the bulk, as explained in the text. **Right**: Electro-osmosis probes an electrical potential ζ which is smaller than the surface potential at $y = 0$.

Eventually we wish to deduce the magnitude of the electrical potential at or near the charged surface; to get the electrical potential into the equation we employ the Poisson equation (discussed in Section 5.3) which relates potential Ψ to the net charge density ρ^*:

$$\varepsilon\varepsilon_0 \frac{d^2\Psi}{dy^2} = -\rho^*(y). \tag{11.44}$$

On substitution of the net charge $\rho^*(y)$ from the Poisson equation (11.44) into (11.43) we obtain a differential equation that now contains the electrical potential profile $\Psi = \Psi(y)$:

$$E\varepsilon\varepsilon_0 \frac{d^2\Psi}{dy^2} = \eta \frac{d^2v_x}{dy^2}. \tag{11.45}$$

A first integration yields

$$E\varepsilon\varepsilon_0 \frac{d\Psi}{dy} = \eta \frac{dv_x}{dy} + (c_1 = 0). \tag{11.46}$$

The integration constant c_1 is zero because moving away from the surface, both derivatives $d\Psi/dy$ and dv_x/dy vanish: in the bulk solution $\Psi = 0$ and the fluid speed has the constant value V. A second integration gives

$$E\varepsilon\varepsilon_0 \Psi = \eta v_x + c_2. \tag{11.47}$$

Since in the bulk $\Psi = 0$ and $v_x = V$, the integration constant in (11.47) must equal $c_2 = -\eta V$, so that we obtain from (11.47):

$$E\varepsilon\varepsilon_0 \Psi = \eta(v_x - V). \tag{11.48}$$

This equation tells us how electrical potential Ψ and fluid speed v_x at altitude y are related. Recalling that the zeta potential ζ is the potential in the slipping plane (or the surface of shear) where $v_x = 0$ we find from (11.48) the relation between zeta potential and the osmotic flow rate V:

$$E\varepsilon\varepsilon_0 \zeta = -\eta V. \tag{11.49}$$

We can now remark that the electrophoretic mobility of the particle emerges when the fluid is kept stagnant and the charged wall in Figure 11.8 is allowed to move. The speed v_{el} of the particle is equal in magnitude to V but has an opposite sign; therefore we have to substitute $v_{el} = -V$ in (11.49) to eventually obtain for the electrophoretic particle mobility

$$\mu_{el} = \frac{v_{el}}{E} = \frac{\varepsilon\varepsilon_0 \zeta}{\eta}; \quad \kappa a \gg 1. \tag{11.50}$$

This is the so-called **Smoluchowski equation** for the mobility of a particle of *arbitrary* shape, valid when the Debye screening length is much smaller than the typical particle dimension a. The remarkable feature of equation (11.50) is that the particle charge does not appear in it, as in the Hückel mobility in (11.30), but that the zeta potential does so directly. So we are not obliged to make any assumption about the charge on the arbitrarily shaped particle and how this charge is related to the particle's surface potential. Recall that for the Hückel limit $\kappa a \ll 1$ we had to assume that the migrating particle is a sphere with a homogeneous distribution of surface charge that is converted to a surface potential via equation (11.29). Also important is the assessment that the Smoluchowski equation does not put restrictions on the magnitude of the surface potential on the arbitrarily shaped colloid; recall that the Hückel equation in (11.34)–(11.39) was derived under the Debye–Hückel assumption of small surface potentials.

Electrophoresis at intermediate κa values

We have computed electrophoretic mobilities for the case $\kappa a \leq 1$ and for the Smoluchowski thin double-layer limit $\kappa a \gg 1$. What about mobilities for κa values intermediate between $\kappa a \approx 1$ and the Smoluchowski limit? The calculation of μ_{el} in this intermediate κa range is a difficult undertaking which has to account for the deformation of the EDL surrounding a colloid that retards its electric migration (see Hunter, 1981). This retardation effect is incorporated in the so-called Henry function (Ohshima, 1994) which approximates the dependence of the electrophoretic mobility on κa in the κa intermediate regime.

In view of the challenging interpretation of mobilities in the intermediate κa range, a pragmatic stance is to perform electrophoresis only at ionic strengths such that $\kappa a \leq 1$ and Hückel's equation applies, or $\kappa a \gg 1$ where Smoluchowski's equation holds. The latter has the advantage of its applicability to particles of arbitrary shape with arbitrary surface potential. When a limiting case is unachievable for a particular dispersion of charged particles, then one should attempt to analyse mobilities in terms of the Henry function found in Ohshima (1994), instead of applying the equations of Hückel or Smoluchowski, which in the intermediate κa regime may lead to very erroneous outcomes for zeta potentials, particularly when they are high, say above 50 mV.

Isoelectric points

We will conclude our introduction to electrophoresis with a straightforward application—one that is not affected by issues of theoretical modelling—namely the determination of isoelectric points by measuring the pH at which the electrophoretic particle mobility is zero. This procedure can be applied to the wide variety of particles, such as biological cells, clay colloids, bacteria, and proteins which have a pH-dependent surface charge density. Charge-stabilized colloids will have a strong tendency to aggregate when brought to their isoelectric point. An often-cited rule of thumb is that stability of charged colloids can be expected when the zeta potential is minimally in the range

$|\zeta| \approx 30-40$ mV. This rule, however, should be taken with a pinch of salt since the occurrence of colloidal (in)stability also strongly depends on the magnitude of the inter-colloidal Van der Waals attractions.

Summary

- Static light scattering (SLS) gauges particle sizes via the angular dependence of the time-averaged intensity of light scattered by non-absorbing particles in translucent, dilute dispersions.
- SLS applied to small enough particles (Rayleigh scattering) can, in principle, yield a mass of particles that have an arbitrary shape.
- By dynamic light scattering (DLS), particle sizes are determined via diffusion coefficients, extracted from the fluctuating scattering intensities caused by Brownian motion.
- Size polydispersity has an important effect on apparent particle sizes from SLS and DLS, that can be quantified for size distributions of colloidal spheres.
- Electrophoretic mobilities of charged colloids yield a model-dependent value of the zeta potential ζ; of the colloid surface potential we can say little more than that it exceeds ζ.
- EDLs that are thin enough have a marked effect on electrophoretic mobility, described by the Smoluchowski equation that is valid for arbitrary particle shape.
- Electro-osmosis is viscous liquid flow driven by an electric field from which a zeta potential ζ can be deduced, via the Smoluchowski equation, for the material that is rubbed by the liquid.

References

R. Berne and R. Pecora (1976), *Dynamic Light Scattering*. John Wiley, New York.

R. J. Hunter (1981), *Zeta Potential in Colloid Science; Principles and Applications*. Academic Press, London.

M. Kerker (1969), *The Scattering of Light and other Electromagnetic Radiation*. Academic Press, New York and London.

G. H. Koenderink et al. (2001), 'Preparation and Properties of Optically Transparent Aqueous Dispersions of Monodisperse Fluorinated Colloids', *Langmuir* **17**, 6086–6093.

J. Ohshima (1994), 'A Simple Expression for Henry's Function for the Retardation Effect in Electrophoresis of Spherical Colloidal Particles', *J. Colloid Interface Science* **168**, 587–590.

R. H. Ottewill (1982), 'Small Angle Neutron Scattering', in *Colloidal Dispersions*, J. W. Goodwin (ed.). Royal Society of Chemistry, London, pp. 143–163.

A. Vrij and R. Tuinier (2005), 'Structure of Concentrated Colloidal Dispersions', in *Fundamentals of Interface and Colloid Science*, vol. 4, J. Lyklema (ed.). Elsevier, Amsterdam, pp. 5.1–5.103.

Appendix 11A: Apparent light-scattering radii

When a dispersion contains particles with a certain size distribution, characterization of the dispersion by SLS and DLS yields (as remarked near the end of Section 11.3) different apparent particle sizes. We will first evaluate the apparent size that is obtained from SLS. Suppose a dispersion contains a mixture of sphere species labelled j = 1,2,3... From (11.8) we infer that the SLS intensity $I_j(Q)$ of particle species j is given by:

$$I_j(Q) = I_{j,0} P_j(Q), \tag{11.51}$$

where $I_{j,0} = I_j(Q = 0)$ is the forward scattering intensity by the j-particles at zero wave vector $Q = 0$; $P_j(Q)$ is the form factor, for spheres given by (11.10). For low concentrations of independently scattering spheres, the measured total scattered intensity $I(Q)$ is the sum of all intensities scattered by the individual species:

$$I(Q) = \sum_j I_{j,0} P_j(Q). \tag{11.52}$$

On division of both sides of equation (11.52) by the total forward scattered intensity $I_0 = \sum_j I_{0,j}$,

$$\frac{I(Q)}{I_0} = \frac{\sum_j I_{0,j} P_j(Q)}{\sum_j I_{0,j}}, \tag{11.53}$$

we see that the average of the form factor is actually an intensity-weighted one: the form factor of each species is multiplied by the intensity scattered by that species. $I_{0,j}$ is proportional to particle weight concentration c_j and particle mass m_j that are, in turn, both proportional to particle volume $(4\pi/3)a_j^3$. Hence the average in (11.53) modifies to

$$\frac{I(Q)}{I_0} = \frac{\sum_j \rho_j a_j^6 P_j(Q)}{\sum_j \rho_j a_j^6} = \frac{\sum_j \rho_j}{\sum_j \rho_j a_j^6} \times \frac{\sum_j \rho_j a_j^6 P_j(Q)}{\sum_j \rho_j} = \frac{\langle a^6 P(Q) \rangle}{\langle a^6 \rangle}, \tag{11.54}$$

where ρ_j is the number density of sphere with radius a_j; the brackets <> denote a number average over the size distribution of the spheres. For small values of wave vector Q such that $Qa \ll 1$ we can substitute in (11.54) the Guinier equation $P(Q) = 1 - (Q,a)^2/3$ to obtain:

$$\frac{I(Q)}{I_0} = 1 - \frac{Q^2}{3} a_{SLS}^2; \quad a_{SLS}^2 = \frac{\langle a^8 \rangle}{\langle a^6 \rangle}, \tag{11.55}$$

where a_{SLS} is the apparent particle radius obtained from SLS, in the Guinier region, on a dilute dispersion of polydisperse spheres. To find the apparent DLS radius, we first note that for spheres labelled j, the correlation function of the scattered light field is, in accordance with (11.19), the single exponential:

$$g_j(\tau) = \exp(-\tau/\tau_{j,CO}); \quad \tau_{j,CO} = \frac{1}{D_j Q^2}, \tag{11.56}$$

where $D_j = k_B T/6\pi\eta a_j$ is the Stokes–Einstein diffusion coefficient of a j-sphere. Just as for the form factor in SLS, the correlation function measured with DLS is an intensity-weighted average:

$$g(\tau) = \frac{\sum_j I_{0,j} \exp(-\tau/\tau_{j,CO})}{\sum_j I_{0,j}} = \frac{\langle a^6 \exp(-\tau/\tau_{CO})\rangle}{\langle a^6\rangle} \tag{11.57}$$

For small delay times τ we can expand $\exp(-\tau/\tau_{CO}) = 1 - \tau/\tau_{CO} + \ldots$ to find:

$$g(\tau) = 1 - \tau Q^2 \frac{\langle a^6 D\rangle}{\langle a^6\rangle} = 1 - \tau Q^2 D_{APP}. \tag{11.58}$$

Here $D_{APP} = kT/6\pi\eta a_{DLS}$ is an apparent diffusion coefficient obtained from the initial decay of the time correlation function measured with DLS, yielding an apparent DLS radius:

$$a_{DLS} = \frac{\langle a^6\rangle}{\langle a^5\rangle}. \tag{11.59}$$

Polydispersity may lead to significant differences between apparent radii and the number-averaged radius $<a>$—determined from particle sizes measured from electron micrographs. For a numerical example, consider a binary sphere mixture composed of equal numbers of spheres with radii $a_1 = 20$ nm and $a_2 = 40$ nm; verify to yourself that the mixture's apparent light-scattering radii are $a_{DLS} = 39.4$ nm and $a_{SLS} = 39.8$ nm—values that are considerably larger than the number average $<a> = 30$ nm.

Exercises

11.1 What is the unit of the Rayleigh ratio R?

11.2 Show that the left-hand and right-hand sides of (11.2) have the same unit.

11.3 Express the refractive index increment dn/dc of a dispersion in terms of the optical contrast $n_p - n_0$ between dispersed particles and solvent. Start with writing n as the volume average of n_p and n_0. What information can be obtained from a measured value for dn/dc?

11.4 Consider two emulsion droplets, each with volume V. By which factor does the scattered light intensity change when the two droplets merge into one with volume $2V$? Assume droplets are small enough for Rayleigh theory to apply.

11.5 Find the location $Q_m a$ of the first minimum in the sphere form factor plotted in Figure 11.3.

11.6 Verify that the radius of gyration of a homogeneous sphere equals $a(3/5)^{1/2}$, by evaluating the root-mean-square distance $<r^2>$, where r is the radial distance of a volume element of matter from the sphere centre at $r = 0$.

11.7 An SLS experiment on a polystyrene solution in cyclohexane ($n_0 = 1.43$) with weight concentration $c = 0.01 \text{g cm}^{-3}$ yields a Rayleigh ratio $R = 3.74 \times 10^{-4} \text{ cm}^{-1}$ at scattering angle $\theta = 90°$. The employed wavelength of incident (unpolarized) light is $\lambda_0 = 546$ nm and the refractive index increment of polystyrene solution is $dn/dc = 0.170 \text{ cm}^3\text{g}^{-1}$.
  (a) Calculate the molar mass M of the polystyrene molecules.
  (b) Suppose polystyrene polymers are polydisperse in mass; what sort of average molar mass is then obtained in (a)?
  (c) Suppose that for some reason polystyrene coils shrink to compact, homogeneous spheres with mass density $\delta_p = 1 \text{g cm}^{-3}$. What would be the refractive index n_p of these spheres?
  (d) Compute the volume fraction φ_p of the polystyrene spheres in (c).

11.8 Measured SLS intensities from a dilute dispersion of silica spheres in ethanol ($n_0 = 1.36$) show a minimum in the angular scattering profile at $\theta = 100°$. The profile has been measured for vertically polarized light with wavelength $\lambda_0 = 546$ nm. How large is the radius a of the spheres?

11.9 Show how Taylor expansions of sine and cosine in (11.10) lead to equation (11.11).

11.10 Why does it not make sense to employ a highly focused beam of sunlight for dynamic light scattering?

11.11 (a) Show that the apparent static light-scattering radius for polydisperse, small Rayleigh scatters equals

$$a_{SLS,R} = \left[\frac{\langle a^6 \rangle}{\langle a^3 \rangle}\right]^{1/3}.$$

(b) Make the verification asked for at the end of Appendix 11A.

11.12 The negative surface charge on a charged colloid is exactly compensated by positive counter-ions in the EDL such that the colloid plus its EDL form as a whole is an electrically neutral object. How can, in an electrophoresis experiment, this neutral object be put into motion by an electric field? Is it not the case that an electric field cannot exert a net force on an overall neutral body?

11.13 Suppose for particles with charge Q and buoyant mass Δm we measure both the electrophoretic mobility μ_{el} and sedimentation mobility μ_{sed}.
   (a) How can we determine the particle's charge-to-buoyant-mass ratio $Q/\Delta m$ from these mobilities?
   (b) Does this determination require any assumption on particle size or shape?

11.14 Suppose a colloidal silica sphere (radius $a = 100$ nm) in ethanol ($\varepsilon = 24.55$) has a zeta potential of $\zeta = 50$ mV.
   (a) Estimate the number σ of charges on the sphere.
   (b) How much surface area (in nm²) is available for each charge?
   (c) What is the average distance between two surface charges?

11.15 RBCs contain haemoglobin (Hb), a protein that carries oxygen. In someone who has sickle-cell disease, the haemoglobin is abnormal, causing RBCs to become hard and sticky and adopt a sickle shape. Using electrophoresis, Pauling discovered a difference in surface charge between Hb (IEP pH = 6.8) and sickle-cell haemoglobin (HbS; IEP = 7.1).
   (a) At which pH values will Hb and HbS migrate in the same direction in an electrophoresis experiment?
   (b) Suppose at pH = 7.0 the mobilities in water for Hb and HbS are, respectively, -0.2×10^{-5} cm²/V sec and $+0.3 \times 10^{-5}$ cm²/V sec. Deduce the charge number for each type. Assume that the proteins are spheres with a diameter of $2a = 5.5$ nm. Determine the zeta potentials of the proteins using the Hückel equation (11.34). Take for water $\eta = 1.0$ mPa s and $\varepsilon = 77.9$.
   (c) The proteins are dispersed in a 10^{-3} M NaCl solution. How large is κa? Is the use of the Hückel equation in (b) justified?

11.16 Solve the PB equation (11.21) to obtain the radial potential profile (11.23).

11.17 Show how the integration constant c_0 in (11.26) follows from the condition of electro-neutrality in equation (11.25).

Find the solutions to these exercises at the end of the book.

Answers to Chapter Exercises

Chapter 1

1.1 The maize protein zein dissolves in ethanol but not in water. So the addition of water enhances the supersaturation of the protein molecules, inducing nucleation and growth of spherical zein particles. This is an example of what is also called the 'poor-solvent method' for preparing colloidal dispersions.

1.2 At alkaline pH, silica has a significant solubility, such that in the sol there is an equilibrium between colloidal silica particles and dissolved silica in solution. Ethanol drastically reduces the silica solubility such that the 'soluble silica' precipitates, gluing particles into a gel.

1.3 (a) $A_s^* = \dfrac{S}{V\delta} = \dfrac{4\pi a^2}{(4/3)\pi a^3 \delta} = \dfrac{3}{a\delta}$.

(b) $A_s^* = \dfrac{3}{a\delta} = \dfrac{3}{(2^{-1} \times 10^{-8}\,\text{m})(2g \times 10^6\,\text{m}^{-3})} = 300\,\text{m}^2\text{g}^{-1}$.

1.4 $\dfrac{S_{\text{nanocubes}}}{S_{\text{big cube}}} = \dfrac{6 \times 10^{24}\,\text{nm}^2}{6 \times 10^{16}\,\text{nm}^2} = 10^8$.

1.5 (a) Total area $A_{\text{tot}} = N4\pi r^2 = N4\pi(N^{-2/3}R^2) = N^{1/3}4\pi R^2$. So the specific area per gram is given by: $\dfrac{N^{1/3}4\pi R^2}{(4\pi/3)R^3\delta} = N^{1/3} \times 3/\delta R = N^{1/3} \times$ specific surface area big sphere.

(b) $\gamma 4\pi R^2 \times (R/r) = \gamma 4\pi R^3 / r$.

1.6 Specific area in $1/\text{m}$: $A_s = \dfrac{\text{surface}}{\text{volume}}$. Sphere: $A_s = \dfrac{4\pi a^2}{(4/3)\pi a^3} = \dfrac{3}{a}$;

Cube: $A_s = \dfrac{6d^2}{d^3} = \dfrac{6}{d}$.

Thin cylinder with length L and diameter d:

$$A_s = \dfrac{2 \times (\pi/4)d^2 + \pi dL}{(\pi/4)d^2 L} = \dfrac{2}{L} + \dfrac{4}{d} \approx \dfrac{4}{d}, \text{ for } L \gg d.$$

Circular disc with radius a and thickness d:

$$A_s = \dfrac{2 \times \pi a^2 + 2\pi a \times d}{\pi a^2 d} = \dfrac{2}{d} + \dfrac{2}{a} \approx \dfrac{2}{d}, \text{ for } a \gg d.$$

Note that the latter result applies to thin discs or thin pancakes of any shape; see Figure (7.11). Note also that for anisotropic particles with high aspect ratio, the specific surface only depends on the smallest dimension d.

1.7

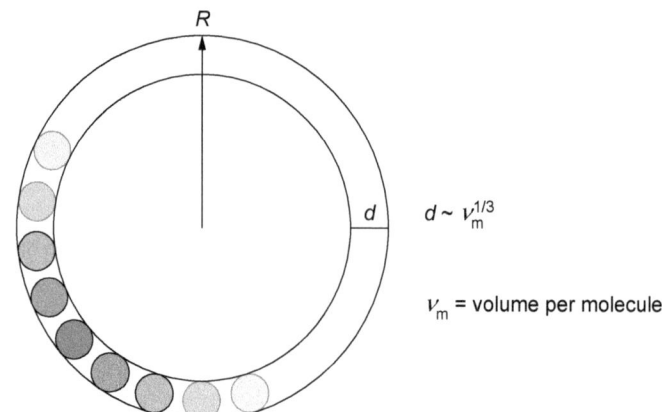

$d \sim v_m^{1/3}$

v_m = volume per molecule

$$\phi = \frac{\text{shell volume}}{\text{total volume}} = \frac{(4/3)\pi R^3 - (4/3)\pi(R-d)^3}{(4/3)\pi R^3}$$

$$= 1 - \left(1 - \frac{d}{R}\right)^3 \approx \frac{3d}{R}, \text{ for } \frac{d}{R} \gg 1.$$

Note: Employ here the Taylor expansion $(1-x)^3 = 1 - 3x + ...$, for $x \ll 1$. For (water) molecules with a diameter of $d \approx 0.1$ nm:

R	ϕ	3d/R
1 mm	3×10^{-7}	3×10^{-7}
1 micron	3×10^{-4}	3×10^{-4}
10 nm	0.030	0.03
1 nm	0.271	0.3

So $3d/R$ yields an accurate value for the fraction of molecules at the particle surface, even for nanoparticles.

1.8 (a) $O = \pi R^2 \Rightarrow R = (O/\pi)^{1/2} \Rightarrow V_h = (2/3)\pi R^3 = \frac{2 O^{3/2}}{3\sqrt{\pi}}$.

(b) Weight $F_G = V_h \times (\delta_{glass} - \delta_{water})g$

$$= \frac{2}{3\sqrt{\pi}} (1\,m^2)^{3/2} \times (10^3\,kg\,m^{-3}) \times (9.8\,m\,s^{-2}) = 3686\,N.$$

(c) Force F_{att} is minus the spatial derivative of *its* associated potential energy A_{att}. Hence:

$$F_{att} = -\frac{dA_{att}}{dh} = -\frac{d}{dh}\left[\frac{-A_H}{12\pi h^2}\right] = -\frac{A_H}{6\pi h^3} (Nm^{-2}).$$

The minus sign expresses that the force is directed towards the origin at $h = 0$:

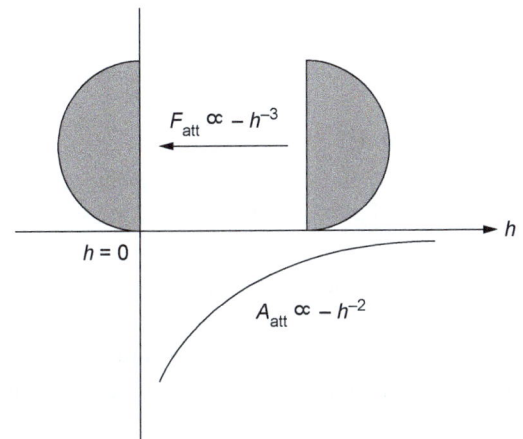

(d) Total Van der Waals attraction = Area × F_{att} = 2×10^5 N, for $h = 1$ nm
                                                                                                                                                   200 N   10 nm
                                                                                                                                                   0.2 N   100 nm

Chapter 2

2.1 $r_k = \dfrac{2\gamma v_m}{k_B T} = \dfrac{2\gamma V_m}{RT} = \dfrac{2 \times (72 \times 10^{-3}\, \text{Nm}^{-1}) \times (18 \times 10^{-6}\, \text{m}^3 \text{mol}^{-1})}{2.48 \times 10^3\, \text{J mol}^{-1}} \approx 1\, \text{nm}.$

2.2 For a cube of side d: $\Delta G = \Delta G_{surf} + \Delta G_{volume} = \gamma 6 d^2 - (d^3/v_m)|\Delta\mu|$. From $\partial \Delta G / \partial d = 0$ we find the critical cube side $d^* = 4\gamma v_m / |\Delta\mu|$, from which it follows that $\Delta G = \gamma 6 d^2 [1 - (2d/3d)^*]$, leading to $\Delta G^* = (1/3)\Delta G^*_{surf}$. This is the same result as for a spherical nucleus.

2.3

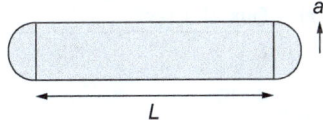

(a) Area $A = 4\pi a^2 + 2\pi a L$; cylinder volume $V = (4/3)\pi a^3 + \pi a^2 L$, so the formation Gibbs energy of the cylinder is

$\Delta G = \gamma A + (V/v_m)\Delta\mu = 4\pi a^2 (1 + L/2a)\gamma + (4/3)\pi a^3 (\Delta\mu/v_m)(1 + 3L/4a)$

$\Delta\mu = -k_B T \ln S.$

(b) $\dfrac{\partial \Delta G}{\partial L} = 2\pi a \gamma - \pi a^2 v_m^{-1} k_B T \ln S = 0,$

so the critical diameter equals $a^* = 2\gamma v_m / k_B T \ln S = r_k / \ln S$ and the growth is spontaneous ($\partial \Delta G / \partial L < 0$) when $a > a^*$.

Note A cylinder can only grow spontaneously in length above a certain critical diameter, due to the competition between surface and bulk effects. If the cylinder is too thin, there is insufficient increase in bulk volume to compensate for the surface area increase caused by lengthwise growth.

2.4 $\int J \int_\infty^a \frac{dr}{r^2} = 4\pi D \int_{c_m}^0 dc(r) \Rightarrow J = 4\pi Da c_m$. This diffusion flux has been derived under the assumptions

(1) There is a steady-state diffusion of molecules towards the sphere; it takes a time of order $t \gg a^2/D$ before the steady-state has been reached.

(2) The sphere is insoluble: $c(r = a) = 0$—which requires that the sphere is big enough.

(3) Infinite bulk with constant bulk concentration $c(r \to \infty) = c_m$ (many growing particles in a finite volume will decrease the bulk concentration).

(4) No interaction between molecules and sphere at $r > a$ (invalidated when, for example, molecules and spheres are charged).

2.5 (a) The pre-factor in the nucleation flux in eq. (2.24) is $4\pi Da^* c_m^2$. For an order-of-magnitude estimate we ignore the size difference between the small critical nuclei and molecules. Hence we take $D \sim k_B T / \eta a^*$ such that the pre-factor is of order $(kT/\eta)c_m^2$.

(b)

c_m/ppm	ln S	a^*/nm	$\Delta G^*/k_B T$	$\exp[-\Delta G^*/k_B T]$
200	0.69	1.6	128	0
500	1.6	0.7	23	10^{-10}
1,000	2.3	0.5	12	10^{-5}

c_m/ppm	$(k_B T/\eta)c_m^2$ (1/L sec)	$\exp[-\Delta G^*/k_B T]$	Flux / (1/L sec)
200	64×10^{26}	0	0
500	4×10^{27}	10^{-10}	4×10^{17}
1,000	2×10^{29}	10^{-5}	2×10^{24}

NB: convert silica concentration c ppm = c mg/L to concentration c_m in molecules per litre using: c_m = (c in ppm) $\times 6.3 \times 10^{18}$ (molecules/L).

Note: What Exercise 2.5 illustrates is that (1) a substantial supersaturation is needed to trigger homogeneous nucleation and (2) homogeneous nucleation is all or nothing (see also Figure 2.4): in a small range of supersaturation the nucleation flux goes from zero to astronomically large, because of the steep rise in magnitude of $\exp[-\Delta G^*/kT]$ that multiplies a very large pre-factor.

2.6 (a)

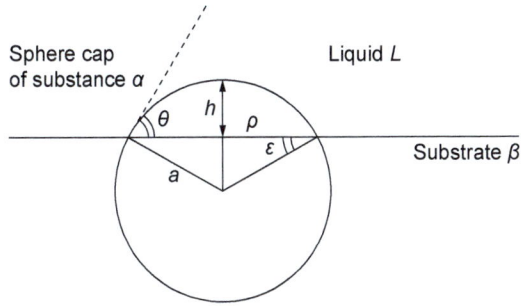

$\cos\theta = \sin\varepsilon = (a - h)/a$.

The areas of the base and the cap are

$$A_{base} = \pi\rho^2 = \pi\left[a^2 - (a-h)^2\right] = \pi a^2(1-\cos^2\theta) = \pi a^2\sin^2\theta$$

$$A_{cap} = 2\pi ah = 2\pi a^2(1-\cos\theta).$$

The volume of the sphere cap equals

$$V_{cap} = (\pi/3)h^2(3a-h) = (\pi/3)a^2(1-\cos\theta)^2(3a - a(1-\cos\theta))$$
$$= (\pi/3)a^3(1-\cos\theta)^2(2+\cos\theta).$$

The function $f(\theta)$ is defined by:

$$f(\theta) = \frac{V_{cap}}{(4\pi/3)a^3} = \frac{1}{4}(1-\cos\theta)^2(2+\cos\theta) = \frac{1}{4}(2-3\cos\theta+\cos^3\theta).$$

The surface Gibbs energy term for the cap is:

$$\Delta G^A_{het} = A_{cap}\gamma_{\alpha L} + A_{base}(\gamma_{\alpha\beta} - \gamma_{\beta L})$$
$$= 2\pi a^2(1-\cos\theta)\gamma_{\alpha L} + \pi a^2\sin^2\theta(\gamma_{\alpha\beta} - \gamma_{\beta L}).$$

The interfacial tensions are related by Young's equation $\gamma_{\beta L} = \gamma_{\alpha\beta} + \gamma_{\alpha L}\cos\theta$ which on its substitution yields

$$\Delta G^A_{het} = 2\pi a^2(1-\cos\theta)\gamma_{\alpha L} - \pi a^2\sin^2\theta(\cos\theta)\gamma_{\alpha L} = 4\pi a^2\gamma_{\alpha L}f(\theta).$$

Hence the total Gibbs energy of cap formation is:

$$\Delta G_{het} = \Delta G^A_{het} + \Delta G^V_{het} = 4\pi a^2\gamma_{\alpha L}f(\theta) + \frac{4}{3}\pi a^3 f(\theta)\frac{\Delta\mu}{V_m} = f(\theta)\Delta G_{hom}.$$

For a superhydrophobic surface $\theta \approx 180°$, $f(\theta) \approx 1$ and $\Delta G_{het} \approx \Delta G_{hom}$. Then the substrate has no effect on droplet nucleation. For a very hydrophilic surface, completely wetted by water: $\theta \approx 0°$, $f(\theta) \approx 0$ and $\Delta G_{het} \approx 0$. In the latter case supersaturation near the surface cannot be maintained.

(b) Sketch of $f(\theta)$ versus θ:

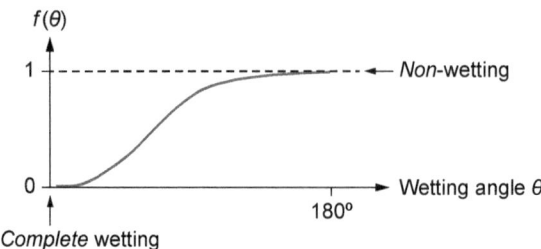

For a contact angle of 180°, the nucleation phase does not wet the substrate at all, which entails $f(\theta) = 1$, meaning that homogeneous nucleation of a sphere occurs, as if no substrate was present: $\Delta G^*_{het} = \Delta G^*_{hom}$. When the nucleating phase completely wets the substrate as a thin film, $f(\theta) = 0$, signifying that the Gibbs energy barrier has vanished: $\Delta G^*_{het} \approx 0$.

2.7 **(a)** The derivation of (2.37) can be found in Section 2.5.

(b) $D \approx 10^{-5}$ cm^2 s^{-1}; $c(\text{sat}) = 100$ ppm $= 1$ mol Si(OH)$_4$ m^{-3}
$V_m = 27.2$ cm^3 mol^{-1}; $r_k \approx 1$ nm (see Exercise 2.1)
$\Rightarrow da/dt = -Dc(\text{sat})r_k V_m a^{-2} \approx -3$ nm s^{-1}.

2.8 In the subsection 'Narrowing distributions' from Section 2.4, you will find the answers to both **(a)** and **(b)**.

2.9 $I_{het} \propto \exp[-\Delta G^*_{het}/k_BT]$; $I_{hom} \propto \exp[-\Delta G^*_{hom}/k_BT]$; $\Delta G^*_{het} = f(\theta)\Delta G^*_{hom}$

$\Rightarrow \dfrac{I_{het}}{I_{hom}} = \exp[(-\Delta G^*_{het} + \Delta G^*_{hom})/k_BT] = \exp[(1 - f(\theta))\Delta G^*_{hom}/k_BT]$.

2.10 $a^* = \dfrac{r_k}{\ln S} \approx 1 - 2$ nm

At neutral pH the bulk solubility $c(\text{sat})$ of silica is about 100 ppm, which implies that in an aqueous waterglass solution containing $c = 300$ ppm of soluble silica the equilibrium radius is in the range $a^* = 0.9$–1.8 nm.

Note: At alkaline pH the bulk solubility of silica rapidly increases to a value of $c(\text{sat}) = 300$ at about pH = 10. Thus, in the 300 ppm waterglass solution at pH = 10 the critical particle size $a = r_k/\ln S$ tends to infinity, so homogeneous precipitation of silica particles will not occur.

2.11 Particle surface area $A = \beta(v_mN)^{2/3}$, volume $V = Nv_m \Rightarrow \beta = A/V^{2/3}$.

Sphere of radius a: $\beta = \dfrac{4\pi a^2}{(4\pi a^3/3)^{2/3}} = 3^{2/3}(4\pi)^{1/3} \approx 4.84$;

for a cube: $\beta = \dfrac{6d^2}{(d^3)^{2/3}} = 6$; for a thin rod with aspect ratio

$L/d \gg 1$: $\beta = \dfrac{\pi dL}{\left[(\pi d^2/4)L\right]^{2/3}} = 16(4\pi)^{1/3}\left(\dfrac{L}{d}\right)^{1/3} \approx 37.2\left(\dfrac{L}{d}\right)^{1/3}$. Note that the sphere has the smallest β-value; this is not a coincidence because a sphere is the shape with the smallest surface area for a given particle volume.

2.12 $\langle \varphi \rangle = \dfrac{3}{4\pi a^3} \int_0^a \varphi(x)\, 4\pi x^2 dx = \dfrac{d_f}{a^3}\left(\dfrac{1}{p}\right)^{d_f-3} \int_0^a x^{d_f-1} dx = \left(\dfrac{a}{p}\right)^{d_f-3}$.

Chapter 3

3.1 (a) $<x^2>^{1/2} = (2Dt)^{1/2} = 0.37$ mm.

(b) In one hour, a grain with a radius of 50 microns diffuses a distance of about $<x^2>^{1/2} \approx 5.9\,\mu$m, so you do not see grains moving when you observe them under a microscope. What Brown viewed was not motions of pollen grains but the erratic motions of tiny micron-size organelles released by the grains into their surroundings.

3.2 (a) Sphere mass
$m = (4\pi/3) \times (10^{-7}\,\text{m})^3 \times (1.5 \times 10^3\,\text{kg m}^{-3}) = 6.28 \times 10^{-18}$ kg

$\Rightarrow \langle v^2 \rangle = \dfrac{3k_B T}{m} = \dfrac{3 \times (1.38 \times 10^{-23}\,\text{JK}^{-1}) \times (298\,\text{K})}{6.28 \times 10^{-18}\,\text{kg}} = 196.5 \times 10^{-5}\,\text{m}^2\,\text{s}^{-2}$

$\Rightarrow \sqrt{\langle v^2 \rangle} = 4.4\,\text{cm s}^{-1}$.

(b) 4.4 cm.

(c) $D = \dfrac{k_B T}{6\pi \eta R} = \dfrac{4.112 \times 10^{-21}\,\text{J}}{6\pi \times (0.89 \times 10^{-3}\,\text{Pa s}) \times (10^{-7}\,\text{m})} = 2.45 \times 10^{-12}\,\text{m}^2\text{s}^{-1}$.

$\Rightarrow \langle x^2 \rangle = 2Dt = 2 \times (2.45 \times 10^{-12}\,\text{m}^2\text{s}^{-1}) \times 1\text{s} = 4.9 \times 10^{-12}\,\text{m}^2$

$\Rightarrow \sqrt{\langle x^2 \rangle} = 2.2\,\mu$m.

3.3 $\tau_{CR} \sim \dfrac{\eta R^3}{k_B T}$; for M-sphere: $\tau_{CR} \sim 2 \times 10^{-13}$ s; N: 5×10^{-7} s; C: 2×10^{-4} s; G: 2×10^8 s.

NB: these are estimates in orders of magnitude; any constant in the expression for the relaxation time is ignored here.

3.4 Each step takes τ_{MR} seconds so the number of steps is

$\dfrac{\tau_{CR}}{\tau_{MR}} \approx \dfrac{2 \times 10^{-4}\,\text{s}}{5 \times 10^{-9}\,\text{s}} = 4 \times 10^4$, corresponding to a distance $4 \times 10^4\,\ell \approx 4\,\mu$m;

40 times the colloid radius.

3.5 $w = \int_0^\infty fv(t)\,dr = f\int_0^\infty v(t)\dfrac{dr}{dt}dt = f\int_0^{\tau_{MR}} v^2(t)\,dt = fv_0^2 \int_0^{\tau_{MR}} \exp[-2t/\tau_{MR}]/dt$

$= fv_0^2\dfrac{\tau_{MR}}{2}(1-e^{-2}) \stackrel{\tau_{MR}=m/f}{=} \dfrac{1}{2}mv_0^2(1-e^{-2})$.

Thus, if a colloid receives at $t = 0$ a kinetic energy of $(1/2)mv_0^2$, this energy is dissipated back to its environment in about τ_{MR} seconds.

3.6

R (nm)	$\tau_{CR} \sim \eta R^3/k_B T$ (s)
10	2×10^{-7}
10^2	2×10^{-4}
10^3	2×10^{-1}
10^4	2×10^2

3.7 $\int_{-\infty}^{+\infty} \frac{d^2 P}{dx^2} x^2 dx, P = P(x,t)$

$= \int_{-\infty}^{+\infty} x^2 \frac{d}{dx}\left(\frac{dP}{dx}\right) dx = \int_{-\infty}^{+\infty} x^2 d\left(\frac{dP}{dx}\right) = \left[x^2 \frac{dP}{dx}\right]_{-\infty}^{+\infty} - \int_{-\infty}^{+\infty} \frac{dP}{dx} dx^2 = 0 - 2\int_{-\infty}^{+\infty} x dP$

$= -2[xP]_{-\infty}^{+\infty} + 2\int_{-\infty}^{+\infty} P dx = 0 + 2 = 2.$

3.8 $l(t) = v_0 \tau_{MR}\left[1 - e^{-t/\tau_{MR}}\right] = v_0 \tau_{MR}\left[1 - \left(1 - \frac{t}{\tau_{MR}} + \ldots\right)\right] = v_0 t$, for $t \ll \tau_{MR}$

For $t \gg \tau_{MR}$: $e^{-t/\tau_{MR}} \approx 0$ and $l(t) = v_0 \tau_{MR}$

3.9 For the standard colloidal C-sphere (radius 100 nm) in water: $\tau_{MR} = 5 \times 10^{-9}$ s.

3.10 **(a)** $l(t) = \int_0^t v(t') dt' = v_0 \int_0^t \exp(-t'/\tau_{MR}) dt' = -v_0 \tau_{MR}\left[\exp(-t'/\tau_{MR})\right]_0^t.$

$= v_0 \tau_{MR}\left[1 - \exp(-t/\tau_{MR})\right]$

The typical momentum relaxation step length ℓ is about

$\ell = v_0 \tau_{MR}[1 - e^{-1}] \approx 0.63 v_0 \tau_{MR}$, for $t = \tau_{MR}$.

(b) For the initial speed v_0 we take the rms-speed which for the colloidal C-sphere equals 3.9 cm s^{-1}; using $\tau_{MR} = 5.10^{-9}$ s from Exercise **3.9**, the relaxation step for the C-sphere turns out to be: $\ell \approx 0.63 \times (3.9 \times 10^{-2} \text{ m s}^{-1}) \times (5 \times 10^{-9} \text{s}) = 0.1$ nm.

So, in its kinetic energy exchange with the surrounding solvent, the colloid executes ballistic steps of molecular size (the radius of an M-sphere). Due to its much larger mass, however, the colloid takes these steps at a very much lower frequency than a molecule.

3.11 $J = -4\pi r^2 D \frac{d\rho}{dr} \Rightarrow J \int_R^\infty \frac{dr}{r^2} = -4\pi D \int_{\rho_0}^{\rho_\infty} d\rho \Rightarrow -J\left[\frac{1}{r}\right]_R^\infty = -4\pi D(\rho_\infty - \rho_0)$

$\Rightarrow J = 4\pi DR(\rho_0 - \rho_\infty).$

Chapter 4

4.1 **(a)** One litre of salt solution contains $n = 2 \times 0.172$ mol $= 0.344$ mol ions. Hence:

$\pi = \frac{n}{V} RT = \frac{(0.344 \text{ mol}) \times (8.314 \text{ J mol}^{-1}\text{K}^{-1}) \times 300 \text{ K}}{10^{-3} \text{ m}^3}$

$= 8.58 \times 10^5$ J m^{-3} $= 8.58$ bar.

(b) Hydrostatic pressure exerted by column of water (with mass density $\delta = 1\,\text{g/ml}$) of height h is $P = \delta g h$. Hence

$$h = \frac{8.58 \times 10^5 \,\text{Nm}^{-2}}{(10^3 \,\text{kg m}^{-3}) \times (9.8\,\text{m s}^{-2})} = 87.6 \frac{\text{N}}{\text{kg s}^{-2}} = 87.6\,\text{m}.$$

(c) $0.9\,\text{wt\% NaCl} = 9\,\text{g NaCl/L} = 0.31\,\text{moles of ions/L}$; 0.31 moles of glucose corresponds to a glucose weight concentration of $56\,\text{g/L}$.

Note: That's a lot of sugar in one litre; for comparison: the solubility of glucose in water at room temperature is about $91\,\text{g/L}$. It is unlikely that Van 't Hoff's law holds for such a high glucose concentration, so this $56\,\text{g/L}$ is at best a rough estimate of the isotonic glucose concentration.

4.2 Coulomb potential energy for charges Q_1 and Q_2 at distance r:

$$V(r) = \frac{Q_1 \times Q_2}{4\pi\varepsilon\varepsilon_0 r}.$$

The charge on one mol of electrons equals the Faraday constant: $96{,}500\,\text{C mol}^{-1}$, so:

$$V(r=1\,\text{m}) = \frac{(96500\,\text{C mol}^{-1} \times 1\,\text{mol})^2}{4\pi \times 78.4 \times (\varepsilon_0 = 8.85 \times 10^{-12}\,\text{F m}^{-1}) \times (r=1\,\text{m})}$$

$$= 1.1 \times 10^{18} \,(\text{C}^2\text{F}^{-1} = \text{VC} = \text{J}).$$

4.3 (a) Disc volume is $V = \pi r^2 h = \pi(4 \times 10^{-6}\,\text{m})^2 \times (2 \times 10^{-6}\,\text{m}) = 32\pi \times 10^{-18}\,\text{m}^3$.

Number density of HB molecules is $\rho_c = \dfrac{250 \times 10^6}{V} = 2.49 \times 10^{24}\,\text{m}^{-3}$;

$\pi = \rho_c k_B (T = 298\,\text{K}) = 0.10\,\text{bar}$.

(b) $\pi \approx (1+z)\rho_c kT = 100\rho_c kT \Rightarrow z \approx 100$. This is an upper estimate because presence of salt will lower the pressure difference across the RBC membrane.

4.4 If H^+ is the only cation: $[H^+]^i[Cl^-]^i = [H^+]^u[Cl^-]^u \Rightarrow \dfrac{[H^+]^u}{[H^+]^i} = 0.69 \Rightarrow \text{pH} = 7.24$.

4.5

in	out
$C_{HB} = 6\,\text{g/L}$	$[Cl^-]^u = 0.012\,\text{mol/L}$
$[Na^+]^i = 0.018\,\text{mol/L}$	$[Na^+]^u = 0.012\,\text{mol/L}$

(a) HB is negative because the concentration of sodium ions in the cell is higher than outside.

(b) Electro-neutrality requires that $[Na^+]^i = z[HB] + [Cl^-]^i$. Salt equilibrium implies: $[Na^+]^i[Cl^-]^i = [Na^+]^u[Cl^-]^u$. Hence $z[HB] = 0.01\,\text{mol/L}$

$$\Rightarrow \frac{M}{z} = \frac{C_{HB}}{z[HB]} = 600\,\text{g/mol}.$$

Note: So instead of the molar mass M we find here from the Donnan equilibrium the much smaller value M/z. This ratio, also called the 'equivalent weight', is the inverse of the charge-to-mass ratio determined via the Donnan equilibrium in the form of a charge sensor; see Exercise 4.13 and equation (4.66).

4.6 From the Boltzmann distribution for cations in the form of protons, see equation (4.43):

$$[H^+]^i = [H^+]^u \exp\left[-\frac{e\Psi_D}{k_B T}\right] \Rightarrow pH^i = pH^u + \frac{e\Psi_D}{2.3 k_B T} = 7.40 + \frac{-9.4\,mV}{59\,mV} = 7.24.$$

4.7 $B_2 = 2\pi \int_0^\infty (1-\beta) r^2 dr; \; \beta = \exp[-w(r)/k_B T] = 0$ for $0 \le r < 2R; \beta = 1$ for $r > 2R$

$$\Rightarrow B_2 = 2\pi \int_0^{2R} r^2 dr = (16/3)\pi R^3 = 4 \times \text{sphere volume}.$$

4.8 $\rho_-^2 + (z\rho_c)\rho_- - \rho_s^2 = 0 \Rightarrow (\rho_-)_{1,2} = -\frac{z\rho_c}{2} \pm \frac{1}{2}\left[(z\rho_c)^2 + 4\rho_s^2\right]^{1/2}$. Discard the negative root that would lead to a negative anion concentration. Hence: $\rho_-/\rho_s = -y + \sqrt{1+y^2}; \; y = z\rho_c/2\rho_s$.

4.9 (a) Expand the square root $\sqrt{1+y^2} = 1 + (1/2)y^2 - \ldots$ and substitute $y = z\rho_c/2\rho_s$ to obtain $\dfrac{\Delta \pi_{ion}}{k_B T} = \dfrac{z^2}{4\rho_s}\rho_c^2.$

(b) $\dfrac{\Delta \pi_{tot}}{k_B T} = \rho_c + \dfrac{z^2}{4\rho_s}\rho_c^2,$ for $0 \le y \ll 1$. Comparing this pressure to (4.27) we see that $z^2/4\rho_s z$ is a second virial coefficient; see for further discussion on this point Philipse and Vrij (2011).

4.10 Employ the logarithmic representation of $\cosh^{-1}(z)$ and $\sinh^{-1}(z)$ from Appendix 4B:

$$\cosh^{-1}[(1+y^2)^{1/2}] = \ln[y + (1+y^2)^{1/2}] = \sinh^{-1}(y)$$

4.11 We measure the total osmotic pressure π_{tot} exerted by a total weight concentration c_{tot}:

$$\frac{\pi_{tot}}{RT} = \frac{c_{tot}}{<M>}.$$

Now each species j exerts a partial pressure

$$\frac{\pi_j}{RT} = \frac{c_j}{M_j},$$

so the total pressure is also given by:

$$\frac{\pi_{tot}}{RT} = \sum \frac{\pi_j}{RT} = \sum \frac{c_j}{M_j}.$$

Hence the average molar mass $<M>$ obtained from measured osmotic pressures is indeed the number-average M_n:

$$\frac{c_{tot}}{<M>} = \sum \frac{c_j}{M_j} \Rightarrow <M> = \frac{c_{tot} = \sum c_j}{\sum c_j/M_j} = \frac{\sum n_j M_j}{\sum n_j} = M_n.$$

Here n_j is the number of moles of species j with molar mass M_j.

4.12 The monovalent anionic colloids expel sodium ions via salt molecules to the reservoir. So the number of depleted sodium ions equals the number of expelled salt molecules. The ion density ratio here is:

$$y = \frac{\rho_c z}{2\rho_s} = \frac{0.1M \times 1}{0.2M} = 0.5.$$

Hence

$$\frac{L_s}{\rho_s} = \frac{\rho_{Na^+} - \rho_s}{\rho_s} = -y + \sqrt{1+y^2} - 1 = -0.382 \implies \rho_{Na^+} = 0.06\,M.$$

4.13 The solubility product for salt in the small sensor (index R) and dispersion (index S) is:

$$\rho_+^R \rho_-^R = \rho_+^S \rho_-^S, \qquad (4.67)$$

where ρ_+, ρ_- are number densities of, respectively, cations and anions. In absence of charged solutes both volumes have the same initial concentration ρ_0 of salt molecules. This initial equilibrium is perturbed by an added weight concentration c of solutes (with mass m) to the dispersion that dissociate into colloidal anions and z counter- (Na^+) ions per anion. The suspension's Na^+ concentration now exceeds its equilibrium value, and salt is depleted to the reservoir to reestablish equilibrium. For large suspension volumes $V_S \gg V_R$, equilibrium ion densities in the dispersion follow from their initial values ρ_0 and the number density zc/m of counter-ions as:

$$\rho_+^S = \rho_0 + zc/m; \quad \rho_-^S = \rho_0, \quad \text{for } V_S \gg V_R. \qquad (4.68)$$

Anion densities in the large suspension volume remain constant, in contrast to the small salt solution (the 'sensor') where the anion density ρ_-^R increases significantly. Because of electro-neutrality $\rho_-^R = \rho_+^R = \rho_s^R$ so the sensor's salt concentration increase follows from equations (4.67) and (4.68) as:

$$\left(\frac{\rho_s^R}{\rho_0}\right)^2 = 1 + \left(\frac{z}{m}\right)\frac{c}{\rho_0}, \quad \text{for } V_R \ll V_S. \qquad (4.69)$$

Here z/m is the ratio of charge number to mass ratio (CMR) of colloids or polyelectrolytes. Equation (4.69) entails that the CMR follows from the quadratic increase of the sensor's relative salt concentration against solute weight concentration.

4.14 The larger the reservoir, the smaller its salt concentration increase due to salt expulsion by the charged colloids in the suspension, and the more difficult it will be to measure accurately the amount of expelled salt.

4.15 Molarity is 0.56 mol of 1:1 electrolyte per litre of seawater.

4.16 An isotonic solution contains 0.15 mol NaCl/L.

4.17 (a) $C\int_{-\infty}^{+\infty} \exp[-mv_x^2/2kT]dv_x \stackrel{y^2=mv_x^2/2k_BT}{=} C(2k_BT/m)^{1/2}\int_{-\infty}^{+\infty} e^{-y^2}dy$

$= C(2k_BT/m)^{1/2} \times \sqrt{\pi} = 1.$

$\Rightarrow C = \left(\dfrac{m}{2\pi k_B T}\right)^{1/2}.$

(b) The average of the square of velocity components v_x is:

$$\langle v_x^2\rangle = \left(\dfrac{m}{2\pi k_B T}\right)^{1/2} \int_{-\infty}^{+\infty} \exp[-mv_x^2/2k_B T]\,v_x^2 dv_x.$$

Rewriting this integral to:

$$\langle v_x^2\rangle = \left(\dfrac{a}{\pi}\right)^{1/2} \int_{-\infty}^{+\infty} e^{-av_x^2} v_x^2 dv_x;\ a = \dfrac{m}{2k_B T},$$

we obtain

$$\langle v_x^2\rangle = -\left(\dfrac{a}{\pi}\right)^{1/2}\dfrac{d}{da}\int_{-\infty}^{+\infty} e^{-av_x^2}dv_x = -\left(\dfrac{a}{\pi}\right)^{1/2}\dfrac{d}{da}\left(\dfrac{\pi}{a}\right)^{1/2} = \dfrac{1}{2a} = \dfrac{k_B T}{m}.$$

Averages are the same in all directions so the outcome for the y- and z-velocity components must also equal $k_B T/m$. Thus the average of the squared speed is:

$$\langle u^2\rangle = \langle v_x^2\rangle + \langle v_y^2\rangle + \langle v_z^2\rangle = \dfrac{3k_B T}{m} \Rightarrow u_{rms} = \left(\dfrac{3k_B T}{m}\right)^{1/2} = \left(\dfrac{3RT}{M}\right)^{1/2}.$$

(c) Rms-speed for N_2 molecules: 515 m/sec; CO_2: 411 m/sec. NB: these are supersonic speeds as they exceed the sound speed of 330 m/sec).

(d) The kinetic energy of a particle equals $E_{kin} = (1/2)mu^2 \Rightarrow$

$$\langle E_{kin}\rangle = \dfrac{1}{2}m\langle u^2\rangle = \dfrac{3}{2}k_B T\ \text{(per particle)};\ \langle E_{kin}\rangle = \dfrac{3}{2}RT\ \text{(per mol)}.$$

4.18 $-\int_0^\infty \exp[-V_{MF}(x)/k_B T]dV_{MF}(x) = k_B T \exp[-V_{MF}(x)/k_B T]|_0^\infty = k_B T(1-0) = k_B T.$

4.19 $\dfrac{w(h)}{\pi_d} = \dfrac{\pi R}{\pi_d}\int_h^\infty w(x)dx \stackrel{\text{insert}(4.56)}{=} -\pi R\int_h^d (d-x)\,dx = -\pi R\left[\dfrac{1}{2}d^2 - dh + \dfrac{1}{2}h^2\right]$

$= -\dfrac{1}{2}\pi R(d-h)^2$

Note that the integration upper limit is d; for $x > d$ the integrand $w(x) = 0$.

4.20 From (4.19) and (4.21) we obtain for the force on a colloid: $K_{MF}(x) = k_B T d\ln\rho(x)/dx$, which is directed away from the membrane in Figure 4.4. This force imparts a speed $u(x)$ to a colloid giving rise to a friction force $fu(x)$. The stationary particle speed is $u(x) = K_{MF}(x)/f$ so the convective flux j_C away from the membrane due to force $K_{MF}(x)$ equals:

$$j_C = \rho(x)u(x) = \rho(x)\dfrac{K_{MF}(x)}{f} = \rho(x)\dfrac{k_B T}{f}\dfrac{d}{dx}\ln\rho(x) = \dfrac{k_B T}{f}\dfrac{d\rho(x)}{dx}.$$

ANSWERS TO CHAPTER EXERCISES

The diffusive particle flux towards the membrane in Figure 4.4 is given by Fick's first law:

$$j_D = -D\frac{d\rho(x)}{dx}$$

In equilibrium the net particle flux is zero:

$$\frac{k_B T}{f}\frac{d\rho(x)}{dx} - D\frac{d\rho(x)}{dx} = 0,$$

which entails for the diffusion coefficient: $D = k_B T/f$.

Chapter 5

5.1 The charged object injects counter-ions in solution, a certain fraction of which (accompanied by a co-ion to form a salt molecule) spontaneously diffuses to the surroundings until thermodynamic salt equilibrium has been re-installed.

5.2 97 nm, 9.7 nm, 0.97 nm, and 0.3 nm.

5.3 Here is a completed version of Table 5.1 (self-ionizing solvents):

	Water	Ethylene diamine	Ethanol
ε	78.3	12.9	24.55
ρ_s/M	10^{-7}	$10^{-7.6}$	$10^{-9.4}$
r_B/nm	0.7	4.3	2.3
$\kappa^{-1}/\mu m$	0.96	0.78	8.9
N_D	54	7	156
E_{kin}/V_{pot}	539	138	1,092

5.4 The potential difference is an equilibrium value and equilibria cannot deliver work. The requirement for a spontaneous process is a decrease in the system's Gibbs free energy: $(\Delta G)_T < 0$; at equilibrium $(\Delta G)_T = 0$. Since $(\Delta G)_T = w_{rev}$ is the *maximal* work a process can deliver, it follows that for an equilibrium $w = 0$.

5.5 Surface potentials have no effect on the screening length.

5.6 25 mV.

5.7 (a) The potential at a distance r from the centre of a sphere carrying a charge of Q coulombs is, according to Coulomb's law,

$$\Psi(r) = \frac{Q}{4\pi\varepsilon\varepsilon_0 r}.$$

Make the surface potential approximately equal to the zeta potential:

$$\Psi(r = R) = \frac{Q}{4\pi\varepsilon\varepsilon_0 R} \approx \zeta.$$

Hence the number of charges on the sphere is approximately

$$\sigma = \frac{Q}{e} \approx \frac{4\pi\varepsilon\varepsilon_0 R\zeta}{e} = \frac{4\pi \times 24.55 \times 8.85 \times 10^{-12} \times 50 \times 10^{-10} C^2 Vm}{1.6 \times 10^{-19} Nm^2 C} = 85.4.$$

(b) 85 charges per sphere; that is not a lot, so every surface charge has plenty of surface room:
$$\frac{4\pi R^2}{85.4} = 1471.5\,\text{nm}^2.$$
(c) Average distance $\sqrt{1471.5\,\text{nm}^2} \approx 38\,\text{nm}$.

5.8 That constant is zero; the single, free EDL expands until ion pressure and Maxwell stress are in balance:
$$\pi_{\text{ion}} = \frac{\varepsilon\varepsilon_0}{2}\left(\frac{d\Psi}{dx}\right)^2.$$

To turn this into an equation for potential profile Φ we have to specify how π_{ion} depends on Φ. The excess ion density at location x in the EDL is given by equation (5.5):
$$\rho_+(x) + \rho_-(x) - 2\rho_s = 2\rho_s[\cosh(\Phi) - 1].$$

Hence:
$$\frac{\pi_{\text{ion}}}{k_B T} = 2\rho_s[\cosh(\Phi) - 1] = \frac{\varepsilon\varepsilon_0}{2k_B T}\left(\frac{d\Psi}{dx}\right)^2,$$

from which it follows that
$$2\cosh(\Phi) - 2 = \frac{\varepsilon\varepsilon_0}{2\rho_s k_B T}\left(\frac{d\Psi}{dx}\right)^2 = \left(\frac{d\Phi}{du}\right)^2; \quad \Phi = \frac{ze\Psi}{k_B T};\quad u = \kappa x.$$

This is identical to equation (5.18) for the square of the electric field in the main text:
$$\left(\frac{d\Phi}{du}\right)^2 = 2\cosh(\Phi) - 2.$$

Integration twice then leads to the GC profile
$$\tanh(\Phi/4) = \tanh(\Phi_0/4)e^{-u},$$

as is further spelled out in Section 5.4 and Exercise 5.20.

5.9 (a) See the sketch of the counter-ion profile in Figure 5.4.

(b) The probability density $P_+(x)$ to find an excess counter-cation at distance x from a negatively charged surface is proportional to
$$P_+(x) \propto \frac{\rho_+ - \rho_s}{\rho_s} = \frac{\rho_+}{\rho_s} - 1 = e^{-\Phi(x)} - 1;\quad \Phi(x) < 0.$$

Assuming a small surface potential we have $e^{-\Phi(x)} \sim 1 - \Phi(x) = 1 - \Phi_0 e^{-\kappa x}$, and $P_+(x) \propto -\Phi_0 e^{-\kappa x}$. Then the average distance between excess ions and the surface is:
$$\langle x \rangle = \frac{\int_0^\infty P_+(x)x\,dx}{\int_0^\infty P_+(x)\,dx} = \frac{\int_0^\infty e^{-\kappa x} x\,dx}{\int_0^\infty e^{-\kappa x}\,dx} = \kappa^{-1}.$$

Note that the integral in the denominator ensures that the probability $P_+(x)dx$ is normalized.

5.10 $\dfrac{L_s}{\rho_s} = \kappa^{-1}\displaystyle\int_{\Phi_0}^{0}\dfrac{e^{\Phi}-1}{d\Phi/dx}d\Phi \overset{\text{Insert}(5.20)}{=} \kappa^{-1}\displaystyle\int_{\Phi_0}^{0}\dfrac{e^{\Phi}-1}{-2\sinh(\Phi/2)}d\Phi = -\kappa^{-1}\displaystyle\int_{\Phi_0}^{0}\dfrac{e^{\Phi}-1}{e^{\Phi/2}-e^{-\Phi/2}}d\Phi$

$= -\kappa^{-1}\displaystyle\int_{\Phi_0}^{0}e^{\Phi/2}d\Phi = 2\kappa^{-1}[e^{\Phi_0/2}-1]$

5.11 Rivers transport clay colloids and other suspended matter that aggregates due to the high sea salinity that settle faster than the more stable colloids in lake water.

5.12 $\tanh(\Phi_0/8) = \tanh(\Phi_0/4)e^{-\kappa x_{1/2}} \Rightarrow \kappa x_{1/2} = \ln\left[\dfrac{\tanh(\Phi_0/4)}{\tanh(\Phi_0/8)}\right]$.

In the Debye-Hückel limit $\Phi_0 \ll 1$: $\tanh(\Phi_0/4) \sim \Phi_0/4$ and $\tanh(\Phi_0/8) \sim \Phi_0/8$.

Hence $\kappa x_{1/2} = \ln 2$.

5.13 When the surfaces are very close together, the counter-ion (proton) density rises, which drives the equilibrium:
Surface–SiOH $\rightleftarrows$ Surface–SiO$^-$ +H$^+$ to the left; upon contact the surfaces will (partly) discharge each other. Counter-ion condensation on silica, incidentally, is only significant at alkaline pH.[1]

5.14 $\pi_{\text{ext}}(h) = 2\rho_s kT \times 32\tanh^2(\Phi_0/4)e^{-\kappa h} \Rightarrow \dfrac{A_{\text{rep}}}{kT} = -\displaystyle\int_{\infty}^{h}\dfrac{\pi_{\text{ext}}(h)}{kT}dh'$

$= -64\rho_s \tanh^2(\Phi_0/4)\displaystyle\int_{\infty}^{h}e^{-\kappa h'}dh' = 64\rho_s\kappa^{-1}\tanh^2(\Phi_0/4)e^{-\kappa h}$.

5.15 (a) $A_{\text{tot}} = 64kT\rho_s\kappa^{-1}\tanh^2(\Phi_0/4)e^{-\kappa h} - \dfrac{\kappa^2 A_H}{12\pi(\kappa h)^2} = \alpha e^{-\kappa h} - \dfrac{\beta}{(\kappa h)^2}$.

(b) $\dfrac{dA_{\text{tot}}}{d(\kappa h)} = 0 \Rightarrow (\kappa h_{\max})^3 e^{-\kappa h_{\max}} = \dfrac{2\beta}{\alpha}$ (1)

$A_{\text{tot}} = 0 \Rightarrow (\kappa h_{\max})^2 e^{-\kappa h_{\max}} = \dfrac{\beta}{\alpha}$ (2)

$\dfrac{(1)}{(2)} \Rightarrow \kappa h_{\max} = 2$ (location maximum)   (3)

(3) in (2) $\Rightarrow \alpha = \dfrac{e^2}{4}\beta$ (4)

Substitute

α, β in (4) $\Rightarrow \dfrac{\kappa^3}{\rho_{\text{floc}}} = \text{constant}_1 \times A_H^{-1}\tanh^2(\Phi_0/4)$ (5)

---

[1] See B. Stojimirovic et al. (2020), 'Experimental evidence for algebraic double-layer forces', *Langmuir* **36**, 47–54.

Since $\kappa^2 \propto z^2 \rho_{floc}$ and $\Phi_0 = ze\Psi_0/kT$ we find for the flocculation concentration the scaling:

$$\rho_{floc} = \text{constant}_2 \times A_H^{-2} z^{-6} \tanh^4(ze\Psi_0/4kT).$$

(c) For high surface potential $\tanh(\Phi_0) \sim 1$ and we obtain the proportionality

$$\rho_{floc} \propto A_H^{-2} z^{-6}, \text{ for } \Phi_0 \gg 1,$$

whereas for small surface potentials where $\tanh(\Phi_0) \sim \Phi_0$

$$\rho_{floc} \propto A_H^{-2} z^{-2} \Psi_0^4, \text{ for } \Psi_0 \ll 1.$$

5.16 In the midplane m between the plates there is no electric field but there is a (dimensionless) potential Φ_m. The sum of counter- and co-ion concentrations in the midplane is a sum of Boltzmann distributions: $\rho_s e^{\Phi_m} + \rho_s e^{-\Phi_m}$ where ρ_s is an ion density in the bulk where $\Phi = 0$. Since $e^{\Phi_m} + e^{-\Phi_m} > 2$, whatever the sign of the potential, it follows that the ion concentration at m exceeds the ion density $2\rho_s$ in the bulk. Hence, the midplane osmotic pressure exceeds the bulk osmotic pressure so the plates repel each other.

5.17 Because the Van der Waals attraction decays more slowly than the EDL repulsion.

5.18 The DH approximation is that potentials are low (< 25 mV). For a charged plate electrical field lines are parallel; around a sphere field lines radially diverge and the potential drops faster such that more ions diffuse around in low-potential regions.

5.19 $\tanh y = \dfrac{\sinh y}{\cosh y} = \dfrac{e^y - e^{-y}}{e^{+y} + e^{-y}} = \dfrac{e^{2y} - 1}{e^{2y} - 1}; \; y = \dfrac{x}{2} \Rightarrow \tanh\left(\dfrac{x}{2}\right) = \dfrac{e^x - 1}{e^x + 1}.$

5.20 From equation (5.20): $\dfrac{d\Phi}{du} = -2\sinh(\Phi/2) = e^{-\Phi/2} - e^{\Phi/2}$.

To obtain the potential profile $\Phi = \Phi(u)$ in the EDL, we employ the substitutions $e^{\Phi/2} = x$ and $d\Phi = (2/x)\, dx$ to get

$$\frac{dx}{1-x^2} = \frac{dx}{2(1-x)} + \frac{dx}{2(1+x)} = \frac{1}{2} du,$$

which upon integration yields

$$\ln\left(\frac{1+x}{1-x}\right) = u + C.$$

Substitution of $x = \exp[+\Phi/2]$ and employing the boundary condition $\Phi(u = 0) = \Phi_0$ leads to

$$\frac{1 - e^{\Phi/2}}{1 + e^{\Phi/2}} = \frac{1 - e^{\Phi_0/2}}{1 + e^{\Phi_0/2}} e^{-u}.$$

Employing the identity (Exercise 5.19)

$$\tanh(x/2) = \frac{e^x - 1}{e^x + 1},$$

we can rewrite the GC potential profile to the more compact form

$$\tanh(\Phi/4) = \tanh(\Phi_0/4)e^{-u}; \quad u = \kappa x,$$

which is equation (5.21) in the text.

5.21 $4\rho_s \kappa^{-1} \Phi_0^2 = 4\rho_s \kappa \left(\dfrac{ze\Psi_0}{k_B T}\right)^2 \times \left(\kappa^{-2} = \dfrac{\varepsilon\varepsilon_0 k_B T}{2\rho_s(ze)^2}\right) = \dfrac{2\varepsilon\varepsilon_0 \kappa \Psi_0^2}{k_B T}.$

5.22 For sufficiently small electrical potentials $\sinh(\Phi)$ can be linearized to $\sinh(\Phi) \sim \Phi$ such that the PB equation simplifies to

$$\dfrac{d^2\Phi}{dx^2} = \kappa^2 \Phi, \text{ for } \Phi \ll 1,$$

which is known as the linearized PB equation; the small-potential approximation is also referred to as the Debye–Hückel approximation. Its solution is

$$\Phi(x) = Ae^{-\kappa x} + Be^{\kappa x}.$$

The boundary conditions on the electrical potential are that it equals the surface potential Φ_0 at $x = 0$ and that it vanishes at infinity, implying that the integration constants equal $A = \Phi_0$ and $B = 0$ such that:

$$\Phi(x) = \Phi_0 e^{-\kappa x}.$$

5.23 $2\cosh(x) - 2 = e^x + e^{-x} - 2 = (e^{x/2} - e^{-x/2})^2 = 4\sinh^2(x/2).$

Chapter 6

6.1 Take discs of equal radii R that diffuse towards a target disc centred at the origin. The diffusive disc flux is:

$$J = 2\pi r D \dfrac{d\rho}{dr}.$$

Here ρ is the disc number density (particles per unit area) at distance r from the origin. Since J is constant (stationary diffusion) we find:

$$J\int_\delta^{2R} \dfrac{dr}{r} = 2\pi D \int_{\rho_{bulk}}^0 d\rho \Rightarrow J = 2\pi D \rho_{bulk} \ln^{-1}[\delta/2R].$$

One cannot, in contrast to the three-dimensional case, take the limit $\delta \to \infty$ because then the diffusion flux would vanish. Thus, for diffusion-controlled processes on a surface one must specify a certain diffusion 'zone thickness' δ, a choice that is not required for the three-dimensional case.

6.2 Integration of $\int_{c_0}^{c(t)} c^{-2} dc = -k_{11} \int_0^t dt$ yields $c(t) = c_0/(1 + c_0 k_{11} t)$; the half time $t_{1/2} = 1/k_{11} c_0$ follows from substitution of $c(t) = c_0/2$ and $t = t_{1/2}$.

6.3 $J(j \to i) = \text{const.} \times (2 + x + x^{-1})$; $x = \dfrac{R_i}{R_j} \Rightarrow \dfrac{d}{dx} J(j \to i) = 0$ for $x = 1$

(is a minimum)

6.4 3.8×10^{-5}, 3.8×10^{3} sec.

6.5 In Section 6.6 we found that the radius R_i of sphere i (with radius R_0 at time t_0) grows by the diffusive uptake of small molecules j as:

$$R_i^2 - R_0^2 = 2D_j \phi_j (t - t_0); \quad \phi_j \approx c_{j,\infty} v_j.$$

Hence: $\dfrac{dR}{dt} = \dfrac{A}{R}$. Here A is a constant; the subscript i has been dropped.

Another sphere has a larger radius $R + \Delta R$, with $\Delta R > 0$:

$$\dfrac{d(R + \Delta R)}{dt} = \dfrac{A}{R + \Delta R}.$$

The change in time of ΔR is therefore:

$$\dfrac{d\Delta R}{dt} = -\dfrac{A}{R}\left[\dfrac{\Delta R}{R + \Delta R}\right]; \quad \Delta R > 0.$$

Hence $d\Delta R/dt < 0$. So, for any arbitrary pair of spheres the relative size difference decreases in time; consequently, the whole particle size distribution sharpens by diffusional growth.

6.6 $R = (k_B T / 4\eta \dot{\gamma})^{1/3} = 1.0 \,\mu\text{m}$. For radii $R > 1\,\mu\text{m}$ the flow-induced flocculation rate will exceed the diffusional rate constant by a factor proportional to the cube of the particle radius. For colloids or clusters (much) larger than a micron, stirring dominates Brownian motion; for dispersions of nanoparticles, in contrast, shaking or shearing has no influence on flocculation rates.

6.7 (a) $J = 4\pi D_{\text{virion}} R_{\text{host}} c_\infty$. Here c_∞ is the virion bulk number density which follows from the weight concentration via: $c_\infty = c_w / m$; $m = $ virion mass $\approx 4\pi a^3 / 3\bar{V}$.

(b)

Table 6.2 Encounter frequencies $J (\text{s}^{-1})$ between virions and host cell

c_w (g/L)	10^{-3}	10^{-2}	10^{-1}
a (nm)			
10	92	920	9,200
20	5.75	57.5	575
40	0.36	3.59	35.9

$\eta_{H_2O} = 10^{-3} \text{kg m}^{-1}\text{s}^{-1}$; $T = 298\,\text{K}$; $R = 2\,\mu\text{m}$; $\bar{V} = 0.7\,\text{cm}^3\text{g}^{-1}$.

6.8 Number density decreases as

$$\frac{dc}{dt} = k_{11}c^2; \quad \frac{c}{c_0} = \frac{1}{1+k_{11}c_0 t}.$$

From the experimental data: $k_{11} \approx 5 \times 10^{-12}\,\text{cm}^3/\text{sec}$. For rapid flocculation the prediction is a higher rate constant: $k_{11} = 8k_B T/3\eta \approx 12 \times 10^{-12}\,\text{cm}^3/\text{sec}$. Van der Waals attraction between the silica particles at contact is possibly not large enough to induce irreversible aggregation for every encounter between the particles.

6.9 $J(2 \to 1) = 18 \times 10^5\,\text{s}^{-1}$.

6.10 Formation of 4-mers:

$$\overset{i=1}{\text{O}} + \overset{j=3}{\text{OOO}} \xrightarrow{k_{13}} \text{OOOO}; \quad \frac{dc_4}{dt} = k_{13}c_1 c_3$$

$$\overset{i=2}{\text{OO}} + \overset{j=2}{\text{OO}} \xrightarrow{k_{22}} \text{OOOO}; \quad \frac{dc_4}{dt} = \frac{1}{2}k_{22}c_2 c_2$$

$$\overset{i=3}{\text{OOO}} + \overset{j=1}{\text{O}} \xrightarrow{k_{31}} \text{OOOO}; \quad \frac{dc_4}{dt} = k_{31}c_3 c_1; \text{ nb: } k_{31} = k_{13}$$

$$\Rightarrow \frac{dc_4}{dt} = \frac{1}{2}(k_{13}c_1 c_3 + k_{31}c_3 c_1) + \frac{1}{2}k_{22}c_2 c_2 = \frac{1}{2}\sum_{i=1}^{4-1} k_{ij}c_i c_j; \; i+j=4; j=4-i$$

Disappearance of 4-mers:

$$\text{OOOO} + \text{O} \xrightarrow{k_{41}}$$

$$\quad\quad\quad + \text{OO} \xrightarrow{k_{42}} \Rightarrow \frac{dc_4}{dt} = -k_{41}c_4 c_1 - k_{42}c_4 c_2 - k_{43}c_4 c_3 - \ldots = -c_4\sum_{i=1}^{\infty} k_{4i}c_i$$

$$\quad\quad\quad + \text{OOO} \xrightarrow{k_{43}}$$

$$\quad\quad\quad + \text{OOOO} \xrightarrow{k_{44}}$$

etc.

$$\Rightarrow \text{Net change in } c_4 : \frac{dc_4}{dt} = \frac{1}{2}\sum_{i=1}^{4-1} k_{ij}c_i c_j - c_4\sum_{i=1}^{\infty} k_{4i}c_i; \; j=4-i.$$

Replace 4 with α to obtain the rate equation for α-mers in equation (6.41).

6.11

$$\frac{dc_\alpha}{dt} = \frac{1}{2}k_{11}\sum_{i=1}^{\alpha-1} c_i c_{\alpha-i} - k_{11}c_\alpha\sum_{i=1}^{\infty} c_i$$

$$\Rightarrow \frac{dc_{tot}}{dt} = \sum_{\alpha=1}^{\infty}\frac{dc_\alpha}{dt} = \frac{1}{2}k_{11}\sum_{\alpha=1}^{\infty}\sum_{i=1}^{\alpha-1} c_i c_{\alpha-i} - k_{11}\sum_{\alpha=1}^{\infty} c_\alpha \sum_{i=1}^{\infty} c_i = \frac{1}{2}k_{11}c_{tot}^2 - k_{11}c_{tot}c_{tot} = -\frac{1}{2}k_{11}c_{tot}^2.$$

6.12 For the derivative of the non-equilibrium concentration profile $c = c(x)$,

$$c = \gamma c_0 \exp[-V/k_B T],$$

with respect to x we can write:

$$\frac{dc}{dx} = c_0 \exp[-V/k_B T]\frac{d\gamma}{dx} + \gamma c_0 \frac{d}{dx}\exp[-V/k_B T]$$

$$= c_0 \exp[-V/k_B T]\frac{d\gamma}{dx} - \frac{\gamma c_0 \exp[-V/k_B T]}{k_B T}\frac{dV}{dx}$$

$$= c_0 \exp[-V/k_B T]\frac{d\gamma}{dx} - \frac{c}{k_B T}\frac{dV}{dx},$$

which on substitution in differential equation (6.31)

$$-\frac{j}{D} = \frac{dc}{dx} + \frac{c}{k_B T}\frac{dV}{dx}$$

yields

$$-\frac{j}{D} = c_0 \exp[-V/k_B T]\frac{d\gamma}{dx} \Rightarrow \frac{d\gamma}{dx} = -\frac{j}{Dc_0}\exp[V/k_B T].$$

6.13 It does: $\frac{dc}{dt} = -k_{11}c^2 \times \frac{1}{2} \times 2$, where the factor ½ corrects for counting each particle twice, and the factor 2 accounts for the fact that every collision removes two particles.

Chapter 7

7.1 (a) The pressure drop Δp over the cube in Figure 7.6 entails a net force on the cube equal to $\Delta p(\Delta x)^2$. The viscous friction the flow exerts on the cube is proportional to

(1) the fluid viscosity η
(2) the average flow speed $<u>$
(3) the total internal area of the cube against which fluid is rubbing, an area that is proportional to the cube volume $(\Delta x)^3$.

Hence the force balance on the cube reads: $\eta <u> (\Delta x)^3 = $ constant $\Delta p(\Delta x)^2$.

Identifying the constant as the liquid permeability k and taking the limit $\Delta x \to dx$ we obtain Darcy's law for x-directed flow in its differential form:

$$\langle u \rangle = -\frac{k}{\eta}\frac{dp}{dx}.$$

NB: The minus sign takes into account that a negative pressure gradient $dp/dx < 0$ drives the flow in the positive x-direction.

(b) $\langle u \rangle = -\frac{k}{\eta}\frac{dp}{dx} \Rightarrow \langle u \rangle \int_0^L dx = -\frac{k}{\eta}\int_{P+\Delta P}^{P} dp \Rightarrow \langle u \rangle = \frac{k}{\eta}\frac{\Delta P}{L}.$

(c) The assumption in **(b)** is that permeability k is constant, signifying that the particle packing's microstructure is independent of spatial coordinate x. When the microstructure or particle sizes vary along the x-axis, then $k = k(x)$ and the integral form of Darcy's law reads

$$\langle u \rangle \int_0^L \frac{dx}{k(x)} = \frac{\Delta P}{\eta} \Rightarrow \langle u \rangle = \frac{\bar{k}}{\eta} \frac{\Delta P}{L}; \quad \bar{k} = \frac{L}{\int_0^L k^{-1}(x)dx}.$$

Here $\bar{k}$ is the so-called harmonic mean of $k(x)$, expressing that regions with low permeability (e.g. those composed of small particles) strongly reduce the flow speed.

7.2 (a) The net force on a particle settling at speed $v(t)$ in a viscous medium is $(\Delta m)g - fv(t)$. Hence from Newton's second law we obtain the equation of motion

$$m\frac{dv(t)}{dt} = (\Delta m)g - fv(t).$$

Here m and Δm are, respectively, the inertial and buoyant particle mass, g is the gravitational acceleration, and f is the Stokes friction factor. Integration of this equation, with the boundary condition $v(t = 0) = 0$, yields:

$$\frac{dv(t)}{(\Delta m)g - fv(t)} = \frac{dt}{m} \Rightarrow \ln\left[(\Delta m)g - fv(t)\right]_0^{v(t)} \times \frac{-1}{f} = \frac{t}{m}$$

$$\Rightarrow v(t) = \frac{(\Delta m)g}{f}(1 - e^{-t/\tau_{MR}}) = v(1 - e^{-t/\tau_{MR}})$$

Here $\tau_{MR} = m/f$ is the momentum relaxation time, and v is the stationary sedimentation speed, reached at a time $t \gg \tau_{MR}$.

(b) $\mu = \frac{\Delta m}{f} = \frac{2R^2(\delta_p - \delta)}{9\eta} = \frac{2 \times (10^{-7} \text{m})^2 \times (1 \times 10^3 \text{kg m}^{-3})}{9 \times 10^{-3} \text{Pa s}} = 2.2 \times 10^{-9} \text{s}.$

7.3 $\rho \times f \times u \ [=] \ \text{m}^{-3} \times (\text{Nm}^{-2}\text{s} \times \text{m}) \times (\text{m s}^{-1}) = \text{Nm}^{-3}$

$\frac{dp}{dx} [=] \frac{\text{Nm}^{-2}}{\text{m}} = \text{Nm}^{-3}.$

7.4 (a) From solving the Stokes equation for a slit pore (see Section 8.4) we obtain for the average fluid speed in one slit:

$$<u>_{slit} = \frac{d^2}{12\eta} \frac{\Delta P}{L},$$

where ΔP is the total pressure drop, going a distance L in the x-direction (Figure 7.9). The flow velocity averaged over the whole porous structure in Figure 7.9 is the volume average:

$$\langle u \rangle = \varphi <u>_{solid} + (1-\varphi)<u>_t = 0 + (1-\varphi)<u>_t = (1-\varphi)\frac{d^2}{12\eta}\frac{\Delta P}{L},$$

taking into account that inside the solid phase (with volume fraction φ) the flow speed is zero, comparison with Darcy's law (7.21) shows that for a microstructure of parallel platelets the permeability is:

$$k = \frac{1-\varphi}{12}d^2.$$

(b) We rewrite this result in terms of the plate thickness D as follows. For N plates in a cubic volume of side L the solid volume fraction is $\varphi = NL^2D/L^3 = ND/L$. The pore volume fraction is $(1-\varphi) = NL^2d/L^3 = Nd/L$ which entails that:

$$d = \frac{(1-\varphi)}{\varphi}D.$$

The specific surface area of the platelets is

$$A_s = \frac{\text{Area}}{\text{Volume}} \approx \frac{2}{D},$$

as explained in Figure 7.11. Hence, we arrive at

$$k = \frac{1}{3}\frac{(1-\varphi)^3}{\varphi^2}A_s^{-2}.$$

This result equals the KC scaling in equation (7.29), with a Kozeny constant $C = 3$. What we see confirmed here is that the scaling $k \propto (1-\varphi)^3(\varphi A_s)^{-2}$ is independent of the internal geometry of the porous medium; this geometry only needs to be specified for the calculation of the KC constant, here $C = 3$ for a structure of parallel plates.

7.5 Average flow speed *in* the pore of Figure 7.10 is $<u>_{\text{pore}} = a^2\Delta P/8\eta L$; hence the permeability of the capillary equals $k = (1-\varphi)a^2/8$, where φ is the solid volume fraction of the capillary in Figure 7.10. The specific surface area of the solid shell of the capillary is:

$$A_s = \frac{\text{solid area}}{\text{solid volume}} = \frac{2\pi aL}{\pi R^2 L - \pi a^2 L} = \left(\frac{2}{a}\right)\frac{(a/R)^2}{1-(a/R)^2}.$$

The pore volume fraction equals:

$$1-\varphi = \frac{\pi a^2 L}{\pi R^2 L} = \left(\frac{a}{R}\right)^2.$$

Thus $A_s = \frac{2}{a}\left(\frac{1-\varphi}{\varphi}\right)^2 \Rightarrow a^2 = \frac{4}{A_s^2}\left(\frac{1-\varphi}{\varphi}\right)^2,$

which leads to

$$k = \frac{1}{2}\frac{(1-\varphi)^3}{\varphi^2}A_s^{-2} \Rightarrow \text{Kozeny constant } C = 2.$$

ANSWERS TO CHAPTER EXERCISES

7.6 $k(\varphi) = \dfrac{\eta}{\rho f_{sa}(\varphi)}$ $[=]\dfrac{\text{Pa s}}{\text{m}^{-3} \times (\text{Pa s} \times \text{m})} = \text{m}^2.$

7.7 For the nano N-sphere: $\mu = \dfrac{2a^2}{9\eta}(\delta_p - \delta) = \dfrac{2 \times (5 \times 10^{-9}\text{m})^2}{9 \times (10^{-3}\text{Pa s})}(1 \times 10^3 \text{ kg m}^{-3})$

$= 5.56 \times 10^{-12} \dfrac{\text{kg m}}{\text{N s}} = 5.56 \times 10^{-12}\text{ s}.$

For the colloidal C-sphere: $\mu = \left(\dfrac{100\text{ nm}}{5\text{ nm}}\right)^2 \times 5.56 \times 10^{-12}\text{ s} = 2.22 \times 10^{-9}\text{ s}.$

For the granular G-sphere: $\mu = \left(\dfrac{10^6\text{ nm}}{5\text{ nm}}\right)^2 \times 5.56 \times 10^{-12}\text{ s} = 0.22\text{ s}.$

7.8 Conservation of (number and volume) of colloids in Figure 7.7 entails:

$x_0\varphi = \varphi_c L(t) + \varphi[x(t) - L(t)]$

$\Rightarrow \dfrac{dx(t)}{dt} = \dfrac{dL(t)}{dt}\dfrac{\varphi - \varphi_c}{\varphi}$

Darcy's law for the average flow rate in Figure 7.7: $\langle u \rangle = -\dfrac{k\Delta P}{\eta L(t)}.$

Since $\dfrac{dx(t)}{dt} = \langle u \rangle$:

$-\dfrac{k\Delta P}{\eta L(t)} = \dfrac{dL(t)}{dt}\dfrac{\varphi - \varphi_c}{\varphi} \Rightarrow L^2(t) = \dfrac{k}{\eta}\dfrac{2\varphi}{\varphi_c - \varphi}\Delta P t$

7.9 $(\Delta m)\omega^2 r = 10 \times \Delta mg \Rightarrow \omega = \sqrt{\dfrac{10g}{r}} = \dfrac{100}{2\pi}\text{s}^{-1} \times 60\text{ s min}^{-1} = 955\text{ rpm}.$

7.10 Force F equals by definition minus the gradient in potential energy $U(r)$:

$F = -\dfrac{dU(r)}{dr}.$ Hence,

$\int\limits_0^{U(r)} dU(r') = -\int\limits_a^r F dr' \Rightarrow U(r) = -(\Delta m)\omega^2 \int\limits_a^r r dr' = (\Delta m)\omega^2 \dfrac{1}{2}(a^2 - r^2).$

7.11 The specific surface area of a sphere is $A_s = 3/R$. Hence for with radius $R = 500$ nm the liquid permeability of the sphere packing equals:

$k = \dfrac{1}{C}\dfrac{(1-\varphi)^3}{\varphi^2}\dfrac{R^2}{9} \stackrel{C=5,\varphi=0.64}{=} 2.53 \times 10^{-2} R^2 \stackrel{R=500\text{nm}}{=} 6.32 \times 10^{-16}\text{ m}^2$

The required pressure drop is therefore:

$\Delta P = \dfrac{\langle u \rangle \eta L}{k} = \dfrac{(2.778 \times 10^{-6}\text{ m s}^{-1})(10^{-3}\text{Pa s})(10^{-2}\text{m})}{6.32 \times 10^{-16}\text{ m}^2} \approx 0.4\text{ bar}.$

For radii $R = 50$ and 5 nm the needed pressures are, respectively, $\Delta P \approx 40$ and $4{,}000$ bar. The enormous pressure of $4{,}000$ bar clearly illustrates the infeasibility of separating (stable, non-aggregated) nanoparticles from solution by filtration.

7.12 (a) The specific surface area of a thin rod with length $L \gg D$ is
$$A_s = \frac{O}{V} = \frac{\pi DL}{(\pi/4)D^2 L} = \frac{4}{D}, \text{ for } L \gg D \Rightarrow k = \frac{1}{C}\frac{(1-\varphi)^3}{\varphi^2}A_s^{-2} = \frac{1}{C}\frac{(1-\varphi)^3}{\varphi^2}\frac{D^2}{16}.$$

(b) If rod length L and diameter D are both doubled the aspect ratio L/D remains the same and the rod volume fraction φ will not change since $(L/D)\varphi = \text{constant} \approx 5$. Doubling the rod diameter will increase k with a factor of four.

(c) $\varphi \approx \dfrac{5}{(L/D)=20} = 0.25 \Rightarrow k \approx \dfrac{1}{5} \times \dfrac{(0.75)^3}{(0.25)^2} \times \dfrac{25\,\text{nm}^2}{16} \approx 2\,\text{nm}^2.$

7.13 (a) $k_0 = \dfrac{\eta}{\rho f_0} = \dfrac{\eta}{(3\varphi/4\pi a^3)(6\pi \eta a)} = \dfrac{2a^2}{9\varphi}.$

(b) If for given φ the sphere radius drops, the sphere area σ (per volume of array) increases and hence the hydrodynamic friction per volume increases and the permeability diminishes. Nb:
$$\sigma = \rho 4\pi a^2 = \frac{\varphi}{(4/3)\pi a^3}4\pi a^2 = \frac{3\varphi}{a}, \text{ so indeed } \sigma \text{ goes up if } a \text{ goes down.}$$

7.14 (a) $D = \dfrac{k_B T}{6\pi \eta a} = 2.14 \times 10^{-11}\,\text{m}^2\text{s}^{-1}.$

(b) $\mu = \dfrac{2a^2}{9\eta}(\delta_p - \delta) = 3.4 \times 10^{-12}\,\text{s}.$

(c) $\sqrt{\langle x^2 \rangle} = \sqrt{2Dt} = 5.1 \times 10^{-5}\,\text{m}.$

(d) Sedimentation speed: $v = \mu g = 3.4 \times 10^{-12}\,\text{s} \times 9.8\,\text{m s}^{-2} = 3.3 \times 10^{-11}\,\text{m s}^{-1}$. Distance sedimented in 1 minute is 2.0×10^{-9} m, a distance that is totally swamped by Brownian motion of the protein.

(e) Sedimentation rate in a centrifuge: $v = \mu \omega^2 r$;
$\omega = (20{,}000\,\text{rpm}/60\,\text{s min}^{-1}) \times 2\pi = 2094\,\text{rad s}^{-1}$
$\Rightarrow v = 3.4 \times 10^{-12}\,\text{s} \times (2094\,\text{rad s}^{-1})^2 \times 0.1\,\text{m} = 1.5 \times 10^{-6}\,\text{m s}^{-1}$
$\Rightarrow$ distance in 1 minute is 8.9×10^{-5} m.

Now displacements by diffusion and sedimentation are of comparable magnitude.

7.15 (a) The RBC is modelled in Exercise 4.3 by a disc with diameter $d = 8\,\mu\text{m}$ and height $h = 2\,\mu\text{m}$. Put the volume V_{eq} of the equivalent sphere with radius R_{eq} equal to the volume of an RBC:
$$\frac{4\pi}{3}R_{eq}^3 = \frac{\pi}{4}d^2 h \Rightarrow R_{eq} = \left(\frac{3}{16}d^2 h\right)^{1/3} = \left(\frac{3}{16}(8\,\mu\text{m})^2 \times 2\,\mu\text{m}\right)^{1/3} \approx 2.9\,\mu\text{m}.$$

(b) Settling speed of the equivalent sphere:

$$v = \frac{2R_{eq}^2}{9\eta}(\delta_p - \delta)g = \frac{2\times(2.9\times10^{-6}\,\text{m})^2}{9\times(10^{-3}\,\text{Pa s})}(0.1\times10^3\,\text{kg m}^{-3})(9.8\,\text{m s}^{-2})$$

$$= 1.8\times10^{-6}\,\frac{\text{kg s}^{-1}}{\text{N m}^{-2}} = 1.8\times10^{-6}\,\text{m s}^{-1}$$

(c) Distance $= v \times t = (1.8\times10^{-6}\,\text{m s}^{-1}) \times (24\,\text{h} \times 60\,\text{min h}^{-1} \times 60\,\text{s min}^{-1}) = 0.16\,\text{m}$.

7.16 (a)

Table 7.1 Sedimentation and diffusion data in water[2]

Substance	T(°C)	$\mu/10^{-13}$ s	$D/10^{-7}$ cm²s⁻¹	$\bar{V}_p$/cm³g⁻¹	M
Proteins					
Bovine serum albumin	25	5.01	6.97	0.734	66,500
Fibrinogen	20	7.9	2.02	0.706	330,000
Lysozyme	20	1.91	11.2	0.703	14,400
Viruses					
Bushy stunt virus	20	132	1.15	0.74	10,700,000
Tobacco mosaic virus	20	198	0.46	0.74	40,590,000

(b) Equal temperatures ensure that μ and D are measured at the same viscosity since η significantly depends on temperature.

7.17 For a number density $\rho_c(x)$ of colloids settling at stationary speed $v = (\Delta m)g / f$ the downward sedimentation flux is $j_{sed} = \rho_c(x)v$; the upwards diffusion flux equals according to Fick's first law $j_{dif} = (k_B T / f)d\rho_c(x)/dx$. From $j_{sed} + j_{dif} = 0$ a differential equation in $\rho_c(x)$ follows that, for the boundary condition $\rho_c(x \to 0) = \rho_0$, has the barometric profile (7.9) as its solution.

7.18 On working out the integral in (7.10) for the barometric profile (7.9) we recover Van 't Hoff's law for uncharged, ideal colloids, as could be expected because the barometric profile is derived on the basis of Van 't Hoff's law; see text after equation (7.8).

7.19 (a) $\varphi_c \dfrac{L}{D} \approx 5 \Rightarrow \varphi_c \approx \dfrac{5}{20} = 0.25$.

(b) The total sphere volume V follows from $V = \varphi_s V_s$, where $V_s = Oh_s$ is the volume of the sphere packing in a column with cross-sectional area O. Hence $V_s = 0.64 \times (1\,\text{m}) \times (O\,\text{m}^2) = 0.64\,O\,\text{m}^3$. The volume fraction of the random cylinder packing is:

$$\varphi_c = \frac{0.64\,O\,\text{m}^3}{h_c(O\,\text{m}^2)} \approx 0.25 \Rightarrow h_c \approx 2.6\,\text{m}.$$ Note the substantial increase by a factor of 2.6 in random packing volume when, for a given total solid volume, spheres are replaced by thin rods.

---

[2] Table source: H. Sober (ed.) (1968), *The Handbook of Biochemistry*. Chemical Rubber Co., Cleveland, Ohio.

7.20 The probability $P(x)$ to find a particle at height x is proportional to the number density $\rho_c(x)$. Hence, the average altitude $<x>$ of particles in the barometric distribution is:

$$\langle x \rangle = \frac{\int_0^\infty P(x)x\,dx}{\int_0^\infty P(x)\,dx} = \frac{\int_0^\infty \rho_c(x)x\,dx}{\int_0^\infty \rho_c(x)\,dx} = \frac{\int_0^\infty e^{-x/l_g} x\,dx}{\int_0^\infty e^{-x/l_g}\,dx} = l_g.$$

So, the average altitude of the particles equals their gravitational length.

Chapter 8

8.1 $\mathrm{Re} \sim \dfrac{\delta U L}{\eta}[=]\dfrac{\mathrm{kg\,m^{-3} \times m\,s^{-1} \times m}}{\mathrm{Pa\,s}} = \dfrac{\mathrm{kg\,m}}{\mathrm{N s^2}} = \dfrac{N}{N} = 1.$

8.2 Taking $L = R$, and for U the Stokes sedimentation velocity $u_{sed} = 2R^2(\delta_p - \delta)g/9\eta$, we have for the Reynolds number: $\mathrm{Re} = \dfrac{\delta u_{sed} L}{\eta}$
$= \dfrac{2(\delta_p - \delta)\delta g}{9\eta^2} R^3 = (2.75 \times 10^{12}\,\mathrm{m^{-3}})(R\,\mathrm{in\,m})^3.$ Here δ_p and δ are the mass density of, respectively, particles and water; $\eta = 0.89\,cP$. For the particles from Table 3.1: N: $\mathrm{Re} = 5 \times 10^{-13}$; C: $\mathrm{Re} = 3 \times 10^{-9}$; G: $\mathrm{Re} = 3 \times 10^3$.

8.3 (a) $\mathrm{Re} \approx 6 \times 10^5$.

(b) By far water displacement; viscous drag is unimportant because you swim at high Re. Your energy input is proportional to U^2 and not to U as would be the case for viscous flow.

8.4 Re will decrease by a factor $D/d = (2D/3L)^{1/3} = 0.28$, for $L/D = 30$, assuming that D^2 is the frontal area facing the flow. One could argue that for flow perpendicular to the rod, Re actually increases with a factor L/D.

8.5 (a) $\dfrac{d^2 u_x}{dy^2} = 0 \Rightarrow u_x = C_1 y + C_2$; employ the boundary conditions
$u(y = D) = u(D)$ and $u_x(y = 0) = 0$: $u_x(y) = u(D)\dfrac{y}{D} \Rightarrow \langle u \rangle = \dfrac{1}{D}\int_0^D u_x\,dy$
$= \dfrac{1}{2}u(D).$

(b) Apply Newton's viscosity law: $\dfrac{F}{\mathrm{area}} = \eta \dfrac{du_x}{dy} = \eta \dfrac{u(D)}{D} = 10^{-3}\,\mathrm{Pa}.$

8.6 (a) $\dfrac{d}{dy}\dfrac{du_x}{dy} = \dfrac{1}{\eta}\dfrac{dp}{dx} \Rightarrow \dfrac{du_x}{dy} = \dfrac{1}{\eta}\dfrac{dp}{dx} y + A; A = 0$ because $\dfrac{du_x}{dy} = 0$ at $y = 0$.

$\Rightarrow u_x = \dfrac{1}{2\eta}\dfrac{dp}{dx}y^2 + B; B = -\dfrac{1}{8\eta}\dfrac{dp}{dx}d^2$ because $u_x(y = d/2) = 0$

(stick boundary) $\Rightarrow u_x = \dfrac{1}{2\eta}\dfrac{dp}{dx}\left(y^2 - \dfrac{d^2}{4}\right).$

(b) Stress has the unit of pressure: Pa.

(c) $\sigma_{xy} = -\eta \dfrac{du_x}{dy} = -\eta\left(\dfrac{1}{\eta}\dfrac{dp}{dx}y\right) = -\dfrac{dp}{dx}y = +\dfrac{\Delta P}{L}y.$

(d) Maximal stress: $\sigma_{xy}(y=\tfrac{1}{2}d) = \dfrac{\Delta P}{2L}d.$

(e) Total viscous force on the two plate surfaces is $\sigma_{xy}(y=\tfrac{1}{2}d)\times 2L^2$
$= \dfrac{\Delta P}{2L}d\times 2L^2 = \Delta P dL$. The total viscous force equals the net external force $\Delta P dL$ that drives the flow (net pressure difference multiplied by area dL).

8.7 $Q = dL\langle u\rangle = dL\dfrac{d^2}{12\eta}\dfrac{\Delta P}{L} = \dfrac{d^3}{12\eta}\Delta P.$

8.8 In the case of a pure-slip boundary condition flowing fluid does not experience any friction from the surface such that no fluid profile develops; the spatially constant fluid speed does not reach a steady state as there is no viscous force to counteract the driving pressure gradient.

8.9 Volume flow rate scales as $Q \propto R^4\Delta P \Rightarrow \dfrac{\Delta P_2}{\Delta P_1} = \left(\dfrac{R_2=5}{R_1=4}\right)^4$; pressure increase is a factor of 2.44.

8.10 EDLs surrounding charged colloids enhance the dispersion viscosity in comparison to that of uncharged particles. Addition of salt shrinks the double layers and, hence, lowers the viscosity towards its value for uncharged colloids.

8.11 The viscosity will drop substantially: fibres aligning to a streamline dissipate less energy than randomly oriented ones that tumble in the flow. An alternative argumentation is that aligned fibres maximally pack at densities (far) above the random rod packing density: thus the gap between a given fibre concentration and the maximal packing density widens so viscosity decreases—in agreement with the KD relation (8.47).

8.12 (a) $\langle u\rangle = \dfrac{2}{dL}\int_0^{d/2}\int_0^L u_x\,dx\,dy \overset{\text{Subst.(8.23)}}{=} \dfrac{1}{d\eta L}\int_0^L\dfrac{dp}{dx}dx \times \int_0^{d/2}(y^2 - \dfrac{d^2}{4})dy$

$= \dfrac{-1}{d\eta L}\int_0^{\Delta P} dp \times \left(\dfrac{-d^3}{12}\right) = \dfrac{d^2}{12\eta}\dfrac{\Delta P}{L}.$

(b) $\langle u\rangle = \dfrac{1}{\pi R^2 L}\int_0^L\int_0^R u_z\, 2\pi r\,dr\,dz \overset{\text{Subst.(8.30)}}{=} \dfrac{-\Delta P}{2\eta R^2 L}\times \int_0^R (r^2 - R^2)r\,dr$

$= \dfrac{-\Delta P}{2\eta R^2 L}\times \dfrac{-R^4}{4} = \dfrac{R^2}{8\eta}\dfrac{\Delta P}{L}.$

8.13 The viscosity of the mixture will be lower than the viscosity of the separate dispersions because the random-packing density (RPD) of the mixture is higher than the RPD of monodisperse spheres. In other words, the gap between volume fraction $\varphi = 0.5$ and the RPD widens which lowers the viscosity (see also Figure 8.10).

8.14 $\eta_r = 1 + 2.5\varphi = 1 + 2.5\,c\bar{v} = 1 + 2.5 \times \dfrac{10\,g}{10^3\,cm^3} \times \dfrac{1\,cm^3}{g} = 1.025$.

8.15 Taylor expansion of the KD equation for low colloid concentration $\varphi \ll \varphi_M$ yields:

$$\eta_r = \left(1 - \dfrac{\varphi}{\varphi_M}\right)^{-[\eta]\varphi_M} = 1 + [\eta]\varphi_M \dfrac{\varphi}{\varphi_M} = 1 + [\eta]\varphi, \text{ for } \varphi \ll \varphi_M, \text{ which}$$

for spheres ($[\eta] = 2.5$) equals Einstein's viscosity equation (8.46).

8.16 It does: viscosities of dilute sphere dispersions only depend on the *total* volume fraction $\varphi = (4\pi/3)\sum_{j=1}^{t} \rho_j R_j^3$, where ρ_j is the number density of spheres with radius R_j.

8.17 $\displaystyle\int_{\sigma_0}^{\sigma'} \dfrac{d\sigma}{\sigma} = -\dfrac{G}{\eta}\int_0^t dt \Rightarrow \sigma = \sigma_0 \exp[-Gt/\eta] + (C=0) = \sigma_0 \exp[-t/t_R]; t_R = \eta/G$.

8.18 (a) $\sigma_{zr} = -\eta\dfrac{du_z}{dr} = -\dfrac{1}{2}\dfrac{dp}{dz}r = \dfrac{1}{2}\dfrac{\Delta P}{L}r$. The stress is zero at $r = 0$ and increases linearly with r to its maximum value at $r = R$ (Figure 8.7).

(b) Since the stress is a viscous force per unit area, the total viscous force exerted on the tube wall is: $F_{vis} = 2\pi R L \times \sigma_{zr}|_{r=R} = \pi R^2 \Delta P$. So the flow in the tube is driven by a net external force $F_{ext} = \pi R^2 \Delta P$, exerted on the inlet of the tube, which for a steady flow is balanced by the total viscous force on the tube's inner surface.

8.19 From equation (8.38) it follows that the capillary radius R_B for which $Pe \approx 1$ is about $R_B \approx 1.5$ and $R_B \approx 4$ microns for, respectively, the colloidal C-sphere and the nano-N sphere from Table 3.1.

Chapter 9

9.1 (a) For the magnetite particle with $d = 1\,nm$: $\lambda = \left(\dfrac{\mu_0 m_{sat}^2}{24 k_B T}\right)\dfrac{\pi d^3}{6}$

$= \dfrac{(4\pi \times 10^{-7}\,JA^{-2}m^{-1})(430 \times 10^3\,Am^{-1})^2}{24 \times 1.38 \times 10^{-23}\,JK^{-1} \times 298\,K} \times \dfrac{\pi (10^{-9}\,m)^3}{6} = 1.23 \times 10^{-3}$.

For $d = 10\,nm$ and $30\,nm$: $\lambda = 1.23$ and 33.21.

(b) $\dfrac{U_{max}}{k_B T} = -1$, for $\lambda = \dfrac{1}{2}$. So $\left(\dfrac{\mu_0 m_{sat}^2}{24 k_B T}\right)\dfrac{\pi d^3}{6} = \dfrac{1}{2}$

$\Rightarrow d = \left(\dfrac{72 k_B T}{\pi \mu_0 m_{sat}^2}\right)^{1/3} = 7.4\,nm$.

9.2 The distance of closest approach due to the oleic acid layers increases from d to $d + 2\Delta$; hence instead of λ from (9.19) we have

$\lambda = \dfrac{\mu_0 m_d^2}{4\pi k_B T(d+2\Delta)^3} = \dfrac{\mu_0 m_{sat}^2}{24 k_B T}\dfrac{V_p}{(1+2\Delta/d)^3}; m_d = m_{sat} V_p = m_{sat}\pi d^3/6$.

So λ in (9.19) is diminished by a factor of $(1+2\Delta/d)^3$.

9.3 (a) $\tau_{RR} = \dfrac{3(\pi/6)(10^{-8}\text{m})^3(2\times 10^{-3}\text{kg.m}^{-1}.\text{s}^{-1})}{(1.38\times 10^{-23}\text{ JK}^{-1})(298\text{K})} = 7.6\times 10^{-7}\text{ s}.$

(b) $\tau_{NR} = \dfrac{1}{f_0}\exp\left[KV_p/k_BT\right] = 7.6\times 10^{-7}\text{ s}$

$\Rightarrow KV_p/k_BT \approx \ln(10^9\text{s}\times 7.6\times 10^{-7}\text{s}) = 6.63$

(c) $K \approx 6.63\times\dfrac{(1.38\times 10^{-23}\text{ JK}^{-1})(298\text{K})}{(\pi/6)(10^{-8}\text{m})^3} \approx 52\text{ kJ m}^{-3}.$

9.4 $M = \rho m_d\Lambda(\alpha) = \rho m_d\dfrac{\alpha}{3},$ for $\alpha\ll 1$; $\alpha = \dfrac{\mu_0 m_d H}{k_B T} \Rightarrow \chi_i = \dfrac{M}{H} = \rho\dfrac{\mu_0 m_d^2}{3k_B T}.$

9.5 Distribution function $P(\theta)$ is properly normalized when $\int_0^\pi P(\theta)d\theta = 1$. Hence:

$2\pi C\int_0^\pi e^{\alpha\cos\theta}\sin\theta d\theta \overset{x=\cos\theta}{=} -2\pi C\int_1^{-1} e^{\alpha x}dx = 1 \Rightarrow C = \alpha/2\pi(e^\alpha - e^{-\alpha})$

$= \alpha/4\pi\sinh(\alpha).$

9.6 $<\cos\theta> = 2\pi C\int_0^\pi \cos\theta\, e^{\alpha\cos\theta}\sin\theta\, d\theta \overset{x=\cos\theta}{=} -2\pi C\int_1^{-1} x\, e^{\alpha x}dx$

$= -\dfrac{2\pi C}{\alpha}[x\, e^{\alpha x} - \alpha^{-1}e^{\alpha x}]_1^{-1}$

$= \dfrac{1}{\sinh(\alpha)}[\cosh(\alpha) - \alpha^{-1}\sinh(\alpha)] = \dfrac{\cosh(\alpha)}{\sinh(\alpha)} - \dfrac{1}{\alpha} = \Lambda(\alpha).$

9.7 $\Lambda(\alpha) = \dfrac{e^\alpha + e^{-\alpha}}{e^\alpha - e^{-\alpha}} - \dfrac{1}{\alpha}$; employ the Taylor expansion $e^x = 1 + x + (1/2)x^2 + (1/6)x^3...$ to find for small α:

$\Lambda(\alpha) = \dfrac{2+\alpha^2...}{2\alpha+(1/3)\alpha^3...} - \dfrac{1}{\alpha} = \dfrac{(2/3)\alpha^3...}{2\alpha^2+(1/3)\alpha^4...} = \dfrac{\alpha}{3},$ for $\alpha\ll 1.$

Note that $e^{\pm\alpha}$ has to be expanded up to terms of order α^3 to find the correct limit $\alpha/3$.

9.8 This problem uses equation (9.17) to calculate values for the pair potential U.

(a) Scaled to the pre-factor λ, the answers are (in sequence): $-2, +2, 0, 1, -1$. The first, so-called head-to-tail, configuration is the most stable.

(b) Here, we have to summate over all six bonds in the cluster. Both the angles and the different contact distances have to be taken into account. The answers are (in sequence):

$U = \lambda\left[4\left(+\dfrac{1}{2}+1\right)+2\dfrac{1}{(\sqrt{2})^3}(2)\right] = \lambda\left[2+\sqrt{2}\right] = \lambda\cdot[7.414]$

$U = \lambda\left[4\cdot(0)+2\dfrac{1}{(\sqrt{2})^3}(-\dfrac{1}{2}+2\dfrac{1}{2})\right] = \lambda\left[\dfrac{1}{(\sqrt{2})^3}\right] = \lambda\cdot[0.3536]$

$U = \lambda\left[2\cdot(-2)+2\cdot(+1)+2\dfrac{1}{(\sqrt{2})^3}(\dfrac{1}{2}-2\dfrac{1}{2})\right] = \lambda\cdot[-2.3536]$

$$U = \lambda\left[2\cdot(-2)+3\cdot(\frac{3}{4}-2\frac{1}{4})+\frac{1}{(\sqrt{3})^3}(\frac{1}{4}-2\frac{3}{4})\right] = \lambda\cdot[-3.4906]$$

$$U = \lambda[1\cdot(-2)+4\cdot(\frac{3}{4}-2\frac{1}{4})+\frac{1}{(\sqrt{3})^3}(+1)] = \lambda\cdot[-0.8075]$$

The fourth configuration is the most stable, lowest in energy.

9.9 (a) $m_d = (\frac{\pi}{6}\times 10^{-24}\,m^3)(430\times 10^3\,A\,m^{-1}) = 2.25\times 10^{-19}\,Am^2$

(b) The answers are 0.054, 0.54, 5.4, and 54. For 1 tesla the calculation reads:

$$\frac{mH}{k_BT} = (2.25\cdot 10^{-19}\,Am^2)(1T)\frac{JA^{-1}m^{-2}}{T}\frac{1}{(1.38\cdot 10^{-23}\,JK^{-1})(300\,K)} = 54.$$

(c) For the horizontal axis, 1 T corresponds roughly to 800 kA/m ($10^7/4\pi$). The vertical axis gives the Langevin factor (equation 9.8), which for the points in (b) equals 0.018, 0.18, 0.81, and 0.98.

(d) The volume fraction is $(100/430) \times 100\% = 23\%$.

(e) From $M = \chi H = M_{sat}\frac{\alpha}{3} = M_{sat}\frac{\mu_0 mH}{3k_BT}$ it follows that $\chi = M_{sat}\frac{\mu_0 m}{3k_BT}$, so that:

$$\chi = (100\cdot 10^3\,A/m)\frac{(4\pi\cdot 10^{-7}\,JA^{-2}m^{-1})(2.25\cdot 10^{-19}\,Am^2)}{3(1.38\cdot 10^{-23}\,JK^{-1})(300\,K)} = 2.3.$$

9.10 (a) $\dfrac{k_BT}{\mu_0 m_s} = \dfrac{(300\,K)(1.38\times 10^{-23}\,JK^{-1})}{(4\pi\times 10^{-7}\,JA^{-2}m^{-1})(2.252\times 10^{-19}\,Am^2)} = 14.63\,kA\,m^{-1}.$

(b) $k_BT/\mu_0 m_d$ has the unit of field strength, and equals the external field $H(x)$ minimally needed to produce a significant SD-particle concentration profile. For $H(x) \ll k_BT/\mu_0 m_d$ particles remain homogeneously distributed, whereas field strengths $H(x) \gg k_BT/\mu_0 m_d$ induce steep concentration profiles.

9.11 (a) The second virial coefficient is defined as

$$B_2 = 2\pi\int_0^\infty (1-\beta)r^2 dr;\quad \beta = \exp[-w(r)/k_BT].$$

The interaction potential to be substituted is the DHS potential:

$$\frac{w(r)}{k_BT} = \begin{cases} = \infty & 0 \leq r < d+2\Delta \\ = -\frac{\lambda^2}{3}\left(\frac{d}{r}\right)^6 & r \geq d+2\Delta \text{ and } \lambda < 1. \end{cases}$$

First, we consider the second virial coefficient for a hard sphere of radius $d+2\Delta$:

$$B_2^{HS} = 2\pi\int_0^{d+2\Delta} (1-0)r^2 dr = \frac{2\pi}{3}(d+2\Delta)^3.$$

The magnetic attraction is weak such that we can linearize the exponent, to obtain for the magnetic part of the second virial coefficient:

$$B_2^{MAG} = 2\pi \int_{d+2\Delta}^{\infty} (1-\exp[-w(r)/k_B T])r^2 dr \approx 2\pi \int_{d+2\Delta}^{\infty} [w(r)/k_B T]r^2 dr$$

$$= -\frac{2\pi\lambda^2}{3} \int_{d+2\Delta}^{\infty} \left(\frac{d}{r}\right)^6 r^2 dr = -\frac{2\pi\lambda^2}{9}\frac{d^6}{(d+2\Delta)^3}$$

So, for the second virial coefficient of the dipolar hard-sphere, scaled on the hard-sphere coefficient, we obtain:

$$\frac{B_2}{B_2^{HS}} = 1 + \frac{B_2^{MAG}}{B_2^{HS}} = 1 - \frac{\lambda^2}{3}\left(\frac{d}{d+2\Delta}\right)^6, \text{ for } \lambda < 1.$$

(b) We see, as could be expected, that the magnetic attraction, that gives rise to the λ^2-term, diminishes the magnitude of DHS second virial coefficient; the implication is that the particle magnetic moment lessens the osmotic pressure of a DHS dispersion in comparison to a dispersion of non-magnetic spheres.

Chapter 10

10.1 (a) Number of oil droplets $= \dfrac{400\times 10^{-6}\,m^3}{(\pi/6)(5\times 10^{-6})^3} = 6.1\times 10^{12}$.

Total oil area: $(6.1\times 10^{12})\pi(5\times 10^{-6}\,m)^2 = 479\,m^2$.

(b) $V_{cylinder} = \pi R^3 \Rightarrow A_{cylinder} = 4\pi R^2 = 4\pi(V_{cylinder}\pi^{-1})^{2/3}$
$= 4\pi(5\times 10^{-4}\,m^2\pi^{-1})^{2/3} = 0.037\,m^2$.

Hence the oil area is larger by a factor of $479/0.037 = 12{,}946$.

10.2 For $a = 3\,nm$: $\dfrac{\Delta G}{k_B T} = \dfrac{\pi a^2 \gamma_{ow}(1-\cos\theta)^2}{k_B T}$

$= \dfrac{\pi(3\times 10^{-9}\,m)^2(0.03\,Nm^{-1})(1-\cos 60°)^2}{(298\,K)(1.38\times 10^{-23}\,JK^{-1})} = 52.$

For $a = 30\,nm$: $\Delta G = 5{,}200\,k_B T$.

10.3 See the answer to Exercise **2.6(a)**.

10.4 The plot will look like Figure 10.6, with a minimum of $\Delta G_A = -687.5\,k_B T$ at $90°$.

10.5 (a) The creaming rate is the Stokes velocity:

$v = \dfrac{2}{9}\dfrac{R^2}{\eta}(\delta_p - \delta)g = \dfrac{2}{9}\dfrac{(1\times 10^{-6}\,m)^2}{10^{-3}\,Pa\,s}(-0.2\times 10^3\,kg\,m^{-3})(9.8\,m\,s^{-2})$

$\approx -4.4\times 10^{-7}\,m\,s^{-1}.$

For $R = 10\,\mu m$: $v = -4.4\times 10^{-5}\,m\,s^{-1}$.

(b) Time for 1 μm oil droplet to rise from bottom to top:
$$\frac{10^{-1}\text{m}}{4.4\times 10^{-7}\text{ m s}^{-1}}\times 3600\text{ hr s}^{-1} \approx 63\text{ hr}.$$ For the 10 μm droplets it takes about 0.6 hr for creaming to complete.

10.6 Area reduction: $4\pi(a-da)^2 - 4\pi a^2 = -8\pi a\,da + 4\pi(da)^2$
$= -8\pi a\,da$, for $da \ll a$
$\Rightarrow dG_{surf} = \gamma \times$ area change $= -\gamma 8\pi a\,da$; $dG_{surf} < 0$: droplet contraction is spontaneous.

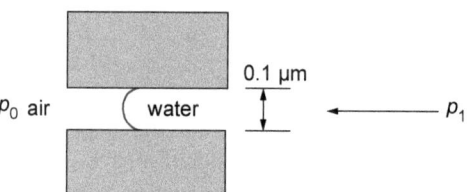

10.7 For a spherical water meniscus in a pore, the Laplace equation predicts a pressure difference Δp across the curved interface with magnitude
$$\Delta p = \frac{2\times 72\times 10^{-3}\text{Nm}^{-1}}{(1/2)(0.1\times 10^{-6}\text{m})} = 2.88\times 10^6\text{Nm}^{-2} = 28.4\text{ atm}.$$
Thus water pressure p_1 minus air pressure p_0 must exceed 28.4 atmospheres to squeeze water out of a pore.

10.8 (a) Equation (10.19) for $n=2$ yields the quadratic
$$2K(c_{mo})^2 + c_{mo} - c_{tot} = 0,\text{ with the solution: } c_{mo} = \frac{[1+8Kc_{tot}]^{1/2}-1}{4K}.$$
(b) See Figure 10.8: the square-root growth of c_{mo} beyond the CMC levels to an almost-constant plateau upon increase of the number of soap molecules in a micelle to $n \gg 1$.

10.9 (a) $\dfrac{\pi}{k_B T} = c_{mo} + c_{mi} = c_{mo} + \dfrac{c_{tot}-c_{mo}}{n} = c_{mo}\left[1-n^{-1}\right] + c_{tot}n^{-1}$;
$$c_{mo} = \frac{[1+8Kc_{tot}]^{1/2}-1}{4K}.$$
(b) In the limit $n \gg 1$: $\dfrac{\pi}{k_B T} = c_{mo}$. Then the osmotic pressure is determined only by the free soap molecules.
(c) When the equilibrium constant K is zero, no micelles are formed, so there are only monomeric soap molecules present, with number density c_{tot}, exerting an osmotic pressure:
$$\frac{\pi}{k_B T} = c_{tot}.$$

For infinite K, all monomers have been incorporated in micelles; hence the solution only contains micelles with concentration $c_{tot}n^{-1}$; then the solution's osmotic pressure is

$$\frac{\pi}{k_B T} = c_{tot}n^{-1},$$

which is a factor of n lower than the pressure exerted by a solution of monomeric soap molecules.

Chapter 11

11.1 The unit of Rayleigh ratio R is m^{-1}.

11.2 $K = \dfrac{2\pi^2 n^2}{N_{AV}\lambda_0^4}\left(\dfrac{dn}{dc}\right)^2 [=] \dfrac{1}{\text{mol}^{-1}\times\text{m}^4\times(\text{kg m}^{-3})^2} = \text{mol m}^2\text{kg}^{-2}$

$\Rightarrow KcM [=] \text{mol m}^2\text{kg}^{-2}\times\text{kg m}^{-3}\times\text{kg mol}^{-1} = \text{m}^{-1}$, which equals the unit of the Rayleigh ratio.

11.3 The volume-average refractive index n of the solution is the sum of particle index n_p and solvent index n_0 weighted by volume fractions of, respectively, particles and solvent:

$$n = n_0\varphi_0 + n_p\varphi_p = n_0 + \varphi_p(n_p - n_0); \quad \varphi_0 + \varphi_p = 1.$$

For the particle volume fraction we can write: $\varphi_p = c\overline{V}_p$, where $\overline{V}_p$ is the specific particle volume, that is, the volume per particle mass. Hence:

$$\frac{dn}{dc} = \frac{n - n_0}{c} = \frac{\varphi_p(n_p - n_0)}{c} = \overline{V}_p(n_p - n_0).$$

Thus a measured value of dn/dc yields, for particles of known specific volume, the particle refractive index n_p. Note that for massive particles the specific volume is simply the reciprocal of particle mass density: $\overline{V}_p = 1/\delta_p$. For polymers or polymer colloids that may swell or shrink in, respectively, a good or poor solvent, this reciprocal relation is invalid.

11.4 Rayleigh scattering intensity of the two droplets is proportional to $2V^2$; that of the single droplet to $(2V)^2$. Hence droplet coalescence *doubles* the total scattering intensity.

11.5 $P(Q) = \left[\dfrac{3(\sin Qa - Qa\cos Qa)}{(Qa)^3}\right]^2 = 0$ for $\dfrac{\sin Q_m a}{\cos Q_m a} = Q_m a \Rightarrow \tan Q_m a = Q_m a.$

The solutions of $\tan Q_m a = Q_m a$ are marked by the dots in the figure below. The first non-zero positive solution is $Q_m a = 4.5$, which is the location of the first minimum in the sphere-form factor in Figure 11.3.

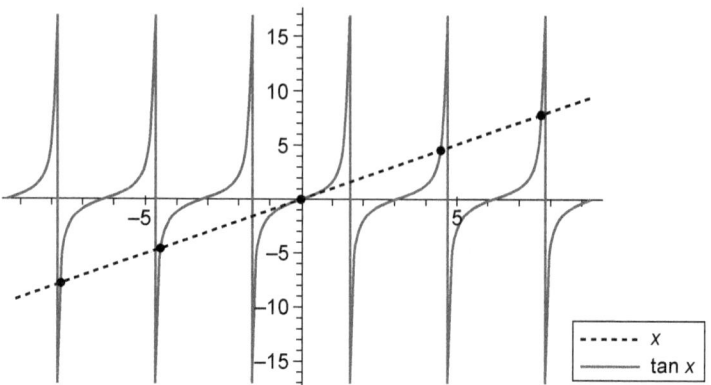

11.6 $R_g^2 = \langle r^2 \rangle = \dfrac{\int_0^a r^2 4\pi r^2 dr}{\int_0^a 4\pi r^2 dr} = \dfrac{(1/5)a^5}{(1/3)a^3} = \dfrac{3}{5}a^2.$

11.7 **(a)** At $\theta = 90°$ scattered light is vertically polarized so we can apply equation (11.2):

$$M = \dfrac{R_V N_{AV}}{Kc}; \quad K = \dfrac{2\pi^2 n^2}{\lambda_0^4}\left(\dfrac{dn}{dc}\right)^2 = \dfrac{2\pi^2 (1.43)^2 (0.170 \times 10^{-3} \text{m}^3\text{kg}^{-1})^2}{(546 \times 10^{-9}\text{m})^4}$$

$$= 6.419 \times 10^{18} \text{m}^2\text{kg}^{-2}$$

$$\Rightarrow M = \dfrac{(3.74 \times 10^{-2}\text{m}^{-1})(6.023 \times 10^{24}\text{ mol}^{-1})}{(6.419 \times 10^{18}\text{m}^2\text{kg}^{-2}) \times (10\text{ kg m}^{-3})} = 3.51 \times 10^3 \text{ kg mol}^{-1}.$$

(b) A weight average:

$$R_V = K\sum_j c_j m_j = K\dfrac{\sum_j c_j m_j}{\sum_j c_j} c = K\langle m \rangle_w c; \quad \langle m \rangle_w = \dfrac{\sum_j \rho_j m_j^2}{\sum_j \rho_j m_j}.$$

Here c is the total weight concentration and $c_j = \rho_j m_j$ is the weight concentration of j-spheres with number density ρ_j and mass m_j.

(c) $\dfrac{dn}{dc} = \bar{V}_p(n_p - n_0) \Rightarrow n_p - n_0 = \delta_p \dfrac{dn}{dc} = 1\text{g cm}^{-3} \times 0.170\text{ g}^{-1}\text{ cm}^3 = 0.170$

$\Rightarrow n_p = n_0 + 0.170 = 1.43 + 0.170 = 1.60.$

(d) $\varphi_p = c\bar{V}_p = \dfrac{c}{\delta_p} = \dfrac{10^{-2}\text{g cm}^{-3}}{1\text{g cm}^{-3}} = 0.01.$

11.8 The wave vector for $\theta = 100°$ equals $Q_m = \dfrac{4\pi(1.36)}{546\text{ nm}}\sin(100°/2) = 0.024\text{ nm}^{-1}.$

Assuming that Q_m is the first minimum of the form factor, the silica sphere radius equals $a = \dfrac{4.5}{0.024\text{ nm}^{-1}} Q_m = 188\text{ nm}.$

11.9 Substitute the Taylor expansions

$\sin x = x - \dfrac{x^3}{3!} + \dfrac{5^3}{5!} \ldots$ for $|x| \ll 1$ and $\cos x = 1 - \dfrac{x^2}{2!} + \dfrac{x^3}{4!} \ldots$ for $|x| \ll 1$

into

$$\frac{1}{3}\sqrt{P(Q)} = \frac{\sin x - x \cos x}{x^3}; \quad x = Qa,$$

to obtain

$$\frac{1}{3}\sqrt{P(Q)} = \frac{1}{3} - \frac{x^2}{30}.$$

Hence

$$P(Q) = \left[1 - \frac{x^2}{10}\right]^2 = 1 - \frac{x^2}{5}, \text{ for } x = Qa \ll 1.$$

11.10 Sunlight is incoherent, meaning that light waves are not in phase with one another. Consequently light scattered from different particles is of random phase so the interference pattern will be smeared out and there are no fluctuating intensity peaks that are needed to make use of DLS.

11.11 (a) $I = \sum_j I_j; \ I_j \propto c_j m_j \Rightarrow I \propto c \dfrac{\sum_j c_j m_j}{\sum_j c_j} = c \dfrac{\sum_j \rho_j a_j^6}{\sum_j \rho_j a_j^3} = c \dfrac{\langle a^6 \rangle}{\langle a^3 \rangle} \equiv c a_{SLS,R}^3$

$$\Rightarrow a_{SLS,R} = \left[\frac{\langle a^6 \rangle}{\langle a^3 \rangle}\right]^{1/3}.$$

(b) $a_{DLS} = \dfrac{\langle a^6 \rangle}{\langle a^5 \rangle} = \dfrac{20^6 + 40^6}{20^5 + 40^5} = 20\left(\dfrac{1+2^6}{1+2^5}\right) = 39.5 \text{ nm}$

$a_{SLS} = \left(\dfrac{\langle a^8 \rangle}{\langle a^6 \rangle}\right)^{1/2} = 20\left(\dfrac{1+2^8}{1+2^6}\right)^{1/2} = 39.8 \text{ nm}.$

11.12 The electrical force QE on the moving colloid is counteracted by a viscous force that is *smaller* than the viscous force jointly experienced by the counter-ions (on which the electrical force $-QE$ is exerted). As a result the charged colloid acquires a net motion, in a direction determined by its charge sign. Nb: for the derivation of (11.39) we employed the concentric shell model to find that the joint friction factor of the ions is a factor $(1+\kappa a)$ larger than the friction factor of one colloidal sphere.

11.13 (a) $\mu_{el} = \dfrac{Q}{f}$ and $\mu_{sed} = \dfrac{\Delta m}{f} \Rightarrow \dfrac{Q}{\Delta m} = \dfrac{\mu_{el}}{\mu_{sed}}.$

(b) The expression $\mu_{el} = Q/f$ assumes that the typical particle size a is small enough such that $\kappa a \ll 1$, but makes no assumption about particle shape; $\mu_{sed} = m/f$ is independent of size and shape. So the result for the particle charge-to-buoyant mass ratio in (a) only requires that $\kappa a \ll 1$.

11.14 (a) The potential at a distance a from the centre of a sphere carrying a charge of Q coulombs is, according to Coulomb's law (11.29)

$$\Psi_0 = \frac{Q}{4\pi\varepsilon\varepsilon_0 a}.$$

Taking the surface potential Ψ_0 approximately equal to the zeta potential ζ, we obtain for the number σ of charges on the sphere a value of about

$$\sigma = \frac{Q}{e} \approx \frac{4\pi\varepsilon\varepsilon_0 a\zeta}{e} = \frac{4\pi \times 24.55 \times 8.85 \times 10^{-12} \times 50 \times 10^{-10}\, C^2 Vm}{1.6 \times 10^{-19}\, Nm^2 C} = 85.4.$$

(b) 85 charges per sphere: that is not a lot, so every surface charge has plenty of surface room:

$$\frac{4\pi a^2}{85.4} = 1{,}471.5\ nm^2.$$

(c) Average distance between two surface charges:

$$\sqrt{1{,}471.5\, nm^2} \approx 38\, nm.$$

11.15 (a) At pH < 6.8 (both proteins positively charged) and at pH > 7.1 (both proteins negative).

(b) $\mu_{el} = \dfrac{2}{3}\dfrac{\varepsilon\varepsilon_0\zeta}{\eta}$

For the Hb proteins: $\zeta = \dfrac{3\eta\mu_{el}}{2\varepsilon\varepsilon_0}$

$$= \frac{3 \times (1.0 \times 10^{-3}\, Pa\ sec) \times (-0.2 \times 10^{-5} \times 10^{-4}\, m^2 V^{-1} sec^{-1})}{2 \times (77.9 \times 8.85 \times 10^{-12}\, J^{-1} C^2 m^{-1})}$$

$$= -0.44\ mV$$

For the HbS proteins: $\zeta = \dfrac{3\eta\mu_{el}}{2\varepsilon\varepsilon_0} = +0.66\ mV.$

These are small zeta potentials; but note that the proteins are migrating at a pH very close to their isoelectric point.

(c) For 10^{-3} M NaCl, $\kappa^{-1} = 9.7$ nm. Hence $\kappa a = \dfrac{2.25\ nm}{9.7\ nm} = 0.23$, which is small enough for the Hückel equation to apply.

11.16 The PB equation in the DH approximation reads:

$$\frac{1}{r^2}\frac{d}{dr}\left(r^2\frac{d\Psi}{dr}\right) = \kappa^2\Psi.$$

On substitution of $\Psi = u/r$ this PB equation simplifies to

$$\frac{d^2 u}{dr^2} = \kappa^2 u.$$

If the second derivative of a function u equals the function u itself, the function must be an exponential one. Therefore the general solution is:

$$u = C_0 e^{-\kappa r} + Ae^{+\kappa r},$$

where c_0 and A are integration constants. So:

$$\Psi = c_0 \frac{e^{-\kappa r}}{r} + A \frac{e^{+\kappa r}}{r}.$$

The constant A must be zero since the potential vanishes to $\Psi = 0$ at large radial distance r. Hence:

$$\Psi = c_0 \frac{e^{-\kappa r}}{r},$$

which is equation (11.23) in the text.

11.17 From the electro-neutrality condition equation (11.25) we obtain:

$$Q = 4\pi\varepsilon\varepsilon_0 \kappa^2 \int_a^\infty \Psi r^2 dr = 4\pi\varepsilon\varepsilon_0 \kappa^2 c_0 \int_a^\infty e^{-\kappa r} r\, dr,$$

where we have substituted the potential $\Psi = c_0 e^{-\kappa r}/r$ from equation (11.23). Working out the integral

$$\int_a^\infty e^{-\kappa r} r\, dr \stackrel{v=\kappa r}{=} \kappa^{-2} \int_{\kappa a}^\infty e^{-v} v\, dv = \kappa^{-2} e^{-\kappa a}(1+\kappa a),$$

we find:

$$Q = 4\pi\varepsilon\varepsilon_0 \kappa^2 c_0 \times \kappa^{-2} e^{-\kappa a}(1+\kappa a),$$

leading to the value of integration constant c_0:

$$c_0 = \frac{Q e^{\kappa a}}{4\pi\varepsilon\varepsilon_0 (1+\kappa a)},$$

which is equation (11.26) in the text.

Glossary

A

Aggregation The phenomenon of **colloids** sticking together as a result of attractive forces between them; see also **coagulation** and **flocculation**.

amphipathic Denotes molecules or particles that have **hydrophilic** as well as **hydrophobic** parts; from *amphi* 'of both kinds' and *pathos* 'experience'.

amphiphilic Literally 'loving both kinds'; synonymous with **amphipathic**.

anisotropy constant The energy barrier, per unit crystal volume, separating two opposite orientations of the **magnetization** of a **magnetic domain** along the easy crystalline axis.

Asakura–Oosawa–Vrij potential The interaction potential between colloids composed of an **excluded volume** repulsion and an attraction stemming from the **depletion force**.

association colloids Structures of colloidal dimensions that spontaneously form in solution by **self-assembly** of **surfactant** molecules; examples are **micelles** and biological membranes.

athermal Describes particles that are too large (exceeding about 10 microns) to exhibit significant **Brownian motion**.

AUC Abbreviation for an analytical **ultracentrifuge**.

Avogadro's constant The number of atoms or molecules in one mole of a substance, equal to $N_A = 6.022 \times 10^{23}$ mol^{-1}.

B

Bingham flow Flow behaviour of a dispersion in which, once the **yield stress** is exceeded, the **shear stress** increases linearly with **shear rate**, the slope being equal to the **plastic viscosity**.

Bjerrum length The separation distance between two ions at which their coulombic interaction energy is equal to their individual **thermal energy**.

Boltzmann's constant Relates a system's temperature to the **Boltzmann distribution** of its components over the available energy levels. It has the value $k_B = 1.3807 \times 10^{-23}$ JK^{-1}.

Boltzmann distribution Specification of the population, the number of molecules or particles, in a state of certain energy in a system at given temperature.

Born repulsion Steeply repulsive force stemming from overlap of electronic shells of two atoms, operative, for example, when two colloidal surfaces are pushed into contact.

Bridging flocculation A process of **flocculation** in which **colloids** become linked to each other by irreversible polymer adsorption onto their mutual surfaces.

Brownian motion The spontaneous, random (angular or positional) displacements executed by small particles in **thermal motion**.

Brownian relaxation Relaxation of the magnetization of **ferrofluids** by rotational **Brownian motion** of particles; compare to **Néel relaxation**.

buoyant mass The mass of a submerged particle minus the mass of the solvent volume that is displaced by the particle.

buoyant weight The product of a particle's **buoyant mass** and the acceleration of gravity; equal to the net attractive force the Earth exerts on the particle.

C

chemical potential The change in **Gibbs free energy** per molecule, a change that equals the **reversible work** needed to bring a molecule to some location or phase.

CMC Abbreviation for **critical micelle concentration**.

coagulant A substance that causes a colloidal dispersion or any other liquid to **coagulate**.

coagulation Transformation of a colloidal dispersion into a **gel** or solid-like state by the action of a **coagulant**. Sometimes also used to denote the process of **flocculation**.

coalescence The process whereby small droplets fuse together to a single large droplet.

coarsening The decrease in specific surface area of particles by **coalescence** or **Ostwald ripening**.

co-ions Ions that have the same charge sign as charged **colloids**; compare to **counter-ions**.

colloidal dispersion Distributions of **colloids** in a continuous solid, liquid, or gas phase.

colloidal filtration The separation of particles from a dispersion by their accumulation on a porous substrate through which solvent, driven by an external pressure, is permeating.

colloids Particles or structures having at least one dimension that is between a few nanometres and a few microns.

colloidal glass A **random packing** of **colloids** in a **metastable** state of collective arrest that precludes the colloids to rearrange—via their **thermal motion**—to crystalline structures of lower **free energy**.

colloidal stability The kinetic stability entailed by a rate of **flocculation** that is imperceptibly low due to a repulsive barrier in the interaction potential between **colloids**.

configuration A three-dimensional distribution or arrangement of **colloids** that fluctuates in space and time due to the colloids' **Brownian motion**.

comminution The reduction of inorganic materials to minute particles or fragments by mechanical actions such as grinding and crushing.

contact angle The angle formed by the free surface of a liquid phase in contact with a solid surface.

counter-ions Ions with a charge opposite to that of charged **colloids**; compare to **co-ions**.

convection The one-way traffic of particles in **stationary** motion, imposed by an external force or flow field; contrary to the random particle movements in **diffusion** or **Brownian motion**.

creaming The phenomenon of upward **sedimentation**, occurring for particles with a lower mass density than the solvent in which they are dispersed.

critical micelle concentration The concentration, in solution, at which **surfactant** molecules spontaneously form multi-molecular aggregates, also referred to as **micelles**.

cryo-TEM Abbreviation for cryogenic **TEM**.

Curie point Another term for **Curie temperature**.

Curie temperature A critical temperature above which **ferromagnetic** substances become **paramagnetic**.

D

DA Abbreviation for **Derjaguin approximation**.

Dalton's law States that the total pressure exerted by a mixture of ideal gases equals the sum of the pressures exercised by the components that make up the mixture.

Darcy's law The equation that links the rate of **viscous flow** in a porous medium to the pressure gradient driving the flow, liquid **viscosity**, and **liquid permeability** of the medium.

DE Abbreviation for the **Donnan equilibrium**.

Deborah number (De) The ratio of a characteristic internal molecular relaxation time of a material, to the time span during which the material is observed.

Debye length A measure for the thickness, or spatial extent, of a diffuse **electrical double layer**, which increases upon lowering the **electrolyte** concentration.

Debye–Hückel approximation The assumption of small **electrical potentials** near a charged surface, for which the **Poisson–Boltzmann equation** entails an exponential decay.

depletants Polymers or small solute particles that may induce a **depletion force** between colloids that are dispersed in the solution.

depletion force Attractive force between closely approaching **colloids** resulting from unbalanced **osmotic pressure** exerted by **depletants**.

Derjaguin approximation An approximate method to convert interactions between two parallel plates to interactions between two spheres—accurate whenever spheres are much larger than the interaction range.

detergent A water-soluble cleansing agent which combines with impurities and dirt to enhance their solubility; **micelles** are instances of such agents.

DHS Abbreviation for **dipolar hard sphere**.

DHS potential Potential of interaction between two uncharged, hard spheres which each have an embedded permanent **dipole moment**.

dialysis The separation of particles in solution on the basis of differences in their ability to pass through a **semi-permeable membrane**.

diamagnetism Tendency of molecular magnetic moments to align, independently of each other, in a direction at 180° to an applied magnetic field; compare to **paramagnetism**.

diffusion Spontaneous random movements executed by molecules and small particles in thermal motion; for colloids, synonymous with **Brownian motion**.

diffusion coefficient The ratio of **thermal energy** driving **diffusion** to the **Stokes friction factor**, accounting for retardation by the viscous force on particles moving in solution.

diffusion equation A differential equation for the concentration of diffusing particles, the solution of which entails how the particles' concentration profile evolves in space and time.

diffusion-limited aggregation The aggregation process in which **diffusion** is the limiting factor in how fast sticky particles combine to complex, branching clusters.

dipolar hard-sphere potential An interaction potential combining the **hard-sphere potential** with the magnetic interaction between dipoles located at the centre of the spheres.

dipole A pair of equal, oppositely charged or magnetic poles separated by a finite distance.

dipole moment Mathematical product of the separation of the ends of a **dipole** and the magnitude of the charges or magnetic poles.

disjoining force Any force that drives **colloids** apart; an example is the **electrical double-layer repulsion** between charged particles, responsible for **colloidal stability**.

dispersion method The process of breaking down a macroscopic phase into progressively smaller parts. For inorganic materials the method is also referred to as **comminution**.

DLA Abbreviation for **diffusion-limited aggregation**.

DLS Abbreviation for **dynamic light scattering**.

DLVO Abbreviation for Derjaguin–Landau–Verwey–Overbeek.

DLVO potential Interaction potential between two colloids, comprising an **electrical double-layer repulsion** and a **Van der Waals force**.

domain wall A region between neighbouring **magnetic domains** in which the **magnetization** direction gradually rotates from the direction that occurs in one domain to that in the other.

Donnan cell A region of solution of zero **electric field** to which charged colloids are confined by a **semi-permeable membrane** or any other external force acting on the colloids.

Donnan equilibrium The thermodynamic salt equilibrium between a **Donnan cell** and a bulk reservoir salt solution.

Donnan potential The constant **electrical potential** in a **Donnan cell**.

Donnan pressure The difference in **osmotic pressure** between a **Donnan cell** and the reservoir solution with which the cell is in thermodynamic equilibrium.

dynamic light scattering Technique to gauge particle **diffusion coefficients** via the fluctuations in light scattering intensities caused by the particles' **Brownian motion**.

E

EDL Abbreviation for **electrical double layer**.

elastic solid A solid able to spontaneously resume its undeformed state upon removal of an external, deforming force; antithesis of a **viscous fluid**.

elasticity modulus The **stress** exerted on an **elastic solid** divided by the deformation or **strain** caused by the stress.

electric field At a given position the electric field equals the force experienced by a **unit charge** placed at that position.

electrical double layer (EDL) The duplex layer of electrical charge comprising charges that are fixed on a surface, and **co-ions** and **counter-ions** that diffuse in solution near that surface.

electrical double-layer repulsion The net repulsive force caused by overlap of two **EDLs**, following from the osmotic ionic pressure in the midplane between the two EDLs.

electrical potential The electrical potential at any point equals the work required to bring a positive unit charge from infinity to the given point.

electrolytes Strong electrolytes are substances fully ionized in solution at all concentrations; weak electrolytes are substances with a degree of ionization that increases when concentrations are reduced.

electro-osmosis Liquid flow under an electric field in a channel or porous medium from which a value for the **zeta potential** is deduced for the material along which the liquid is flowing.

electrophoresis The movement of charged particles in a fluid or **gel** under the influence of an applied electric field.

electrophoretic mobility The speed of **electrophoresis** of a charged particle per unit of applied **electric field**.

elementary charge The electrical charge carried by a proton; its value is $e = 1.602 \times 10^{-19}$ C.

emulsifier A substance employed to stabilize an **emulsion**; often a **surface-active agent** is used, but fine particles may also be used in the case of a **Pickering emulsion**.

emulsion A dispersion of minute droplets of one liquid in another continuous liquid in which the dispersed liquid is not miscible or soluble.

equation of state An equation relating a solution's osmotic pressure to temperature and solute concentration; for **ideal particles** this equation equals **Van 't Hoff's law**.

excluded volume The volume surrounding a centre of a particle that cannot be accessed by the centre of another particle.

F

ferrimagnet A material composed of **magnetic domains** in which a minority of atomic magnetic dipoles are oriented opposite to the other parallel dipoles; compare **ferromagnet**.

ferrofluids Concentrated, stable dispersions of permanently magnetic particles that respond to a magnetic field gradient much like a liquid **ferromagnet** would do.

ferromagnet A material composed of **magnetic domains** each containing individual atomic dipoles that all point in the same fixed direction; compare **ferrimagnet**.

Fick's first law An equation that defines a particle's **diffusion coefficient** as the ratio of the diffusion flux of particles to the concentration gradient that drives the flux.

Fick's second law Another designation for the **diffusion equation**.

flocculation The process in which attractive particles stick together into structures ('flocs') in which particles preserve their size and shape.

Flow-induced flocculation The process in which **aggregation** occurs due to attractive colloids driven together in the liquid-flow field of a stirred dispersion.

form factor A function that quantifies the angular dependence of a particle's light scattering intensity which arises from interference of light emitted by different locations in the particle.

free energy See **Gibbs free energy**.

G

gas constant The gas constant is defined as the product of **Boltzmann's constant** k_B and **Avogadro's constant** N_A. It has the value $R = 8.3145$ J K^{-1} mol^{-1}.

Gauss's flux theorem The equation that relates the gradient of the **electric field** at a given location to the net space charge density at that location.

gel A jelly-like substance composed of sticky polymers or attractive **colloids** that results from a **gelation** process.

gelation The process in which attractive particles in a **colloidal dispersion** stick together to form a **gel**.

Gibbs free energy The part of an energy decrease that is maximally available as work: the **reversible work** delivered in a **reversible process** at constant temperature and pressure.

Gibbs–Kelvin equation The relation between particle solubility and the particle's surface curvature. Convex curvature of, for example, spheres increases their solubility in proportion to the inverse sphere radius.

granular particles Another term for **athermal particles**.

gravitational length The **thermal energy** of a particle divided by its **buoyant weight**; a measure for the spatial extension of a **sedimentation–diffusion equilibrium**.

Guinier equation An equation that shows how gyration radii of arbitrarily shaped particles follows from their **static light scattering** intensities at sufficiently small scattering angles.

H

Hagen–Poiseuille law An equation relating the volume rate of **viscous flow** through a capillary to fluid **viscosity**, capillary radius, and the pressure gradient driving the flow.

half-life time The time needed in **flocculation** for the total particle number to become half of its initial value at time zero; inversely proportional to the flocculation's **rate constant**.

hard-sphere potential The interaction potential between two spheres that is infinitely repulsive to exclude sphere overlap, and that is zero for finite surface-to-surface distances.

heterogeneous nucleation Formation of small particles on a substrate composed of another material than the small particles.

homogeneous nucleation Formation of small particles in the bulk of a **supersaturated** solution containing one type of solute molecules only.

Hooke's law The linear increase, for an **elastic solid**, of shear stress with shear strain; the constant of proportionality being the **elasticity modulus**.

Hückel equation The equation that shows how a **zeta potential** follows from a measured **electrophoretic mobility** of a charged sphere that is much smaller than the **Debye length**.

hydrodynamic radius The radius of a colloidal sphere that includes a layer of solvent adhering to the sphere surface; this immobilized solvent constitutes a **slipping plane**.

hydrophilic Having the tendency to dissolve in, mix with, or be wetted by water; opposite of **hydrophobic**.

hydrophobic Tending to repulse or fail to mix with water; the opposite of **hydrophilic**.

I

ideal gas law States that for a gas of **ideal particles**, gas pressure is the product of the number density of the molecules and their **thermal energy**.

ideal packing law States how for a **random packing** of uncorrelated particles, packing fraction depends on particle shape.

ideal particles Particles that occupy negligible space, that do not interact with each other, and that consequently possess only **kinetic energy**.

inertia Property of matter by which it continues in a state of rest or uniform, linear motion, unless that state is changed by an external force.

interface A boundary between two phases such as two immiscible liquids, or a liquid and a solid phase. The term **surface** is sometimes reserved for the cases when one phase is a gas.

interfacial tension The tension of an **interface**, which equals the Gibbs free energy that is required to form a unit interfacial area. See also **surface tension**.

intrinsic viscosity The limit, at zero particle concentration, of the **reduced viscosity**.

irreversible flocculation The process of particles sticking together by contact attractions much larger than the **thermal energy**, such that it cannot be reversed by **Brownian motion**.

K

kinetic energy The energy a body has on account of its motion; compare with **potential energy**. Average kinetic energy of particles is proportional to their **thermal energy**.

kinetic unit A collective of molecules or particles that diffuses around as one entity, with an average **kinetic energy** that is proportional to the **thermal energy**.

Kozeny–Carman relation A relation that quantifies how the **liquid permeability** of porous media depends on **specific surface area** and volume fraction of the media's solid phase.

Krieger–Dougherty equation A relation between a dispersion's **Newtonian viscosity** and colloid concentration, entailing viscosity divergence on approach of **random close packing**.

L

laminar flow Fluid flow composed of fluid layers that move parallel to each other, at constant but different speeds.

Langevin function The mathematical function that generates the **magnetization curve** for a **superparamagnetic** dispersion of non-interacting magnetic particles.

Langmuir equation An equation that informs how the **disjoining force** between two charged surfaces follows from the **electrical potential** midway between the surfaces.

Laplace pressure The pressure increase in a material when that material is bent to a convex curvature; for a sphere the increase is inversely proportional to the sphere radius.

liquid permeability The term that, in **Darcy's law**, multiplies the pressure gradient to obtain the speed of **viscous flow** in a porous medium.

lyophilic Adjective pertaining to colloids formed by spontaneous dispersal of a substance in a solvent; compare to **lyophobic**.

lyophobic Adjective pertaining to colloids composed of substances that are next to insoluble in the solvent in which the colloids are dispersed; compare to **lyophilic**.

M

magnetic dispersions Solutions of **colloids** that possess a magnetic **dipole moment** that is either permanent or induced by an external magnetic field.

magnetic domains Regions in a **ferromagnetic** material that are each spontaneously magnetized to **saturation** in a definite crystalline direction.

magnetic micro-beads Micron-sized spheres with embedded **superparamagnetic** particles that endow the spheres with a high **saturation magnetization** in an external field.

magnetic relaxation The return of aligned moments by thermal rotations to the state of random orientations after the aligning external field has been switched off.

magnetic susceptibility The **magnetization** per unit of applied magnetic field, equal to the initial, linear slope of a **magnetization curve**.

magnetization Net magnetic **dipole moment** per unit volume of material.

magnetization curve A plot of the average **magnetization** of a sample as function of the applied magnetic field.

magnetophoresis The movement of particles under the influence of an applied magnetic field.

magnetophoresis–diffusion equilibrium The concentration profile resulting from the balance between particle **magnetophoresis** in a field gradient and the opposing particle **diffusion**.

magnetotactic bacteria Bacteria with an internal magnetite crystal chain that orients them in the Earth's magnetic field.

Maxwell model The portrayal of **viscoelasticity** by a spring connected to a piston immersed in a fluid representing, respectively, elastic and viscous response of the fluid.

Maxwell stress A negative pressure stemming from the electrostatic attraction between **counter-ions** and a charged surface, which tends to contract the **electrical double layer**.

MDE Abbreviation for **magnetophoresis-diffusion equilibrium**.

mean-squared displacement The average of the square of displacements a particle is making in a diffusion process; this process can be modelled by a **random walk**.

metastable adjective used for systems that can transform to a more stable **phase** of lower **Gibbs free energy**. For example, **supersaturated** water vapour is metastable with respect to liquid water.

micelles Aggregates of surface-active molecules formed in solution when the solubility limit for single molecules has been reached. See **critical micelle concentration**.

monodisperse Adjective used for particles or colloids that are of uniform size.

MRI Abbreviation for magnetic resonance imaging.

MSD Abbreviation for **mean-squared displacement**.

mutual diffusion coefficient Coefficient describing diffusion of a particle relative to another diffusing particle; for independent diffusers the sum of particles' **diffusion coefficients**.

N

Néel relaxation Relaxation of **magnetization** by thermal rotation of magnetic moments inside a magnetic material; compare **Brownian relaxation**.

Newtonian fluid A fluid with a constant **viscosity** that is independent of **shear rate**.

Newton's viscosity law An equation stating that fluid **viscosity** is the **shear stress** applied to the fluid divided by the resulting **shear rate**.

non-Newtonian fluids Fluids with a **viscosity** that depends on the **shear rate**.

normal stress The component of stress at right angles to the area considered.

nucleation The formation of small particles in a **supersaturated** solution that subsequently grow by the diffusional uptake of solute molecules from their surroundings.

O

optical contrast The difference between the average refractive index of a colloid and the refractive index of the solvent in which the colloid is dispersed.

osmosis The **spontaneous process** of solvent molecules diffusing from a concentrated solution to a less concentrated one; the antithesis of the process is **reverse osmosis**.

osmotic pressure The pressure exerted by solute particles on account of their **thermal energy**; for **ideal particles** this pressure is given by Van 't Hoff's law.

Ostwald ripening Process whereby large particles in a solution grow at the expense of smaller ones, owing to the greater solubility of the smaller particles.

P

paramagnetism Behaviour resulting from the tendency of molecular magnetic moments to align, independently of each other, with an applied magnetic field; compare **diamagnetism**.

Péclet number (*Pe*) A dimensionless quantity expressing the ratio between mass transfer by flow and mass transfer by **diffusion**; for small *Pe*, diffusion predominates the transfer.

phase A continuous region of matter that is homogeneous with respect to composition and the thermodynamic variables pressure and temperature.

phase transition The spontaneous molecular transfer from one **phase** to another, co-existing phase; for example the vapour–liquid transition that occurs in a **supersaturated** vapour.

Pickering emulsion An **emulsion** in which the dispersed droplets are covered by adsorbed particles that retard or prevent droplet **coalescence**.

plastic viscosity The slope of the linear increase of **shear stress** with **shear rate** for a dispersion that exhibits **Bingham flow**.

plug flow Stationary motion of a substance through a tube where all parts of the substance move at the same speed.

Poiseuille flow This is **laminar flow** in a cylindrical tube, with a parabolic flow velocity profile.

Poisson equation A differential equation that relates the **electrical potential** at a location to the net electrical charge density at that location.

Poisson–Boltzmann equation A differential equation, the solution of which is the **electrical potential** profile in the **electrical double layer** in the solution near a charged surface.

polydispersity The phenomenon that a particle property is a distributed quantity; examples are colloids with distributions in size or shape.

porous media Materials composed of a solid phase with high **specific surface area**, and an open, usually tortuous pore space filled with either gas or liquid.

potential energy The energy possessed by a body by virtue of its position, or its physical properties such as magnetic moment and electrical charge; compare with **kinetic energy**.

precipitation The process whereby dissolved **solute** molecules unite in solution to solid particles, often via a process of **nucleation** and growth.

pressure The force per unit plane area when that force is perpendicular ('normal') to the area; compare **stress**.

R

random close packing The highest achievable packing fraction for a particle packing that comprises no long-range (translational or orientational) crystalline ordering.

random rod packing Particle compact composed of randomly oriented rods or fibres.

random sphere packing Disordered sphere packing in which long-range positional order is absent; nevertheless, due to the spheres **excluded volumes**, short-range order does occur.

random walk Sequence of uncorrelated steps taken by a particle in totally arbitrary directions; employed to evaluate the **mean-squared displacement** a particle traverses by **diffusion**.

rate constant The factor that, in the kinetics of **flocculation**, determines the speed at which colloidal aggregates form and disappear in time; see also **half-life time**.

Rayleigh ratio A measure for the intensity of **static light scattering** from a unit volume of solution relative to the intensity of the incident light beam.

Rayleigh scattering Light waves radiated by oscillating electrical dipoles in a particle that are all in phase, since the particle is much smaller than the wavelength of light.

RCP Abbreviation for **random close packing**.

reduced viscosity The **relative viscosity** minus one, divided by the particle volume fraction.

relative viscosity Ratio of the **viscosity** of a dispersion to that of the solvent.

repeptization The reversal of the **flocculation** of a colloidal dispersion by washing out the salt that induced the flocculation, as a result of which **colloidal stability** is restored.

reverse osmosis The forced expulsion, via a **semi-permeable membrane**, of solvent out of solution by a hydrostatic pressure that exceeds the solution's **osmotic pressure**.

reversible process A very slow process that limits towards a sequence of equilibrium states, characterized by a minimal dissipation, or frictional loss, of energy as heat.

reversible work The maximal amount of work delivered via a **reversible process**.

Reynolds number (*Re*) A dimensionless number that expresses the ratio of inertial forces to viscous forces.

rheology The science of the deformation and flow of matter.

RMSD Abbreviation of the square root of the **mean-squared displacement** of a particle in **Brownian motion**.

RO Abbreviation for **reverse osmosis**.

Rosensweig instability The shape change of a flat **ferrofluid** surface to a periodic spike pattern when the fluid is exposed to a sufficiently strong magnetic field.

S

salt depletion The ejection of neutral salt molecules by charged colloids or interfaces to a bulk reservoir salt solution.

saturation concentration The concentration of material dissolved in solution that is in thermodynamic equilibrium with a flat bulk phase composed of that material.

saturation magnetization The maximal **magnetization** of a volume of material, attained when every magnetic **dipole moment** in that volume points in the same direction.

SD Abbreviation for magnetic **single domain**.

SDE Abbreviation for **sedimentation–diffusion equilibrium**.

second virial coefficient Entails the first-order correction, due to particle–particle interactions, to **Van 't Hoff's law** for the **osmotic pressure** exerted by ideal particles.

sedimentation Settling of particles in a gravitational or centrifugal field.

sedimentation–diffusion equilibrium The concentration profile that results from the balance between particle **sedimentation** and the oppositely directed particle **diffusion**.

sedimentation mobility The **sedimentation** speed of a particle per unit of applied gravitational or centrifugal field.

self-assembly The spontaneous association of molecules into multi-molecular aggregates or structures. An instance of self-assembly is the formation of **association colloids**.

SEM Abbreviation for scanning electron microscope.

semi-permeable membrane A thin sheet or film of porous material that is permeable only to solvent molecules and that blocks passage of solute particles.

settling radius The radius of a spherical particle determined by measurement of its rate of **sedimentation**.

shear Motion of a layer of a material relative to adjacent layers.

shear plane A synonym for **slipping plane**.

shear rate The change of **shear strain** per unit time.

shear strain The deformation that occurs in a material in response to a **shear** applied to that material.

shear stress The component of **stress** tangential to the area considered.

shear thickening The property of viscosities increasing with increasing **shear rate**.

shear thinning The property that viscosities reduce with increasing **shear rate**.

single-domain The adjective referring to particles composed of a single **magnetic domain**.

slipping plane Imaginary plane between solvent adhering to a surface, and freely flowing solvent further away from that surface; also called **shear plane**.

slow flocculation The process of **flocculation** in which encounter frequencies by **Brownian motion** are retarded by a repulsive barrier in the interaction potential between two colloids.

SLS Abbreviation for **static light scattering**.

Smoluchowski equation The equation that converts a measured **electrophoretic mobility** of arbitrary shaped colloids that are much larger than the **Debye length** to a **zeta potential**.

soap The alkali metal salt of a fatty acid; often generically used to denote any **surface-active agent**.

solubilization The uptake of non-polar, fatty substances in the hydrocarbon-like, hydrophobic core of **micelles**.

solute The minor component in a solution, being dissolved in the solvent which is the solution's major compound.

specific surface area The area of an object or solid phase per unit volume of that object or phase; sometimes also defined as area per unit weight instead of per unit volume.

spontaneous process In thermodynamics any process that occurs by itself without the help of external agents, and that is characterized by a decrease in the system's **Gibbs free energy**.

static light scattering (SLS) Method by which particle sizes are obtained from the angular dependence of the intensity of monochromatic light scattered by the particles.

stationary flow A flow in which the flow velocity at any point is constant in time.

stationary speed The constant, time-independent speed a particle sustains as long as the net force exerted on the particle is zero.

stick boundary The condition that, for fluid flowing along a surface, the speed for fluid in immediate contact with the surface is zero.

Stokes equation An equation resulting from the balance between pressure and viscous forces in **viscous flow**, the solution of which are flow velocity profiles.

Stokes flow Flow in which viscous forces totally dominate inertia forces.

Stokes friction factor The factor that determines the magnitude of the resistive force exerted on a particle moving in a viscous medium. See also **Stokes' law**.

Stokes' law The equation stating that a particle migrating at **stationary speed** in a viscous medium experiences a frictional force equal to this speed times the **Stokes friction factor**.

Stokes radius The radius of a sphere with the same rate of **sedimentation** as settling particles that are of unknown size or shape.

streaming potential The **zeta potential** of a charged **interface** obtained from measurements of the rate of **electro-osmosis** along that interface.

stress A force per unit plane area, with the force lying ('tangential') in the plane; compare **pressure**.

superparamagnetism Tendency of magnetic particle moments in a ferrofluid to line up in a magnetic field; similar to **paramagnetism** except that the **magnetization** is much larger.

supersaturation A solution is said to be supersaturated when the concentration of dissolved material exceeds the **saturation concentration**.

surface The boundary between a solid or liquid and a gas phase. The term **interface** is preferred for the boundary between two condensed phases that are either solid or liquid.

surface-active agent A compound containing both **hydrophilic** and **hydrophobic** moieties that, as a consequence, make the compound readily adsorb at an **interface**.

surface tension The tension of the surface film of a liquid, in magnitude equal to the **Gibbs free energy** needed to create a unit area of **surface**.

surfactant Another term for **surface-active agent**.

Svedberg relation An equation showing how the **buoyant mass** of a particle can be obtained from measurements of the particle's **diffusion coefficient** and **sedimentation mobility**.

T

TEM Abbreviation for transmission electron microscope.

thermal energy Product of the **Boltzmann constant** k_B and absolute temperature T; the average **kinetic energy** of particles is proportional to $k_B T$.

thermal motion The spontaneous motion of particles that is a manifestation of their **thermal energy**.

thixotropy Property of becoming less viscous when agitated, shown, for example, by **gels** which become temporarily fluid when shaken or stirred.

U

ultracentrifuge Instrument in which a cell containing a solution or **colloidal dispersion** is rotated at very high speeds (typically 10,000–40,000 rpm).

unit charge Another term for **elementary charge**.

V

Van der Waals forces Electrostatic attractive forces between molecules, arising from interaction between permanent or transient electric **dipole moments** of these molecules.

Van 't Hoff's law States that a solution of **ideal particles** exerts an **osmotic pressure** equal to the particle number density times their **thermal energy**. Compare to the **ideal gas law**.

viscoelasticity Property of a substance of exhibiting elastic behaviour for sufficiently rapid distortions, and viscous behaviour at sufficiently slow deformations of the substance.

viscosity Property of a material to resist deformation more when deformation rates increase.

viscous flow Another term for **Stokes flow**.

viscous fluid A fluid in which all energy invested to maintain its flow is dissipated as heat; the antithesis of an **elastic solid**.

Y

yield stress A threshold value of the **shear stress** below which a dispersion does not flow.

Young's equation An equation for the force equilibrium between the three surface tensions acting along the perimeter of a liquid cap on a solid surface submersed in a gas phase.

Z

zero point of charge The pH at which **colloids** have a zero zeta potential such that they do not undergo **electrophoresis**.

zeta potential Considered to be the **electrical potential** in the **slipping plane** near charged **colloids** that are submitted to **electrophoresis**.

Index

Note: Tables and figures are indicated by an italic *t* and *f* following the page number.

A

activation free energy, 14–15
adsorption
   at oil–water interface, 185
   of surfactant molecules at colloid–solution interface, 15
adsorption free energy, 185
   for a hydrophilic sphere, 186
aggregation/aggregates, 2, 3, 8, 93, 116, 175
   charge-stabilized colloids, 211
   containing α spheres, 113
   diffusion-limited aggregation (DLA), 26
   ferrofluids, 166
   flow-induced, 117
   of fractal clusters, 27
   linear, 190
   of lyophilic colloids, 6
   of lyophobic colloids, 6, 75
   by magnetite particles, 167f, 190
   migration of, 174
   prevention of, 6
α-mers, 113, 114
aluminium hydroxide, precipitates of, 76f
alumino-silicate (imogolite) fibres, 121f
amphipathic molecules, 187
analytical ultra-centrifuge (AUC), 123, 124
anisotropic colloids, examples of, 4f
anisotropic interactions, 171
anisotropic particles, packings of, 132
anisotropy constant, 171
antiferromagnetism, 163f
apparent light-scattering radii, 213–214
Archimedes' principle, 121
Asakura–Oosawa–Vrij (AOV) potential, 67–68
association colloids, 75, 180–181
athermal particles, 31
atomic dipole ordering, 176
atomic magnetic dipoles, 162
axial flow, in a straight tube, 146f

B

baby nappies, urine uptake by, 194
Bachelor coefficient, 122–123
back diffusion, 175
bacteria, magnetotactic, 166
barometric height distribution, 124
Bernal sphere packing, 132, 133
Bingham flow, 153
bio-organic stabilizers, 185
Bjerrum length, 81, 82, 97
blood, 4
blood capillaries, 148
Boltzmann distributions, 18, 56, 63, 64, 65, 111, 204
   counter-ion-density, 95
   for ideal particles, 125
   and ion distributions near a charged surface, 79–80
Boltzmann exponent, 58
Boltzmann factor, for dipolar energy in an external field, 168
Boltzmann-weighted averages, 173
bone, 4
Born repulsion, 96, 171
buoyant particle mass, 121, 122, 125
buoyant particle weight, 121
bridging flocculation, 69
Brownian collision(s)
   frequencies, 105, 106
   time, 36–37
Brownian motion, 2, 30–49
   Einstein equations for, 41–45
   Fick's first law and the diffusion coefficient, 39–41
   for mastic spheres, 37f
   quadratic displacements from the diffusion equation, 46–48
   random walk, 37–39
   of sedimenting particles, 120
   time scale, 36
   time scales in colloidal dispersions, 32–37
Brownian relaxation, 170

C

C-spheres, 33, 40f, 148
   relaxation of momentum, 40f
   step length and relaxation time, 55
capillaries
   blood, 148
   flow through, 131
   fluid viscosity profiles in, 140
   permeability, 129, 147
carbon, 6
carbon black, 6, 75
carbon colloids, 4
casein micelles, 92f
catalysts, 8, 23, 24
cell membranes, 181
cells, water management of, 50
centrifugal equilibrium profiles, 125
ceramics, 4
   from monodisperse colloids, 5
   from polydisperse colloids, 5
chain friction factor, 175
charged colloids, 125–126
   dispersions of, 153
   electrophoretic mobilities of, 212
   ionic pressures caused by, 69
   osmotic pressure exerted by, 59
chemical potential, 78
   of ions, 78
   of micelles, 189
   of a solvent, 52
clay colloids in soil, 4
   colloidal instability of, 95
CMC (critical micelle concentration), 187
co-ions, 77, 79
coagulation
   early phase of, 106
   late-stage, 113–114, 117
coalescence, 7, 180, 183, 184f
coarsening, 7
collision frequencies, 106
colloid–colloid interactions, 123
colloidal C-sphere, 31, 32t, 148
colloidal charge, 193
colloidal dispersions, 6–8, 156
   applications of, 3, 4
   Brownian collision time, 36–37
   configurational relaxation time, 36
   diffusive or Brownian time scale, 36
   filtration of, 127
   hydrodynamic decay time, 35
   molecular collision time, 33
   momentum relaxation time, 34–35
   timescales in, 32–37
colloidal domain, 2f, 8, 31
colloidal filtration, 127–128
   process, 120–121
   rates, 131
colloidal glasses, 132, 133
colloidal sedimentation *see* sedimentation
colloidal stability, 76, 166, 193
colloids
   definition, 1
   dressed with a double-layer, 206–207
   features of matter in colloidal state, 1–2
   force-induced flux of, 111
   responses to applied magnetic fields, 161, 162f
comminution, 3, 165

concentric-shell model, 206–207
condensation, 22
    of silanol groups, 26
configurational relaxation time, 36, 37
contact depletion attraction, 69
continuity condition, 144
convection, diffusion vs, 147–148
convective particle flux, 175
convex interfaces, contraction due to surface tension of, 191
core–shell colloids, 23
coulombic interaction energy, 79
Coulomb's law, 81, 205
Counter-ions-only limit, 60, 63, 65, 90
    concentration, 79
    diffusion of, 76–77
creaming, 66, 184f
'creeping flow' equation, 144
critical micelle concentration (CMC), 187
crystalline salt formation, 12
cubic crystals, 19
Curie temperature, 161
cylinders, axial flow of a Newtonian fluid in, 146–147

D

DA (Derjaguin approximation), 68, 70–71, 93, 98
Darcy's law, 126–127
    for unidirectional flow, 145
    for viscous flow in particle packings and porous media, 121
DE *see* Donnan equilibrium (DE)
DE (Donnan equilibrium), 59–65, 69, 88
de Donnan model, 193
Deborah number, 139
Debye cube, 97–98
Debye–Hückel (DH) approximation, 84, 91–92, 204
Debye–Hückel potential profile, 205
Debye length, 81, 82, 93, 204
Debye micellization mode, 188
deformation, of a fluid, 140
depletants, 66
    ghost depletants, 68–69
    volume fraction, 69
    Vrij's depletant model, 68
depletion forces, 66–69
    Asakura–Oosawa–Vrij (AOV) potential, 67–68
    depletion attraction between spheres, 68
    reversibility of destabilization by, 69
Derjaguin approximation (DA), 68, 70–71, 93, 98
Derjaguin–Landau–Verwey–Overbeek (DLVO) interaction potential, 77, 93–96
    repulsive barrier in DLVO interaction potential between two colloids, 103
detergency, 188

DHS (dipolar hard spheres), 173
diamagnetism, 164
diffraction patterns, fluctuating, 200–201
diffuse condenser, 85–86
diffusion, 31
    vs convection, 147–148
    in a force field, 110–111
    position vectors of two diffusing particles, 105f
    stationary, 39–41
    to a target sphere, 104–106
diffusion coefficients, 58
    for a Brownian particle or molecule, 39
    dependence on particle size, 193
    derivation of, 66
    effective, 112–113
    Fick's first law and, 39–41
    mutual, 104, 105–106
    Stokes–Einstein, 115, 117, 175, 202, 214
diffusion equation, quadratic displacements from, 46–48
diffusion flux, 106
diffusion-limited aggregation (DLA), 26
diffusional growth, 114–115
diffusive flux, 175
diffusive time regime, 36
dipolar hard spheres (DHS), interaction potential, 173
dipolar magnetic colloids, interactions between, 171–174
dipolar pair potential, 173
dipole moment, 161, 168f
dipole orientations, relaxation of, 170–171
dipoles, head-to-tail configuration, 172
disjoining force, 76, 86
dispersion method, 3
dispersions, 8
    examples of, 3t
    magnetic, 161
    non-Newtonian, 152–156
    and their applications, 3–5
displacement time, 142
distance decay, 173
distribution function, 47
    for dipole orientations, 168
    normalized angular, 169
DLA (diffusion-limited aggregation), 26
DLS *see* dynamic light scattering (DLS)
DLVO potential *see* Derjaguin–Landau–Verwey–Overbeek (DLVO) interaction potential
domain walls, 164
Donnan cell, 60, 61
Donnan equilibrium (DE), 59–65, 69, 88
Donnan potential, 61, 63–64, 88
Donnan pressure, 61–62, 64
    at high-salt limit, 63
    at low-salt limit, 62–63
drift time, 142
dynamic light scattering (DLS), 193, 200–203, 212

apparent particle sizes/radius, 214
application to a dispersion of identical spheres, 202
intensity fluctuations, correlations between, 201–202
particle sizes from, 203

E

EDL *see* electrical double layers (EDLs)
effective diffusion coefficients, 112–113
Einstein equations for Brownian motion, 41–45
    Einstein I, 32, 42–43
    Einstein II, 32, 43–45, 46, 105
    flux balance, 42–43
    force balance, 42
Einstein's viscosity equation, for sphere dispersions, 150
elastic solids, 139, 153
elasticity modulus, 154
electric field strength, 83, 87
electrical charges, of colloids, 193
electrical double layers (EDLs), 75, 96, 212
    diffuse condenser model, 85–86
    diffuse part, 77
    potential profile in, 77
    two interacting, 86–92
electrical potential, 60, 79, 88
electrical potential profile, 77, 78f, 96, 210
    around a charged sphere, 204–205
    near a charged surface, 80f
    in a single, flat EDL, 82–83
    electrolyte reservoir, 86
electro-osmosis, 212
    experiment, 209f
    and the Smoluchowski equation, 208–211
electrophoresis, 92, 203–212
    at intermediate κ a values, 211
    of an isolated charged sphere, 205–206
electrophoretic mobilities, of charged colloids, 212
electrophoretic mobility, 203
emulsification, 180, 183
emulsions, 3, 4, 180, 190
    coarsening by Ostwald ripening, 183
    kinetic stabilization of, 8
    oil-in-water, 180
    stabilized, 183–186
    water-in-oil, 180
energy dissipation, by rotating particles in a shear flow, 156
ensemble average, of diffusive displacements, 46
equation of state, 124
equilibrium colloid profiles, 111
equilibrium concentration profile, 42f
equilibrium dipole orientations, 168–169
equilibrium distribution, of ideal ions in an EDL, 77

equilibrium magnetization, of
　superparamagnetic dispersions, 167–171
equilibrium solubility
　of colloids, 21
　of a sphere, 21f
excluded volume, 67, 133
exponential decay, of electrical
　potentials, 84
external field, dipole moment of beads
　in, 174

F

Faraday, Michael, gold sols, 23, 76
ferrimagnetic materials, atomic magnetic
　dipoles, 162, 163
ferrimagnetism, 163f
ferrofluids, 161–162, 164–166, 176
　applications of, 166
　magnetization curves, 168f
　molten magnet mimicry, 165
　preparation of, 165
　small-scale synthesis of, 177–178
ferromagnetic metals, 163
　atomic magnetic dipoles, 162
　melting points of, 161
　multi-domain structure in, 164
fibre-dispersion viscosity, 152
Fick's first diffusion law, 20, 104, 110
　and the diffusion coefficient, 39–41
filtration rates, 134, 139
flocculation, 7, 75, 95, 96, 103, 184f
　bridging, 69
　electrolyte concentration, 96
　flow-induced, 116–117
　half-life time, 109, 117
　irreversible, 103, 117
　rapid early-phase, 106–109
　rate constants for diffusion-controlled,
　　108, 117
　retardation of, 117
　salt concentration, 94–95
　shear-induced, 116f
　slow, 109–113, 115
flow
　in a cylinder, 146–147
　-induced flocculation, 116–117
　near a flat surface, 144f
　in particle packings and Darcy's law,
　　126–128
　in simple geometries, 144–147
　in a slit, 145–146
fluid viscosity, 156
fluids, deformation of, 140
form factor, 197–198
fractal clusters, 26
fractal dimensionality, 26, 27
free energy, of interaction, 172–173
frequency responses, 155–156
friction factor, 205, 207 see also Stokes
　friction factor

G

gas, thermal particles in, 51f, 52
Gauss's flux theorem, 80, 81
gels, 26, 27, 124
gelation, 26–27
ghost depletants, 68–69
Gibbs adsorption energy, 185–186, 187f
Gibbs free energy, 6–7
　maximum, in nucleation, 14, 16
Gibbs–Kelvin effect, 21
Gibbs–Kelvin equation, 21
gibbsite minerals/platelets, 4f, 185
glass, 30, 155
　window, 6
gold colloids, 4, 23
gold sols, 23, 76
Gouy–Chapman (GC) potential profiles, 82, 83
gradient,
in fluid velocity, 140, 146
in hydrostatic pressure, 140f, 141, 145
in osmotic pressure, 66, 80f
in electrical potential, 80f, 86
in ion number density, 87
granular G-sphere, 31–32
granular particles, 31
gravitational acceleration, 121
gravitational length, 124
gravity, 2
growth rate
　diffusion-limited growth, 20
　reaction limited, 19
Guinier equation, 198–200, 213
Guinier plots, 199
gyration radius, 199

H

haematite, 4f, 6
Hagen–Poiseuille equation, 147
Hagen–Poiseuille law, 147
half-life time, 109
Hamaker constant, 93, 100
hard-sphere repulsion, 171
Henry function, 211
heterogeneous nucleation, 13
　catalysed, 23–25
　rate, 24–25
heterogeneous structures, 26
heterogenous precipitation, 25
homogeneous nucleation, 13, 18f
homogeneous precipitation, 25
　thermodynamics of, 13–16
Hooke's law, 153, 154
Hückel equation, 205–206, 207
human hair, thickness of, 31
hydrodynamic decay time, 35
hydrodynamic friction, 207
　of a particle in a sphere array, 126
hydrodynamic friction factor, of a sphere, 123
hydrodynamic radius, of a sphere, 122
hydrophilic particles, 185

hydrophilic solid spheres
　Gibbs adsorption energy, 185, 186
　in water, 185f
hydrophobic particles, 185
hydrostatic pressure, 53–54
hyperbolic functions, 71–72
　inverse, 72

i

ideal ions, 78–79
ideal packing law, 133
ideal particles, 97
industrial crystallization processes, 12
inertia
　of a colloid, 142
　of a fluid, 141
inhomogeneous particle growth, 26–27
inks, 6
inorganic colloids, 4
inorganic precipitates, 7
inorganic Pickering stabilizers, 185
interaction energy
　repulsive, 91
　between the two spheres, 70
interaction free energy, 172–173
interaction potential, 81
interaction potential curves, for limits of
　flocculation and repeptization, 96
interface, 181
interfacial tension, 180, 181
intrinsic viscosity, 149, 150, 156
inverse hyperbolic functions, 72
ion density ratio, 65
ion distributions, near a charged surface,
　77–80
ionic osmotic pressure gradient, 66
ionic pressures, caused by charged
　colloids, 69
iron
　Curie point of, 161
　melting point of, 161
iron hydroxide, 6
　precipitation of, 2, 76f
iron oxides, 4
irreversible flocculation, 103, 117
isoelectric points, determination of, 211–212
isotropic interactions, 171

K

kaolinite, 4
Kelvin length, 16
kinetics
　of irreversible flocculation of colloids, 117
　of late-stage coagulation, 113–114, 117
　retardation of flocculation, 117
Kozeny–Carman scaling relation, 128–131
Kozeny constant, 130
Krieger–Dougherty equation, 151–152,
　157–158

L

Laminar flow, 140
La Mer's scheme, 18–19
Langevin function, 169
Langmuir equation, 88
Laplace pressure, 21, 191
    under concave surfaces, 182–183
    interfacial tension and, 181–183
Lifshitz, Slezov, and Wagner (LSW) theory, 23
light-scattering, experiments, measuring geometry for, 195f
light scattering
    intensity of, 197f
    by larger particles, 197–198
    by small particles, Rayleigh scattering, 195, 196
    of unpolarized light, 197
liquid flow velocity, 126
logarithmic representations of hyperbolic functions, 72
LSW (Lifshitz, Slezov, and Wagner) theory, 23
lyophilic colloids, 6, 75
lyophobic colloids, 96
    examples of, 75
    preparation of, 6
    thermodynamic instability of, 75, 76f

M

maghemite, 165, 177
magnesium hydroxide colloids, 4, 5f
magnetic attractions, versus Van der Waals attractions, 173–174
magnetic density separation, 166
magnetic dispersions, 161
magnetic microbeads, 162, 174
magnetic moments, determination of, 170
magnetic relaxation, 170–171, 176
magnetic relaxation time, 170
magnetic resonance imaging (MRI), 166
magnetic susceptibility, 170
magnetism, in solids and solutions, 162–164
magnetite, 3, 165
    oxidation of, 177
    precipitation, 177
    preparation of, 176
    single domains (SDs), 164
magnetizable dispersions, 174–176
magnetization curves, 162, 169–170
magnetophoresis, 174–175
    –diffusion (MD) equilibrium concentration profile, 175
    equilibrium profiles, 175–176
    rapid micro-bead, 176
    speed, 175
magnetotactic bacteria, 166
margarine, 185
Maxwell–Boltzmann distribution, 56
Maxwell model for viscoelasticity, 154–155
Maxwell stress, 87

mayonnaise, 153, 180, 183, 185
mean-squared displacement (MSD), 38–39
membrane pressure probes, 56–59
metallic colloids, 4
metastable systems, 12
micellar size distributions, 189–190
micellar structures, examples of, 187f
micelles, 75, 180, 187
    linear, 190
    –monomer equilibria, 188–189
micellization, 187–190
microbeads, magnetic, 162, 174
microscopy, for particle sizing, 194
milk, casein micelles, 92f
molecular collision time, 33
molecular M-spheres, 31, 32
molecular mass determination, 59
molecular Van der Waals attractions, 173
momentum relaxation time, 34–35, 55, 142
monodisperse colloids, 5
monomers, 191
    weakly attractive monomers in equilibrium, 190f
montmorillonite, 4
MRI (magnetic resonance imaging), 166
MSD (mean-squared displacement), 38–39
mutual diffusion coefficient, 104, 105–106

N

nano N-sphere, 31, 32t, 148
Néel relaxation, 171
Newtonian fluids, 141, 146, 153
Newton's second law, 34, 52, 55, 141, 142
Newton's third law, 57
Newton's viscosity law, 140, 141, 143, 146, 149, 156
no-slip condition, 146
non-equilibrium colloid profiles, 111–112
non-ionic surfactants, 187
non-Newtonian dispersions, 152–156
normalized angular distribution function, 169
normal force or pressure, 43
nucleation, 6, 7, 16
    of carbon-dioxide gas bubbles on a glass surface, 13
    classical nucleation theory, 13–14
    and growth of a spherical precipitate, 14f
    heterogeneous, 13, 23–25
    homogeneous, 13, 17, 18f
    rate, 17
    from water vapour phase, 12

O

oil droplets, immersed in water, 7, 8, 181f, 182
oil films, immersed in water, 181, 182
oil-in-water emulsions, 180
oleic acid
    -coated magnetite particles, 167f
    coatings, 177–178

olive oil, 183
    mixture of vinegar and, 180
opals, 5
optical contrast, 196
osmosis, 50–51
    examples of, 193–194
    reversed, 55
    spontaneous, 52–53
osmotic cell, 60f
osmotic equation of state, of interacting colloids, 124
osmotic equilibrium, 54–55
osmotic force, 86
osmotic pressure, 50, 51
    from depleted polymers, 69
    effect of salt on, 60
    exerted by charged colloids, 59
    exerted by uncharged colloids, 59
    measurement of, 56
    from osmotic equilibrium, 52–55
    total osmotic pressure, 87
Ostwald ripening, 7, 8, 13, 21–22, 25, 181, 184f
    emulsion coarsening by, 183
    kinetics, 22–23
    particle solubility and, 20–21
overlap volume, for two spheres, 68

P

paint, 'non-drip,' 153
paper-making, colloidal filtration in, 121
papermaking, 4
paramagnetism, 161, 164
particle growth and polydispersity, 18–20
    diffusion-limited growth, 20
    La Mer's scheme, 18–19
    narrowing distributions, 19
    reaction limited growth, 19
particle quartet, 32t
particle shape, 131, 158
    relation between particle concentration and, 133
particle sizes, 193, 200
    apparent, 212, 213
    from DLS and SLS, 203
particle(s)
    friction factor, 39
    quartet, 31
    solubility, 16, 20–23
PB (Poisson–Boltzmann) equation, 77, 80–82, 204
Péclet number, 148
permeability, 121, 126–127, 129, 130, 131
    of a capillary, 147
    of a slit pore, 146
    using RCP density to evaluate, 133
    vacuum, 172
phase inversion, 184f
phase transition, 12
Pickering emulsions, 184–185

Pickering stabilization, 184–185
pigments, 4
plastic viscosity, 153
plug flow, 145
Poiseuille flow, 148
Poiseuille law, 144
Poisson–Boltzmann (PB) equation, 77, 80–82, 204
Poisson equation, 81
pollen grains, thickness of, 31
polydispersity, 4–5, 106, 214 see also size polydispersity
polymers
    exertion of pressure on colloids, 66
    stress relaxation time, 156
porous media, 126
    flow through, 120
potential energy, 172
potential interaction energy, 171–172
potential profile, 81
precipitation, 2–3, 12
    heterogenous, 25
    magnetite, 177
    silica, 26
    size polydispersity as consequence of, 5
precipitation kinetics, 16–18
    of colloids in a supersaturated solution, 16
    homogeneous nucleation rate, 17
    pre-exponential kinetic factor, 18
pressure
    in the counter-ions-only limit, 60
    exerted by ideal particles, 69
pressure gradient, 66, 145
pressure waves, 35
primary minimum, 94
Prussian blue, precipitates of, 76f

R

radial clusters, that grow by the uptake of rod-like particles, 26f
radial stationary diffusion, 41
radius of gyration, 69, 199
random close packing (RCP), 131
    density, 132, 133
random particle packings, 131–134
random rod packings, 132
rate constants, 107, 113
    and colloidal shape, 108–109
    and colloidal size, 107–108
    for diffusion-controlled flocculation, 108, 117
Rayleigh–Gans–Debye (RGD) theory, 197–198
Rayleigh light scattering, 180, 212
    particle mass from, 197
    and particle volume, 196–197
    by small particles, 195–196
Rayleigh ratio, 194, 195, 197
Rayleigh's equation, 195
RCP (random close packing), 131, 132, 133
reaction limited growth, 19

reaction rates, 106
    second order in concentration, 107
red blood cells (RBC), 50, 194, 206
reduced viscosity, 150
refractive indices, and optical contrast, 196
relative viscosity, 150, 151f
repeptization phenomena, 95–96, 177
repulsion
    Born repulsion, 96
    between colloids, 6, 7
    electrical double layers (EDLs), 75–76, 77, 78, 90, 92
    exponential decay of tail, 96
    hard-sphere, 171
    inter-colloid, 8
    weak overlap (WO), 93
repulsive colloids, 59
repulsive interaction energy, 91
reservoirs, of salt solution 60f, 64
retardation, of flocculation kinetics, 117
reversed osmosis, 55
reversible electrical work, 78
Reynolds number, 141–143
RGD (Rayleigh–Gans–Debye) theory, 197–198
rheology, 139, 140
ring radius, 70
ripening, 7, 8
root-mean-squared (RMS) speed, 33, 45
root of the MSD (RMSD), 45
Rosensweig instability, 165

S

salad dressing, 180
salt
    concentration, 64
    depletion, 64–65, 69, 79, 86
    effect on osmotic pressure, 60
    solution reservoir 60f
saturation concentration, 15
saturation magnetization, 163, 169
scattering
    small-angle, 198–200
    of unpolarized light, 197
Schulze–Hardy rule, 95
SDE see sedimentation–diffusion equilibrium
seawater, extraction of potable water from, 55
second virial coefficient, 58–59
secondary minimum, 94
sedimentation, 120, 134, 184f
    mobility, 122
    single-particle, 121–123
sedimentation–diffusion equilibrium, 123–126, 176
self-assembly of colloids, 3, 5
semi-permeable membranes, 50, 57f
shear, 140
shear flow, 116f
shear plane, 206
shear rate, 140, 141

shear strain, 140
shear stress, 141, 152
shear thickening, of dispersions, 152
shear thinning, 153
shear waves, 35
silica, 76
    cubes, 4f
    ellipsoids, 121f
    needles, 121f
    particles, 185
    precipitation, 26
silica nanoparticles, aqueous sols of, 21, 22f
silica spheres, 5, 5f
    bi-disperse, 121f
silver halide colloids, 19
single domains, 163
single magnetic domain (SMD), 162
single-particle sedimentation, 121–123
    sedimentation mobility, 121–122
size determination, by microscopy, 194
size distribution, self-sharpening, 19
size of colloids, determination of, 193
size polydispersity, 178, 212
slipping plane, 206, 208, 210
slow flocculation, 109–113, 115
SLS see static light scattering (SLS)
SMD (single magnetic domain), 162
Smoluchowski diffusion model, 104, 115
Smoluchowski equation, electro-osmosis and the, 208–211
Smoluchowski limit, at high salt concentrations, 208
SO (strong-overlap) regime, 88–90
soaps, and surfactants, 187
solubilization, 188
solvent mass density, 121
spatial decay, 84
specific particle volume, 121
specific surface area, 1–2
speckle fluctuation times, 201
spheres
    collision radius of solid, 106
    equilibrium solubility of, 21f
    hydrodynamic radius, 122
    mixtures of spheres and increasing RCP density, 152
    radial stationary diffusion from scented, 41
spherocylinders, random dense packings of, 134f
springs, 154
stabilization, Pickering, 184–185
stabilizers, 180
    bio-organic, 185
    inorganic, 185
    for Pickering emulsions, 184–185
static light scattering (SLS), 193, 194–200, 212
    applications of, 200
    particle sizes from, 203, 213–214
stationary convective colloid flux, 58

stationary diffusion, 39–41
   flux, 104
   radial, 41
steady-state colloid flux, 112
stick boundary, 146
Stokes–Einstein diffusion coefficient, 115, 117, 175, 202, 214
Stokes–Einstein equation, 193
Stokes equation, 128, 145, 146, 156
   for viscous flow, 140, 143–144
Stokes factor, for spheres in Brownian motion, 127
Stokes flow, 141
Stokes friction, 33, 122
Stokes friction factor, 34, 111, 120, 121, 142, 174
   particles, 39, 43
   for a sphere, 122, 123, 127, 207
Stokes' law, 111
Stokes radius, 122
stress, 141
stress relaxation time, 155
strong-overlap (SO) regime, 88–90
sugar crystallization, 12
superparamagnetic dispersions, equilibrium magnetization of, 167–171
superparamagnetic ferrofluids, magnetization by external fields of, 176
superparamagnetism, 163
supersaturation, 12, 15, 165
surface, 181
   of liquid water, 181
surface-active agents, 183
surface-active molecules, 191
surface charge, 193
   density, 85, 211
   surface potential versus, 205
surface potential, 84–85
   versus surface charge, 205
surface roughness, of colloids, 96
surface tension, 181
   tensile force of, 181–182
   thermodynamic definition of, 181
surfactants, 180, 183
   ionic, 187
   non-ionic, 187
   and soaps, 187
   stabilizing, 183
Svedberg relation, 122, 123, 206

T

tangential force, or stress, 141, 143f
Taylor expansions, of hyperbolic functions, 72
temporal correlation function, of the scattered light field, 201
tensile force, of surface tension, 181–182
thermal energy, 31
   of ions, 79
thermal motion, 2, 31
thermal equilibrium, 32, 35
thermodynamic equilibrium, between monomers and micelles, 191
thermodynamics of homogeneous precipitation, 13–16
   activation free energy, 14–15
   classical nucleation theory, 13–14
   degree of supersaturation, 15
   Gibbs energy of spherical nuclei, 16
time-independent sedimentation–diffusion equilibrium (SDE) concentration profile, 121
timescales, in colloidal dispersions, 32–37
titania, 6
toluene–water system, 186
toothpaste, 152
torque, on a sphere rotating in a flow 149
total chemical potential change, for water moving from pure water into solution, 54
total osmotic pressure, from anions and cations, 61, 87

U

ultracentrifugation, 123f
uncharged colloids, 60, 61, 124, 125
   osmotic pressure exerted by, 59
uncharged dipolar colloids, interaction between, 176
unidirectional flow, 127

V

Van der Waals forces, 2, 93, 171, 173
   algebraic decay of, 94
   magnetic versus, 173–174
   between two parallel plates, 98–100

van 't Hoff, J. H., 52
Van 't Hoff's law, 52, 58, 61, 63, 66, 68, 69, 124
   from particle kinetics, 55–56
   for a solution of ideal counter-ions, 90
velocity gradient, 116, 146
vesicles, 180
viscoelasticity, 139
   Maxwell model for, 154–155
viscosity, 139–140, 141
   concentration-dependent, 149–152
   reduced viscosity, 150
   relative, 150, 151f
viscous flow, 141
   frictional forces between sliding fluid layers, 156
   parabolic viscous flow profile, 145f
   in porous media, 120
viscous fluids, 139
viscous friction, 31, 120, 203
viscous strain, 154
viscous stress, 146
Vrij's depletant model, 68

W

water, 50
   exhibition of elasticity in, 155–156
   surface of liquid water, 181
   total chemical potential change for water moving from pure water into solution, 54
water vapour phase, nucleation from, 12
wave vector, 198
weak-overlap regime, 90–91
window glass, 6

Y

yield stress, 153
Young's equation, 186

Z

zero-field approximation, 60–61
zero magnetic field, magnetic moments of embedded nanoparticles in, 174
zeta potentials, 92, 206, 207, 210, 211, 212